Yearbook on Space Policy

More information about this series at http://www.springer.com/series/8166

Yearbook on Space Policy

Edited by the
European Space Policy Institute
Director: Peter Hulsroj

Cenan Al-Ekabi • Blandina Baranes •
Peter Hulsroj • Arne Lahcen
Editors

Yearbook on Space Policy 2014

The Governance of Space

Springer

Editors
Cenan Al-Ekabi
Blandina Baranes
Peter Hulsroj
Arne Lahcen
ESPI - European Space Policy Institute
Vienna, Austria

ISSN 1866-8305 ISSN 2197-9405 (electronic)
Yearbook on Space Policy
ISBN 978-3-7091-1898-6 ISBN 978-3-7091-1899-3 (eBook)
DOI 10.1007/978-3-7091-1899-3

Springer Wien Heidelberg New York Dordrecht London

Cover illustration: Artist's impression of Herschel, ESA - C. Carreau

Printed on acid-free paper

Springer-Verlag GmbH Wien is part of Springer Science+Business Media (www.springer.com)

Preface ESPI Yearbook 2014

"The Governance of Space"

The increasing number and diversity of players in space in recent years, and the onset of new technologies as well as the residue of past space activities, has brought the issue of space governance to the fore. While some aspects of space governance are specifically legal and technical, others raise broader questions about the entire current space regime. In order to clarify and shed additional light on these issues, ESPI has decided to focus on the topic of governance from a wide variety of angles in its Yearbook on Space Policy 2014.

Traditionally, the first part of the Yearbook sets out a comprehensive overview of the economic, political, technological and institutional trends that are affecting space activities. It is prepared in-house in ESPI, and while its perspective is European, it also provides a comparative analysis of space developments around the world.

The second part of the ESPI Yearbook approaches the overall theme from an analytical perspective. This year it includes nine external contributions that bring together the views of various eminent professionals in the space field. This part of the Yearbook opens with a contribution by Professor Stephan Krasner and ESPI Resident Fellow Marco Aliberti on the theoretical link between space and international relations theory. Thus, they analyse the two major paradigms of international relations theory, realism and liberalism, to describe and explain the different and often contending approaches followed by states in the governance of space activities. Thereafter, Professor Emeritus Jacques Blamont assesses the effects of globalisation on the governance of space activities in the USA and NASA in particular. The political and institutional aspects of governance are explored in a contribution by Xavier L.W. Liao from the Ghent Institute for International Studies, focusing on the impacts of regionalisation processes on space governance, and in a contribution by Hokkaido University Professor Kazuto Suzuki on how governance models affect geopolitics in Asia. Gérard Brachet, former President of CNES and former

Chairman of UNCOPUOS, contributes an article on the optimal role for governments in space, and Amy Kaminski, Senior Policy Advisor at the Office of the Chief Scientist of NASA, sets out some of the findings of her research on the roles of the public and NGOs in space governance and how this public engagement has been evolving over time. Rajeswari Rajagopalan, who is a Senior Fellow at the Observer Research Foundation, New Delhi, assesses the International Code of Conduct for Space Activities and describes its importance for the long-term sustainable use of the outer space environment. Finally, the hot topic of governance in the European setting is addressed. Professor Thomas Hoerber offers a historic perspective describing how the processes of European integration have shaped the creation and governance of ESA, while Professor Emmanuel Sigalas writes about the role of the European Parliament in the legitimisation of Europe's space policy.

The third part of the Yearbook continues the character of the Yearbook as an archive of space activities. Again prepared in-house by ESPI, a bibliography, chronology and data about institutions are provided where readers of the now eight volumes of the Yearbook can identify statistical developments and evolutions.

In closing, we would like to thank the contributors of the articles in Part Two for their engagement in this publication. Moreover, we are very grateful to Frances Brown, former editor-in-chief of Space Policy and current member of the ESPI Advisory Council, for her support and inspiration as we prepared the ESPI Autumn Conference 2014. The contributions in Part 2 of the Yearbook reflect the presentations made by the authors at the Autumn Conference.

Vienna, Austria

Cenan Al-Ekabi
Blandina Baranes
Peter Hulsroj
Arne Lahcen

List of Acronyms: Acronym Explanation

A

AARSE	African Association of Remote Sensing of the Environment
ABS	Asia Broadcast Satellite
ACE	Advanced Composition Explorer
ACS	Alcântara Cyclone Space
ADF	Aerospace Defence Forces
AEB	Agência Espacial Brasileira (Brazilian Space Agency)
AG	Aktiengesellschaft
AIA	Atmospheric Imaging Assembly
Airbus D&S	Airbus Defence and Space
ALASA	Airborne Launch Assist Space Access programme
ALOS	Advanced Land Observing Satellite
APSCO	Asia-Pacific Space Cooperation Organization
ARD	Atmospheric Re-entry Demonstrator
ARM	Asteroid Redirect Mission
ASAT	Anti-Satellite
ASI	Agenzia Spaziale Italiana (Italian Space Agency)
ATV	Automated Transfer Vehicle

B

BDC	Business Development Bank of Canada
BMD	Ballistic Missile Defence
BMO	Bank of Montreal

C

CAA	Civil Aviation Authority
CAGR	Compound Annual Growth Rate
CASBAA	Cable and Satellite Broadcasting Association of Asia
CASC	China Aerospace Science and Technology Co.
CASIC	China Aerospace Science and Industry Co.
CAST	China Aerospace Science and Technology Corp.
CBERS	China–Brazil Earth Resources Satellite
CCDev	Commercial Crew Development programme
CCP	Commercial Crew Programme
CCiCap	Commercial Crew integrated Capability initiative
CCtCap	Commercial Crew Transport Capability
CD	Conference on Disarmament
CEPT	European Conference of Postal and Telecommunications Administrations
CHF	Swiss franc
CHIRP	Commercially Hosted Infrared Payload
CLA	Alcântara Launch Center
CMB	Cosmic Microwave Background
CME	Coronal Mass Ejections
CNES	Centre National d'Études Spatiales (French Space Agency)
CNSA	China National Space Administration
COP	Conference of the Parties
COPUOS	Committee on the Peaceful Uses of Outer Space
COROT	COnvection, ROtation and planetary Transits space telescope
CPC	Certification Products Contract
CRS	Commercial Resupply Services
CSA	Canadian Space Agency
CST-100	Commercial Space Transportation-100
CSO	Czech Space Office

D

DARPA	Defense Advanced Research Projects Agency
DARS	Digital Audio Radio Service
DBS	Direct Broadcast Services
Dextre	Special Purpose Dexterous Manipulator
DLR	Deutsches Zentrum für Luft- und Raumfahrt (German Aerospace Center)
DND	Department of National Defence
DoD	Department of Defence
DoT	Department of Telecom
DRDO	Defence Research and Development Organisation

DSCOVR	Deep Space Climate ObserVatoRy
DTH	Direct To Home

E

EADS	European Aeronautic Defence and Space
EAP	Environmental Action Programme
EBITDA	Earnings Before Interest, Taxes, Depreciation and Amortization
ECA	Evolution Cryotechnique type A
EDA	European Defence Agency
EDM	Entry, Descent and Landing Demonstrator Module
EDRS	European Data Relay Satellite System
EELV	Evolved Expendable Launch Vehicle
EFT	Exploration Flight Test
EKV	Exoatmospheric Kill Vehicle
EO	Earth Observation
EPIC	Earth Polychromatic Imaging Camera
EPS-SG	European Polar System Second Generation
ERA	European Research Area
ESA	European Space Agency
ESM	European Service Module
EU	European Union
EUMETSAT	The European Organisation for the Exploitation of Meteorological Satellites
EUTELSAT	European Telecommunications Satellite Organisation
EVE	EUV Variability Experiment

F

FAA	Federal Aviation Administration
FCC	Federal Communications Commission
FSS	Fixed Satellite Services

G

GCF	Green Climate Fund
GCSP	Global Collaborative Space Programme
GDP	Gross Domestic Product
GEO	Geostationary Earth Orbit
GEOSS	Group on Earth Observation System-of-Systems
GERD	Gross Domestic Expenditure on Research and Development
GLONASS	Globalnaya Navigatsionnaya Sputnikovaya Sistemya (Russian GNSS Constellation)

GM	General Motors
GmbH	Gesellschaft mit beschränkter Haftung
GMD	Ground-based Midcourse Defence
GMES	Global Collaborative Space Programme Monitoring for Environment and Security
GNSS	Global Navigation Satellite Systems
GPIM	Green Propellant Infusion Mission
GPS	Global Positioning System
GSA	European GNSS Agency
GSLV	Geosynchronous Satellite Launch Vehicle
GTO	Geosynchronous Transfer Orbits

H

HCN	Hydrogen Cyanide
HDTV	High-Definition Television
HFI	High-Frequency Instrument
HiRISE	High-Resolution Imaging Science Experiment
HMI	Helioseismic and Magnetic Imager
HTV	H-2 Transfer Vehicle

I

IAC	International Astronautical Congress
IAG	International Association of Geodesy
ICG	International Committee on Global Navigation Satellite Systems
ICoC	International Space Code of Conduct for Outer Space Activities
ICS	Information and Communication Systems
ICT	Information and Communication Technology
IGS	International GNSS Service
IHMC	Institute for Human and Machine Cognition
ILN	International Lunar Network
ILS	International Launch Services
IMAAFS	Information for Meeting Africa's Agricultural Transformation and Food Security
IMF	International Monetary Fund
INDC	Intended Nationally Determined Contributions
IOV	In-Orbit Validation
IR	Infrared
IRGC	Islamic Revolution Guards Corps
IRIS	Interface Region Imaging Spectrograph
IS	Islamic State or ISIS
ISEF	International Space Exploration Forum
ISPS	Innovative Space Propulsion Systems

ISRO	Indian Space Research Organisation
ISS	International Space Station
ITU	International Telecommunication Union
IXV	Intermediate Experimental Vehicle

J

JADE	Jovian Auroral Distributions Experiment
JAXA	Japan Aerospace Exploration Agency
JEDI	Jupiter Energetic Particle Detector Instrument
JIRAM	Jovian Infrared Auroral Mapper
JUICE	JUpiter ICy moon Explorer

K

K2	Kepler 2
KSLV	Korea Space Launch Vehicle

L

LADEE	Lunar Atmosphere and Dust Environment Explorer
LDEX	Lunar Dust Experiment
LEO	Low Earth Orbit
LFI	Low Frequency Instrument
LIRIS	Laser InfraRed Imaging Sensors
LMC	Large Magellanic Cloud
LRO	Lunar Reconnaissance Orbiter
LWS	Living With a Star programme

M

MAG	Magnetometer
MAVEN	Mars Atmosphere and Volatile EvolutioN
MDA	Missile Defense Agency
MDA Ltd.	MacDonald, Dettwiler and Associates Ltd.
MDIS	+Mercury Dual Imaging System
ME	Mid-life Evolution
Melco	Mitsubishi Electric Co.
MESSENGER	MErcury Surface, Space ENvironment, GEochemistry and Ranging
Metop	Meteorological Operational Satellite
Metop-SG	Metop Second Generation
MEXT	Ministry of Education, Culture, Sports, Science and Technology

MFF	Multiannual Financial Framework
MFG	Meteosat First Generation
MIRO	Microwave Instrument for Rosetta Orbiter
MLA	Mercury Laser Altimeter
MMO	Mercury Magnetospheric Orbiter
MOIRE	Membrane Optical Imager for Real-Time Exploitation
MOM	Mars Orbiter Mission
MPCV	Multi-Purpose Crew Vehicle
MPO	Mercury Planetary Orbiter
MRO	Mars Reconnaissance Orbiter
MSG	Meteosat Second Generation
MSL	Mars Science Laboratory
MTG	Meteosat Third Generation
MTM	Mercury Transfer Module
MWR	Microwave Radiometer

N

NASA	National Aeronautics and Space Administration
NATO	North Atlantic Treaty Organisation
NDAA	National Defense Authorization Act
NEC	Nippon Electric Company
NEO	Near-Earth Object
NEOWISE	Reboot of the Wide-field Infrared Survey Explorer mission to find NEOs
NGA	National Geospatial-Intelligence Agency
NISTAR	National Institute of Standards and Technology Advanced Radiometer
NOAA	National Oceanic and Atmospheric Administration
NRO	National Reconnaissance Office
NSOAS	National Satellite Ocean Application Service

O

OECD	Organisation for Economic Co-operation and Development
OHB	Orbitale Hochtechnologie Bremen
ONSP	Office of National Space Policy)
OPEC	Organization of the Petroleum Exporting Countries
OPSAT	OPtical SATellite
OSIRIS	Optical, Spectroscopic and Infrared Remote Imaging System

P

PCW	Polar Communications and Weather mission
PDV	Prithvi Defence Vehicle
PLA	People's Liberation Army
PND	Portable Navigation Devices
PNT	Positioning, Navigation and Timing
PPP	Public–Private Partnership
PPWT	Treaty on Prevention of the Placement of Weapons in Outer Space and of the Threat or Use of Force Against Outer Space Objects
PRIDE	Programme for Reusable In-orbit Demonstrator in Europe
PROBA	PRoject for OnBoard Autonomy
PRS	Public Regulated Service
PSLV	Polar Satellite Launch Vehicle

Q

QZSS	Quasi-Zenith Satellite System

R

R2	Robonaut 2
Radio LANS	Radio wireless broadband devices
RCM	Radarsat Constellation Mission
RCS	Reaction Control System
RHESSI	Ramaty High Energy Solar Spectroscopic Imager
RLV	Reusable Launch Vehicle
ROSINA	Rosetta's Orbiter Spectrometer for Ion and Neutral Analysis
RRM	Robotic Refuelling Mission
RROxiTT	Robotic Oxidizer Transfer Test

S

SABRE	Synergistic Air-Breathing Rocket Engine
SAM	Sample Analysis at Mars instrument
SAP	Space Applications Programme
SAR	Synthetic Aperture Radar
SDO	Solar Dynamics Observatory
SE	Societas Europae
SecTelSat	Secure Telecom by Satellite
SES	Société Européenne des Satellites
SHEFEX	Sharp Edge Flight Experiments

SIA	Satellite Industry Association
SIP	Swiss Innovation Park
SLATS	Super Low Altitude Test Satellite
SLS	Space Launch System
SLV	Satellite Launch Vehicle
SM	Standard Missile
SMEX	Small EXplorer
SNC	Sierra Nevada Corporation
SOHO	SOlar and Heliospheric Observatory
SpaceX	Space Exploration Technologies
SPC	Space Programme Committee
SS/L	Space Systems/Loral
SSA	Space Situational Awareness
SSCO	Satellite Servicing Capabilities
SSN	Space Surveillance Network
SST	SpaceShip Two
SSTO	Single-Stage-To-Orbit
STEREO	Solar TErrestrial RElations Observatory
STFC	Science and Technology Facilities Council
SUMER	SOHO's Solar Ultraviolet Measurements of Emitted Radiation instrument

T

TAI	Turkish Aerospace Industries
TCBM	Transparency and Confidence Building Measures
TDM	Technology Demonstration Mission
TFEU	Treaty on the Functioning of the European Union
TGO	Trace Gas Orbiter
THEMIS	Thermal Emission Imaging System

U

UAV	Unmanned Aerial Vehicle
UK	United Kingdom
UKSA	UK Space Agency
ULA	United Launch Alliance
UN	United Nations
UNCTAD	United Nations Conference on Trade and Development
UNDSS	United Nations Department of Safety and Security
UNFCCC	United Nations Framework Convention on Climate Change
UNGA	United Nations General Assembly
UNGIWG	United Nations Graphic Information Working Group

UNIDIR	United Nations Institute for Disarmament Research
UNREDD	United Nations Programme on Reducing Emissions from Deforestation and Forest Degradation in Developing Countries
UNOOSA	United Nations Office for Outer Space Affairs
UNSDI	United Nations Spatial Data Infrastructure
UNSPIDER	United Nations Platform for Space-based Information for Disaster Management and Emergency Response
URSC	United Rocket and Space Corporation
US	United States of America
USAF	US Air Force
USAT	Ultra Small Aperture Terminals
UVS	Ultraviolet Imaging Spectrograph

V

VIPIR	Visual Inspection Poseable Invertebrate Robot
VIRTIS	Visible, Infrared and Thermal Imaging Spectrometer
VLM	Brazil's Microsat Launch Vehicle
VLS-1	Brazil's Satellite Launch Vehicle
VSAT	Very Small Aperture Terminals

W

WGP	World Gross Product
WISE	Wide-field Infrared Survey Explorer
WRC	World Radiocommunication Conference
WRS	World Radiocommunication Seminar
WSF	Weather System Follow-on

X

X1	Robotic Exoskeleton

Contents

Part I
The Year in Space 2014

Chapter 1
European Space Activities in the Global Context

Cenan Al-Ekabi

1.1 Global Political and Economic Trends

1.1.1 Global Economic Outlook

The United Nations Annual Report "World Economic Situation and Prospects" noted a slight improvement in global growth in 2014, which was expected to continue at only a moderate level across most regions and major economic groups.[1]

In 2014, the growth of World gross product (WGP) was estimated to be 2.6 %, thus marginally better than the growth of 2.5 % in 2013, but less than the 2.9 % that had been anticipated mid-year. This was due partially to a number of unexpected shocks such as the heightened geopolitical conflicts in different parts of the world, in addition to unfinished post-crisis adjustments in the recovery process from the global financial crisis.[2]

In the Euro zone, while monetary policy measures led to a significant improvement in the sovereign debt crisis, the economic situation remained fragile. Western Europe continued to struggle with GDP growth estimated to be only 1.2 % in 2014, as growth in the region had yet to reach pre-recession levels. However, Germany, Spain, Portugal, Ireland, and the UK returned to positive growth. Also, Eastern Europe gained further ground due partially to recovering domestic demand, the gradual abandonment of fiscal austerity, and a turnaround in the inventory cycle. And while labour markets continued to improve, progress was uneven across the member states.[3]

[1] "World Economic Situation and Prospects 2015." 10 Dec. 2015. United Nations 7 Mar. 2015. http://www.un.org/en/development/desa/policy/wesp/wesp_archive/2015wesp_full_en.pdf.

[2] Ibid. at 1.

[3] Ibid. at 8.

C. Al-Ekabi (✉)
ESPI, Schwarzenbergplatz 6, 1030 Vienna, Austria
e-mail: cenan.al-ekabi@espi.or.at

C. Al-Ekabi et al. (eds.), *Yearbook on Space Policy 2014*, Yearbook on Space Policy, DOI 10.1007/978-3-7091-1899-3_1

The economy of the United States of America had some fluctuations in 2014 with GDP growth estimated at 2.3 %; however that growth was expected to increase to 2.8 % in 2015. While the fiscal drag on growth was diminishing, the possibility of volatility in financial markets in response to the normalization of monetary policy could lead to adverse effects on the real economy. And with an overall decline in employment rates in developed economies since the beginning of the financial crisis in the United States the labour force participation rate was near its lowest level in the past 10 years due to population ageing, an increase in skill requirements, and a higher number of discouraged workers.[4]

In Japan, the momentum generated by the fiscal stimulus package and monetary easing introduced in 2013 receded due to a rise in inflation expectations and the further increase of the consumption tax by the central bank in late-2014. Exports were expected to eventually benefit from the depreciation of the Japanese yen triggered by the monetary easing, while the planned cut in corporate taxes would support fixed investment. And while GDP grew by 0.4 % in 2014, it was expected to increase to 1.2 % for 2015.[5]

While the global economy has struggled to gain momentum grappling with the legacy of the global financial crisis, developed economies are expected to see growth in the oncoming years, thanks to gradually recovering labour markets, ebbing fiscal consolidation, and low financing costs. While GDP growth was estimated at 1.8 % in 2014, it was expected to grow to 2.2 % in 2015. In developing countries growth was projected to gradually accelerate, rising from 4.4 % in 2014 to 4.8 % in 2015.[6]

Overall, WGP was forecast to grow at a pace of 3.1 % and 3.3 % in 2015 and 2016, respectively. Unemployment figures remain elevated in several developed countries, particularly in the euro zone, and wage stagnation has, to some extent, eroded the benefits of salaried employment. However, in developing economies, unemployment rates have remained relatively stable since 2013, partly owing to lower labour force growth. Nevertheless, high unemployment levels continue to persist in South-Eastern Europe, Northern Africa, and Western Asia.[7]

---

[4] Ibid. at 11.

[5] Ibid. at 8.

[6] "Global Economic Prospects | Having Fiscal Space and Using It." 13 Jan. 2015. The World Bank 7 Mar. 2015: 21. http://www.worldbank.org/content/dam/Worldbank/GEP/GEP2015a/pdfs/GEP15a_web_full.pdf.

[7] "World Economic Situation and Prospects 2015." 10 Dec. 2015. United Nations 7 Mar. 2015. http://www.un.org/en/development/desa/policy/wesp/wesp_archive/2015wesp_full_en.pdf.

1.1.2 Political Developments

1.1.2.1 Geopolitics

The year 2014 saw its own share of significant world events, some of which remained unresolved by year's end. In the Middle East Abdel Fattah el-Sisi's success in the May 2014 Egyptian presidential elections ushered in increased military and political relations with Russia and deteriorating relations with the United States; in Syria and Iraq the emergence of the 'Islamic State' (IS or ISIS) brought renewed concern to the region. In contrast to the Arab Spring that arose through political demonstrations and unrest within the Arab world, ISIS had already existed as 'al-Qaeda in Iraq' and gained notoriety when it began to send forces to fight in Syria's civil war. Denounced even by al-Qaeda for its brutality toward its enemies, its scale as an international threat remains uncertain. While the U.S. has sent troops to Iraq to advise its local army on how to regain territory, and the U.S. and Iran have launched airstrikes on IS territory to slow the growth of the group, the IS continues to draw in foreign jihadists (including some Europeans and Americans), and pledges of allegiances from other jihadi groups.[8]

In Iran, President Hassan Rouhani was unable to rally the country to scale back its nuclear ambitions in exchange for the rollback of Western sanctions. In January 2014 a 6-month joint plan of action between Iran and the U.S., UK, France, Germany, Russia, and China, was launched in an attempt to conclude the negotiations process, yet an agreement was not reached by July—nor was one reached in November, following a 4 month extension in talks. A new deadline was set for 1 July 2015.[9]

Russia's annexation of Crimea significantly impacted EU-Russia relations, especially following the downing of Malaysia Airlines passenger Flight 17 over rebel-held territory in Ukraine. Tensions began at the end of 2013, when the pro-Russia former Ukrainian President Viktor Yanukovich decided not to sign a much anticipated trade deal with the EU. Protests continued into February 2014, when Yanukovich resigned from office and fled the country, and pro-Russian militants seized the Crimean capital in return. Following a questionable referendum, wherein 95 % of Crimeans voted in favour of rejoining Russia, by May pro-Russian separatists in other regions of eastern Ukraine declared independence, and held their own elections; while the rest of the country elected Petro Poroshenko as the pro-Western Ukrainian president. Following the downing of the Malaysia Airlines flight on 17 July 2014, the EU and United States responded by increasing sanctions against Russia, which did not do much to abate tensions by the end of the year.[10]

---

[8] Lindsay, James M. "Top Ten Most Significant World Events in 2014." 15 Dec. 2014. Council on Foreign Relations 10 Mar. 2015. http://blogs.cfr.org/lindsay/2014/12/15/top-ten-most-significant-world-events-in-2014/.

[9] Ibid.

[10] Ibid.

In West Africa, Ebola outbreaks in Liberia, Guinea, and Sierra Leone resulted in a death toll of over 6000 people by the end of 2014. While several outbreaks had occurred since the virus' discovery in 1976, those deaths ranged between several dozen to hundreds, having emerged in less densely populated areas. The opposite was the case this time; and with the inadequate, slow and uneven international response, some isolated cases reached the U.S. and Europe. With economic growth rapidly decreasing, and food shortages becoming common, there was concern that the economic costs of the Ebola outbreak in West Africa would persist into the future.[11]

In Europe, austerity measures accompanying the financial crisis drove a wedge between Euro zone members in the north and south. And while there was some sign of growth, its impact on the high unemployment rates in the countries of the 'periphery' was minimal—youth unemployment rates in particular remained very high, more than 50 % in Greece and Spain.[12]

1.1.2.2 Environment

Space applications have an important role in the monitoring and protection of the environment. Space assets are uniquely positioned to offer a global perspective on climate change. They help to better manage disaster situations around the world, and are often a common multinational platform for collecting relevant meteorological and environmental data. These characteristics make them ideal promoters of international understanding and cooperation in this field. Satellite based systems are being used to gather information on climate change indicators e.g. the melting of the ice-caps, changes in the global sea level, and gathering data on the regions most affected by global warming. Remote sensing technologies can also be used to monitor deforestation and land use, and are important for better utilization of fresh water sources. There is no doubt that space technologies will play an important role in human and environmental security in the future, hence technical development of their capabilities is necessary.

Climate change remains one of the commonly recognized agenda topics within the global political debate. In the days leading up to the 20th UN Framework Convention on Climate Change Conference of Parties (UN FCCC/COP) which took place in Lima, Peru, from 1 to 12 December 2014, the EU had pushed for the legally binding mitigation target of a 40 % reduction in emissions by 2030; whereas the U.S. preferred a more buffet-like approach that included some binding elements, but allowed countries to determine the scale and pace of their emissions reductions even if this could mean that the aim of keeping global temperature rises

[11] Ibid.

[12] "Youth Unemployment Could Prolong Eurozone Crisis, Christine Lagarde Says." 10 Dec. 2013. The Guardian 19 May 2014. http://www.theguardian.com/business/2013/dec/10/youth-unemployment-eurozone-crisis-christine-lagarde-imf.

below 2 °C would not be met.[13] However, rather than establishing a new global agreement on climate change, the Lima meeting provided a set of procedural steps, which consisted of loose arrangements for parties bringing forward their intended nationally determined contributions (INDCs); these INDCs and the remaining issues and options put forward by the parties will be included in the elements for a draft negotiating text that will be produced mid-2015. Containing the raw material for the next meeting, the elements paper covers such issues as mitigation, adaptation, finance, technology transfer, transparency, the legal nature of parties' commitments, the use of market mechanisms, and procedures to periodically update commitments. However, amid the compromises being made by industrialized and emerging nations, unresolved major issues remained by the end of the conference, such as whether INDC's are to focus only on mitigation, or also adaptation and finance, and whether quantifiable information was expected on a voluntary or mandatory basis. Moreover, the method for assessing the aggregate effect of the INDCs, and whether the responsibilities of and financing from industrialized and emerging nations should remain differentiated, remained points of discussion.[14]

The Lima conference featured the first "multilateral assessment" of mitigation efforts by 17 industrialized countries, as part of the transparency procedures established under the Cancun Agreements of 2010; showing their progress toward achieving the 2020 emission pledges as agreed in the "Copenhagen Accord" of 2009. Moreover, by the end of the Lima conference, the Green Climate Fund (GCF) established under the Cancun Agreements of 2010 surpassed an informal initial goal of $10 billion in pledges; at the Cancun conference, industrialized nations committed to mobilizing $100 billion per year in public and private finance by 2020, to assist developing countries in handling the effects of global warming and climate change. And in terms of aid to developing countries for "loss and damage" resulting from climate change, the meeting decided on the composition of the executive committee for the Warsaw International Mechanism for Loss and Damage, established in the Warsaw conference of 2013, and adopted an initial 2-year work plan to mitigate climate impact and risks.

Aside from increased transparency through the development of an online Information Hub on the UN FCCC website, not much progress was made regarding the UN Programme on Reducing Emissions from Deforestation and Forest Degradation in Developing Countries (UN REDD), where governments had previously agreed on a set of measures to reduce emissions from deforestation and forest degradation.

The new global agreement on climate change is expected to be concluded at the 21st UN FCCC/COP in Paris, France, running from 30 November to 11 December 2015.[15]

[13] Nelsen, Arthur. "UN climate change deal must have legally binding targets, says EU." 27 Nov. 2014. The Guardian 31 Mar. 2015. http://www.theguardian.com/environment/2014/nov/27/un-climate-change-deal-must-have-legally-binding-targets-says-eu.

[14] "Outcomes of the U.N. Climate Change Conference in Lima." Dec. 2014. C2ES.org 31 Mar. 2015. http://www.c2es.org/docUploads/cop-20-summary.pdf.

[15] Ibid.

On 20 November 2013 the European Parliament and European Council adopted a decision to establish the EU's 7th Environmental Action Programme (EAP) (running until 2020) "living well, within the limits of our planet", which entered into force in January 2014. It identifies three key objectives: e.g. protection, conservation and enhancement of the EU's natural capital; turning the EU into a resource-efficient, green, and competitive low-carbon economy; and safeguarding the Union's citizens from environment-related pressures and risks to health and wellbeing. It also lists four 'enablers' that will facilitate these goals: better implementation of legislation; better information by improving the knowledge base; increased and wiser investment for environment and climate policy; and full integration of environmental requirements and considerations into other policies. Its two overarching horizontal priority objectives are to make the Union's cities more sustainable, and to help the Union address international environmental and climate challenges more effectively.[16]

1.1.2.3 Energy

Instability in parts of the Middle East, the major source of low-cost oil, and tension between Russia and Ukraine, sparked concerns over natural gas security, demonstrating that the energy system may be in danger of falling short of meeting global energy needs in the coming decades. While global energy demand should grow by 37 % by 2040, the change in the distribution of that demand will be more dramatic, with consumption increasing substantially in China and the rest of Asia (60 % of the global total), along with other emerging regions in Africa, the Middle East, and South America. On the other hand, energy demand will remain flat at current rates in Europe, Japan, South Korea, and North America.[17]

China is expected to become the largest oil-consuming country by the 2030s, surpassing the United States whose own oil production and consumption will start to fall back in the 2020s. As oil demand will increase in emerging countries, the turmoil in the Middle East is a major concern given the increasing reliance on this region for oil production growth. Additionally, the long-term supply of natural gas is another concern, with the key uncertainty being whether it can be produced at an attractive price to consumers across Asia and Europe, while still providing an incentive to invest in gas exploration and exploitation. An investment of around $900 billion per year in upstream oil and gas production is needed by the 2030s to meet the projected demand.[18]

[16] European Union. Decision of the European Parliament and of the Council on a General Union Environment Action Programmes to 2020 "Living Well, Within The Limits Of Our Planet". 20 Nov. 2013, European Parliament and Council Decision PE-CONS 64/1/13 REV 1 of 20 November 2013. Strasbourg: European Union. http://ec.europa.eu/environment/newprg/pdf/PE00064_en.pdf.

[17] International Energy Agency. World Energy Outlook 2014—Executive Summary. IAE: Paris, 2014.

[18] Ibid.

On the other hand, coal production will remain abundant, constrained mainly by measures to tackle pollution and reduce CO2 emissions. India is expected to surpass the United States as the world's second biggest coal importer before 2020, and will overtake China shortly thereafter. Relieving pressure on energy supplies and mitigating the impact of price disparities between regions will require a focus on energy efficiency. In 2013, fossil-fuel subsidies totalled $550 billion, whereas subsidies for renewable energy were nearly a quarter of that amount, i.e. $120 billion. According to the United Nations Environment Programme report 'Global Trends in Renewable Energy Investment 2015', investment in developing countries rose by 36 % to $131.3 billion in 2014, with China investing $83.3 billion (up 39 % from 2013), while Brazil ($7.6 billion), India ($7.4 billion), and South Africa ($5.5 billion) were also among the top 10 of investing countries. Additionally, more than $1 billion was invested in Indonesia, Chile, Mexico, Kenya and Turkey. On the other hand, in industrialized countries, the U.S. invested $38.3 billion (up 7 % from 2013), while Europe overall rose by less than 1 % to $57.5 billion, and Japan rose by 10 % to $35.7 billion.[19] By 2040, the world energy supply mix is expected to be split almost evenly between oil, gas, coal, and renewable/low-carbon sources.[20]

1.1.2.4 Resources

Space applications and Earth monitoring technologies play an important role in the area of resource management, as they can provide better control of and support for the utilisation of scarce natural resources. Likewise, satellite based technologies perform indispensable tasks for international trade, e.g. by streamlining global business transactions and payments. Global navigation satellite systems (GNSS) are already relied on as an integral part of transportation and utilisation of natural, agricultural and industrial resources. And the use of meteorological and imaging satellites is making agricultural output bigger and more reliable, along with greater precision. For many developing countries the rationale for investment in space is improvement of the management of their agricultural and natural resources.

The growth of international trade continued at a rate of 2.2 % in 2013, remaining subdued from the 13.8 % growth experienced in 2010, and was expected to remain between 2.5 % and 3 % during 2014.[21] Moreover, the volume of trade in nearly all regions appeared to have decelerated, excluding some developing countries in Asia, and Sub-Saharan Africa, with varying impact on developed, developing and transitioning economies. Developed economies experienced the greatest slowdown, with 0.4 % reduction of imports in 2013, primarily the result of weak intra-EU

---

[19] "Global Trends in Renewable Energy Investment 2015." 31 Mar. 2015. FS-UNEP 26 Aug. 2015. http://fs-unep-centre.org/publications/global-trends-renewable-energy-investment-2015.

[20] International Energy Agency. World Energy Outlook 2014—Executive Summary. IAE: Paris, 2014.

[21] According to the latest data available from UNCTAD.

trade, along with continued contractions in Japan and the U.S. However, EU exports to countries outside the region resulted in a 1.4 % growth increase, whereas growth in U.S. exports shrank to 2.6 %, while Japan's exports further contracted by 1.8 %. Another example was the situation in the economies in transition, where a slowdown was even more noticeable, due mainly to weak European demand for their exports, and a slump in their own GDP growth, resulting in the halving of the growth rate of imports. Growth in exports in transition economies had slowed even further from 1.3 % in 2012 to 1.0 % in 2013, while imports dropped from 5.0 % in 2012 to 2.7 % in 2013. Developing economies also experienced further weakened growth in exports for 2013, dropping to 3.4 % from 4.6 % in 2012, mainly due to weak external demand from developed countries. However, growth in developing countries' imports remained resilient in 2013, increasing to 5.5 % from 5.3 % in 2012, due to increased demand from some of the larger economies.[22]

In mid-2014, crude oil prices were similar to previous years.[23] According to the United Nations Conference on Trade and Development (UNCTAD), crude oil market prices remained high between the second half of 2013 and the first half of 2014, with prices fluctuating in a $102.3–$108.8 band, and averaging $105.5 per barrel for the 12-month period. However, there were signs of increased volatility in oil prices in the second quarter of 2014 as geopolitical tensions between Ukraine and Russia continued to escalate, raising risk perceptions in energy markets. However, as there was no significant disruption in production associated with those rising tensions by mid-2014, the volatility can be attributed more to market sentiment than to real production effects. In fact, the oil market was well supplied in this period mainly due to increased production by the U.S. linked to its shale oil and gas boom, and other large non-OPEC country supplies that had helped to buffer oil supply disruptions that had occurred in other oil producing countries such as Iran, Libya, Nigeria, and South Sudan. And while growth in the oil trade decreased by roughly 1 % in 2013, it marginally increased again by 1 % in the first half of 2014.[24]

Yet crude oil prices dropped radically in the second half of 2014, reaching the nadir of $63.72 (Brent) and $67.53 (West Texas Intermediate) on 1 December 2014.[25] Oil prices were clearly dampened by the November 2014 decision of the Organization of the Petroleum Exporting Countries (OPEC) to keep the market oversupplied, in order to protect market share against that of shale oil. By the end of the 2014, the price of oil had thus dropped by nearly 40 percent compared to mid-2014.[26]

---

[22] United Nations Conference on Trade and Development. Trade and Development Report, 2014. Geneva: UNCTAD, 2014. 1-14. 2014 data not yet available.

[23] Based on the United Nations Conference on Trade and Development (UNCTAD) mid-year reporting period.

[24] Ibid. at 8–10.

[25] "Oil prices fall to near five-year low again." 5 Dec. 2014. The Financial Times 26 Aug. 2015. http://www.ft.com/intl/fastft/246631/oil-prices-fall-near-five-year-low-again.

[26] Tully, Andy. "OPEC Fires First Shot In Global Oil Price War." 30 Nov. 2014. OilPrice.com 26 Aug. 2015. http://oilprice.com/Energy/Oil-Prices/OPEC-Fires-First-Shot-In-Global-Oil-Price-War.html.

As in previous years there continued to be a high degree of volatility in non-oil commodity prices, which were strongly dependent on variable weather conditions that influenced the growth or stagnation in production of certain commodities. In general, there was continued growth in demand for commodities, at even step with the sluggish growth of the world economy. Moreover demand coming from rapidly growing developing countries such as China continued to increase, due partly to government stimulus measures. Nevertheless, the percentage change of non-oil commodity prices appeared to be bottoming out, following an 8.3 % reduction in 2012, and a 6.7 % reduction in 2013, it showed a 3.9 % reduction in 2014.[27] Non-oil commodities prices were forecast to begin increasing by 2016.[28] As in previous years, uncertainty and instability were the major distinguishing features of commodity markets, which was also reflected in the greater volatility of commodity prices. However, from 2012 to 2014, most commodity prices stayed at substantially higher levels than the average levels between 2003 to 2008, and many commodity prices were still at levels close to their peaks prior to the financial crisis. Nevertheless, in line with previous years, vegetable oil seeds and oils, agricultural raw materials, minerals and metals and crude petroleum all showed higher levels of volatility.[29] Whereas overall, metals and minerals prices decreased by 6.8 % for the reporting period, nickel and zinc were the exception with growth of 10.2 % and 7.4 % respectively.[30] The indexes for gold, silver, and platinum declined for a second year, by 10.3 %, 19.7 %, and 6.9 % respectively in 2014.[31]

1.1.2.5 Knowledge

There is no doubt that sustained education and knowledge improvement is one of the necessary conditions for successful space activities, as well as for the full exploitation of their societal benefits. In general, space technology and development, drawing on multiple scientific disciplines, is one of the most difficult and challenging fields in scientific and technical research. Therefore, coherent and sustainable strategies aimed at improving higher education and supporting technical and scientific activities are particularly relevant and necessary for space sector activities.

For Europe, the expansion of its pool of highly skilled and specialised scientists and professionals should be a constant priority if it is to remain a leading actor in the field of space-related scientific and technological R&D. With other developed and emerging

[27] United Nations Conference on Trade and Development. Trade and Development Report, 2014. Geneva: UNCTAD, 2014. 10.

[28] The World Bank. Global Economic Prospects—Coping with policy normalization in high-income countries. Volume 8 / January 2014. Washington DC: World Bank, 2014.

[29] United Nations Conference on Trade and Development. Trade and Development Report, 2014. Geneva: UNCTAD, 2014. 10.

[30] Ibid.

[31] World Bank Group. Commodity Markets Outlook. January 2015. Washington, DC: World Bank, 2015: 23.

economies increasing investment in R&D in response to the financial crisis, for Europe to remain a leader in the global race for knowledge and excellence in space R&D, current levels of financial spending and political commitment in this area will have to be increased, and not only simply maintained. For example, in 2014 30 % of the European working age population (not including Bulgaria, Cyprus, Malta, Latvia, Lithuania, Romania, and Croatia) held a higher education degree compared to 43 % in the U.S., 47 % in Japan and 53 % in Canada.[32] While OECD figures tend to show that many European states can expect a significant increase in the proportion of their population that attains a higher education degree, countries including Spain, France, and Germany, and to a lesser extent Portugal, Austria, Slovakia, Hungary, Czech Republic, and Italy are at risk of falling further behind the OECD average of 33 % of the working age population (between 25-to-64 years of age) with a higher education degree.[33]

Typically, employment rates are highest among people with high qualifications, while people with the lowest educational qualifications are at greater risk of being unemployed.[34] According to the European Centre for the Development of Vocational Training, while employment is projected to grow by 2 % over the period 2013–2020, the European skills forecast indicates that the EU will miss its target of reaching 75 % employment by 2020, reaching instead 68.5 % due to a weak demand for labour following the financial crisis and the modest recovery and average GDP growth rates. Prior to the financial crisis, between 2000 and 2008, the employment rate increased by 3.7 %. Between 2013 and 2020, the number of high qualification job openings should exceed 55,000 (combining replacement needs and expansion demand), whereas medium qualification openings will be over 40,000 (coming mainly from replacement needs), and low qualification jobs will be lower than 10,000 (due to shrinking demand offsetting replacement needs). In fact, the overall share of the labour force with low qualifications is forecast to fall from 22 % in 2013 to 16.8 % in 2020. And while the EU is on course to surpass its education goal of 40 % of 30–34 year-olds having completed higher education by 2020, possibly reaching 45 % by that time, weak high-skilled labour demand could mean that there will be an increased risk of skill mismatch due to the over-qualification of highly skilled labour that has no alternative than to accept lower skilled employment.[35]

Encouragingly, the population of university and higher education students in Europe has been constantly increasing in recent years. There are around 4000 universities and other kinds of higher education institutions in Europe, with over 20 million students in 2011.[36] Unfortunately, this quantitative increase has not been

---

[32] OECD. Education at a Glance 2014: OECD Indicators, OECD Publishing, 2014: 44. http://www.oecd.org/edu/Education-at-a-Glance-2014.pdf.

[33] Ibid. at 46.

[34] Ibid. at 102.

[35] "Cedefop's latest skill supply and demand forecasts highlight Europe's employment challenge." 1 Apr. 2014. CEDEFOP 7 May 2015. http://www.cedefop.europa.eu/node/12684.

[36] "Tertiary education statistics | Data from September 2013." Eurostat Statistics Explained 8 May 2015. http://ec.europa.eu/eurostat/statistics-explained/index.php/Tertiary_education_statistics.

accompanied by qualitative improvements in governance structures and proposed academic curricula, or by increases in funding. Although increasing in size, Europe's higher education system has not yet achieved an academic curricula distribution that will train scientists and professionals with the right kinds of skills to support economic growth and scientific excellence in new technologies. This is especially true for the space sector, which has a relatively limited human resources supply and demand chain.[37]

According to a report to the European Commission on new modes of learning and teaching in higher education, there remains a culture of conservatism within European higher education that impedes its potential to fulfil its role in society and contribute to Europe's prosperity. While an uncoordinated bottom-up approach has been used to a large degree to spur a broad range of good practices across Europe, there is a need for governments and institutions to develop comprehensive strategies at both the national and institutional level for the adoption of new modes of learning and teaching within higher education. New models of provision, such as open online courses, provide the opportunity for lifelong learning, continuing professional development and internationalisation, and also provide the ability to collect and analyse learner data that could allow for personalised learning and enhanced retention by students.[38] Another form of learning beyond the traditional lecture hall that is being utilized in the European space sector comes in the form of industry school labs, such as the DLR's School Lab programme in several universities throughout Germany, which has helped in the long-term furtherance of interest in the natural sciences.[39]

The EU budget dedicated to funding programmes for education, training, youth, and sport was about 14.774 billion euros for the period 2014–2020, which means an increase of 67.9 % compared to the 2007–2013 period.[40] These programmes feature three key actions which address: (1) the learning mobility of individuals; (2) cooperation for innovation and the exchange of good practices; and (3) support for policy reform. The main activity of the first key action will involve youth exchanges, and European voluntary services. The second key action will focus on strategic partnerships for the purpose of implementing innovative practices that lead to high quality teaching, training, learning and youth work, institutional modernisation and societal innovation, and capacity building to foster cooperation and exchanges in the field of youth between participating countries. The third key action will be in the form of a structured dialogue in various meetings between

---

[37] Europe 2020 Target: Tertiary Education Attainment. Europa.eu 11 May 2014. http://ec.europa.eu/europe2020/pdf/themes/28_tertiary_education.pdf.

[38] High Level Group on the Modernisation of Higher Education. "REPORT TO THE EUROPEAN COMMISSION ON New modes of learning and teaching in higher education." Oct. 2014. Europa.eu 8 May 2015. http://ec.europa.eu/education/library/reports/modernisation-universities_en.pdf.

[39] "Study confirms long-term furtherance of the interest in the natural sciences." 4 Oct. 2010. DLR 8 May 2015. http://www.dlr.de/schoollab/en/desktopdefault.aspx/tabid-1708/13696_read-34709/.

[40] "Erasmus+ | The new EU programme for Education, Training, Youth, and Sport for 2014-2020." Europa 8 May 2015. http://ec.europa.eu/programmes/erasmus.-plus/discover/index_en.htm.

young people and decision-makers in the field of youth.[41] The European Commission has also proposed a 46 % rise in research funding under the planned Horizon 2020 strategy that will bring funding up to 80 billion euros.[42]

1.1.2.6 Mobility

Maritime commerce accounts for the bulk of global trade, whereas air traffic carries most of the world's passenger traffic. Space assets are indispensable to both, as they provide meteorological, navigation and communication services that make sea and air transport safer and cheaper. Maritime navigation and mobile communications are two of the upcoming sectors with significant financial interests in the development of new generations of satellite-based applications.

Maritime transport is the most commonly used form of transport for international trade, representing about 80 % of global merchandise trade. With prospects for the world economy, along with trade and shipping, possibly slowly improving, there are still risks in the fragile recovery of developed economies, especially in the event that geopolitical tensions escalate. By 2013, vessel orders had stopped their downward trend, marking a tentative end to the significant order decline that had followed the financial crisis and terminating one of the longest uninterrupted growth periods in recorded maritime history. In early 2014, order books showed slight increases in terms of bulk carriers, tankers, and container vessels; however the general cargo ship category continued diminishing. Even with the small uptick in orders, there was still a high number of ship demolitions (29,052)—yet, total demolitions in 2013 were 20.0 % lower than in the previous year (36,293). In 2013, dry-bulk carriers retained the greatest share of the total quantity of tonnage sold for demolition at 44 %, followed by oil tankers at 20 %, and container ships at 18 %.[43] However, while new ship orders were made in 2014, it will take several years before the maritime industry enters a new shipbuilding cycle.

Another challenge for the maritime industry is the heightened exposure and vulnerability of international shipping to piracy, armed robbery, and other crimes. Between 2003 and 2012, around 3436 acts of piracy were reported worldwide.[44] And in that period, piracy incidents in East Africa, particularly off the Somalia

---

[41] "Erasmus.+ | The new EU programme for Education, Training, Youth, and Sport for 2014-2020." 24 July 2014. SALTO-YOUTH EASTERN EUROPE AND CAUCASUS RESOURCE CENTRE 8 May 2015. http://msy.gov.ge/files/EaP%20Conference%20Tbilisi%202014/July%202/SALTO_presentation.pdf.

[42] "An EU Strategy for Modernising Higher Education—Questions and Answers." 20 Sept. 2011. European Commission—Press Release. 1 Mar. 2012. http://europa.eu/rapid/pressReleasesAction.do?reference=MEMO/11/615&format=HTML&aged=0&language=EN&guiLanguage=fr.

[43] United Nations Conference on Trade and Development. Review of Maritime Transport 2014. Geneva: UNCTD, 2014. 46.

[44] United Nations Conference on Trade and Development. Maritime Piracy | Part 1: An Overview of Trends, Costs and Trade-related Implications. Geneva: UNCTAD, 2014. 3.

coast, the Gulf of Aden and the Indian Ocean were particularly alarming, especially since they exceeded those in some of the traditional regions for piracy including Indonesia, Malaysia, Singapore, and the Philippines. Indeed, between 2005 and 2012, piracy off the coast of Somalia accounted for nearly 50 % of all recorded hijackings. This spurred an increased international military presence in this region, and preventative measures by merchant ships that have privately contracted armed personnel, reduced Africa's share of piracy incidents from 50 % in 2008 to 17.3 % in 2012.[45] This downward trend has continued, with the number of reported piracy incidents in 2013 having decreased by 11.1 % to 264, followed by another reduction of 7.2 % to 245 in 2014. Thus, the number of Somalia-based piracy incidents had fallen from 49 in 2012 to 3 in 2014, although it continued to increase in East Asian waters.[46] In its heyday, the surge in piracy in the Gulf of Aden generated considerable costs especially for Europe as 80 % of shipments that pass through the area come from or are going to this continent. Re-routing shipments around the Cape of Good Hope alone is estimated to have created over $7.5 billion of additional shipping costs annually with associated increased energy consumption.

1.1.3 Main Science and Technology Indicators Relevant for Space Activities

The space sector requires the continued input of cutting-edge technologies, facilitated by a public sector that occasionally bets against unfavourable financial market conditions. In order to compete on a world scale while mitigating uncertain market conditions, a web of private and public technology development schemes interact with and complement each other at various levels and to different extents in the space industry and associated business domains. This interdependence likewise exists at the European Union level where knowledge and innovation are the key drivers of future economic and societal development in a broader context. This is also reflected in the continued development of the European Research Area (ERA), which consists of activities, programmes and policies that are designed and operated at regional, national, and European levels, and in which the space sector occupies an important place. Technological development and innovation is in fact a quantifiable activity, based on its input and output.[47]

---

[45] Ibid.

[46] ICC International Maritime Bureau. Piracy and Armed Robbery Against Ships: Report for the Period 1January—31 December 2014. Jan. 2015. ICC-CSS 11 May 2015. http://www.hellenicshippingnews.com/wp-content/uploads/2015/01/2014-Annual-IMB-Piracy-Report-ABRIDGED.pdf.

[47] Godin, Benoit. "Science, Accounting and Statistics: the Input-Output Framework." Canadian Social Sciences and Humanities Research Council (2005): 5. http://www.csiic.ca/PDF/Godin_31.pdf.

1.1.3.1 Science and Technology Inputs

Science and technology inputs at the European level are constantly measured against the benefits they generate. In 2014, R&D investment continued to push for increased efficiency and effectiveness and stronger R&D integration in broader industrial and macro-economic policies. The result in the reporting period is measured by gross domestic expenditure on R&D (GERD), a statistical tool showing nominal changes in those expenditures.

According to the latest values published by Eurostat, the overall GERD as a percentage of GDP reached 2.02 % in 2013, from 2.01 % in 2012.[48] While the EU's 2013 GDP increased to 13.52 trillion euros in current price terms, its 2013 GERD was 273.46 billion euros, amounting to a 1.5 % increase in GERD from the previous year[49]; in 2012, EU GERD had grown by 3.9 %. In contrast, this performance remained below Japan's and the United States' indicators whose data available remains limited to 2011 and 2012 respectively. Yet there are signs that the intensity of investment of those frontrunners might be diminishing. Japan's 2011 GERD was estimated to be 3.38 %, amounting to about 143.7 billion euros; and the U.S. saw an increase in its GERD from 2.77 % in 2011 to 2.81 % in 2012, amounting to around 353.01 billion euros. On the European national level, while the top three countries in terms of GERD maintained their positions in 2013, they converged somewhat with Finland (3.32 %), Sweden (3.21 %) and Denmark (3.05 %); while in 2013 ten European Member States, mainly in Central and Eastern Europe, continued to have expenditures below 1 %.[50]

A further breakdown of statistical data helps to underscore the impact of private R&D investment in Europe, when paired to publicly funded projects. In 2013, countries with higher public investment in R&D tended to show a generally high level of innovation investment within the private business sector, the top three again being Finland (2.29 %), Sweden (2.19 %) and Denmark (1.99 %). From the data provided, countries such as Germany and France, with larger GDPs tended to field similarly large government led investments in R&D, mostly through higher education research. Overall, private sector funded R&D accounted for 63.9 % of the EU's total GERD, whereas Japan's private sector funded R&D accounted for 76.9 % in

---

[48] "Gross domestic expenditure on R&D (GERD)—% of GDP." Eurostat 28 Mar. 2015. http://ec.europa.eu/eurostat/tgm/table.do?tab=table&init=1&language=en&pcode=t2020_20&plugin=1; *see also* "Total intramural R&D expenditure (GERD) by sectors of performance—per Million euro." 17 Nov. 2014. European Commission—eurostat 28 Mar. 2015. http://appsso.eurostat.ec.europa.eu/nui/submitViewTableAction.do.

[49] "Gross domestic product at market prices | At current prices." Eurostat 28 Mar. 2015. http://ec.europa.eu/eurostat/tgm/refreshTableAction.do?tab=table&plugin=1&pcode=tec00001&language=en; Total intramural R&D expenditure (GERD) by sectors of performance—per Million euro." 17 Nov. 2014. European Commission—eurostat 28 Mar. 2015. http://appsso.eurostat.ec.europa.eu/nui/submitViewTableAction.do

[50] "Gross domestic expenditure on R&D (GERD)—% of GDP." Eurostat 28 Mar. 2015. http://ec.europa.eu/eurostat/tgm/table.do?tab=table&init=1&language=en&pcode=t2020_20&plugin=1.

2011 and the United States rested at 69.8 % in 2012. The proportion of government expenditures in this field also varied significantly from one European region to the other, with the highest levels observed in Central and Eastern European countries.[51] The mixed approaches to R&D funding remain a challenge when planning appropriate policies at the European level.

1.1.3.2 Science and Technology Outputs

In contrast to R&D input, the outputs achieved from investments in innovation are harder to measure, due to the complex market dynamics within the private sector. While innovation offers a competitive edge to companies that are willing to invest in and develop R&D projects, statistics such as the number of patents lodged by country, industrial sector and individual companies are merely indicative of output, as their effect on financial performance is also heavily dependent on external market competition and other factors. Based on the European Commission's 2014 EU Industrial R&D Investment Scoreboard, in 2013 approximately 2500 companies worldwide each invested more than 15.5 million euros in R&D with growth slowed down across the scoreboard from 2012.[52] Total investment in R&D by leading companies amounted to 538.5 billion euros; i.e. European companies accounted for 30.2 % (162.4 billion euros) of that figure. Together with the U.S. at 36.0 % and Japan at 15.9 %, they constituted 82.1 % of the companies surveyed. Companies realise they can gain an edge by investing heavily in R&D when operating in highly competitive environments, as this action appears to have a linear link between investment and market performance. This trend might be the case with top performing companies, but it is highly dependent on the nature of the market, and would likely not hold true in every sector. What is clear however is that in 2012 the EU experienced increased growth (6.3 %) and that following the 2008 financial crisis the 2000 companies surveyed began increasing their investments in R&D by an average of 6.2 % on a yearly basis, underlining the strategic importance of R&D during times of economic uncertainty.[53]

The uneven benefits of corporate innovation on a European scale are further demonstrated in the European Commission's biennial competitiveness assessment conferences, CONCORDi. The 4th CONCORDi meeting, held on 26–27 September 2013, focussed on financing innovation for company growth, and included themes on financial sources, growth strategies, and the role of public policies. It was concluded that financial constraints, such as internal lack of liquidity paired with

---

[51] Ibid.

[52] According to the latest data available.

[53] European Commission. "EU R&D Scoreboard | The 2014 EU Industrial R&D Investment Scoreboard." 16 Jan. 2015. EU R&D Scoreboard 12 Aug. 2015. http://iri.jrc.ec.europa.eu/scoreboard14.html; *see also* European Commission. "EU companies must boost R&D investment to stay globally competitive." 4 Dec. 2014. Europa—Press releases database 12 Aug. 2015. http://europa.eu/rapid/press-release_IP-14-2342_en.htm.

high competition and lack of demand, are a major obstacle to innovation. Here, the reduction of information asymmetries can considerably lower the barriers to access financial resources, since open financial disclosures, patenting, and publication are positive information signals to investors, and have been proven to positively affect innovation and the growth rate. Access to funding can be supplemented through bank loans, venture capital investment, and perhaps crowd-funded external financing mechanisms. Here, a balance in financing needs to be established, since financial institutions might not have the scope and financial products to deal with long term investment in young firms, and venture capitalists might not see the desired short term returns. Remedial policy measures should be accurately designed and coordinated to fully utilize complementarities and synergies to maximize their effectiveness, considering that the impact of policies may depend on the size of firm, its growth phase, differences in low- and high- tech industry, and whether additional characteristics are to be taken into account. Especially in the space sector it would seem that innovation-fuelled corporate performance greatly depends on the market sector's level of demand for cutting-edge technology. In general, the more high-tech a sector is, the greater and more robust the investment returns that can be expected from innovation. European policy reforms to increase efficiency and effectiveness began in 2011; that said, at a European level any horizontal measures taken to improve competitiveness through innovation would have to be accompanied by appropriate custom-made policies for the different market segments, corporate sizes and geographical regions of Europe.[54] The 5th biennial CONCORDi meeting, to be held on 1–2 October 2015 in Seville, Spain, will cover the role of scientific analyses in the design and implementation of an EU industrial policy based on research, innovation and technological advancement, taking place against the policy backdrop of a renewed agenda to boost jobs, growth and investment.[55]

1.2 Worldwide Space Policies and Strategies

1.2.1 The United Nations System

Various institutions within or associated with the United Nations are relevant for space policy. In this section, the UN General Assembly, its Committees, and other UN bodies and organs that deal with space activities are discussed.

[54] European Commission Joint Research Centre, Institute for Prospective Technological Studies. CONCORD-2013—4th European Conference on Corporate R&D and Innovation: Financing R&D and innovation for corporate growth in the EU: Strategies, drivers and barriers—Summary Report. Dec. 2013. Seville, Spain. JRC Scientific and Technical Reports, 2013.

[55] "5th European Conference on Corporate R&D and Innovation CONCORDi 2015." IRI 14 Apr. 2015. http://iri.jrc.ec.europa.eu/concord/2015/index.html.

1.2.1.1 United Nations General Assembly

The United Nations General Assembly (UNGA) passed four Resolutions in its 69th (2014–2015) session pertaining to the use and exploration of outer space. The first was Resolution A/RES/69/31 adopted on 2 December 2014, on the 'Prevention of an arms race in outer space'.[56] The second, adopted on the same day, was Resolution A/RES/69/32 on 'No first placement of weapons in outer space'.[57] The third resolution, also adopted on 2 December, was Resolution A/RES/69/38 on 'Transparency and confidence-building measures in outer space activities'.[58] On 5 December 2014, the UNGA adopted its annual Resolution A/RES/69/85 on "International cooperation in the peaceful uses of outer space".[59]

First, Resolution A/RES/69/31 on the 'Prevention of an arms race in outer space' echoed similar resolutions tabled in previous years. The instrument reaffirmed the importance of transparency and confidence building measures (TCBM) to avoid an arms race in space and additionally recognised, as stated in the report of the Ad Hoc Committee on the Prevention of an Arms Race in Outer Space, that the legal regime applicable to outer space by itself does not guarantee the prevention of an arms race in outer space, whilst acknowledging that the regime plays a significant role in the prevention of an arms race in that environment, and that there is a need to consolidate and reinforce the regime and enhance its effectiveness. It also recognized the importance of strict compliance with existing bilateral and multilateral agreements. The Conference on Disarmament (CD) was seen as the relevant multilateral disarmament forum for further discussions, and the instrument noted the updated 2014 version of the draft treaty on Treaty on Prevention of the Placement of Weapons in Outer Space and of the Threat or Use of Force Against Outer Space Objects (PPWT), as introduced by China and the Russian Federation during the CD. In general, it also recognised the complementary nature of multilateral and bilateral efforts in this area.

Similarly, Resolution A/RES/69/32 on 'No first placement of weapons in outer space' built on the Report of the First Committee (A/69/438) on the 'Prevention of

---

[56] United Nations General Assembly. Resolution adopted by the General Assembly—Prevention of an arms race—Sixty-Ninth Session. UN Doc. A/RES/69/31 of 11 Dec. 2014. Vienna: United Nations. http://www.un.org/en/ga/search/view_doc.asp?symbol=A/RES/69/31.

[57] United Nations General Assembly. Resolution adopted by the General Assembly—No first placement of weapons in outer space—Sixty-Ninth Session. UN Doc. A/RES/69/32 of 11 Dec. 2014. Vienna: United Nations. http://www.un.org/en/ga/search/view_doc.asp?symbol=A/RES/69/32.

[58] United Nations General Assembly. Resolution adopted by the General Assembly—Transparency and confidence-building measures in outer space activities—Sixty-Ninth Session. UN Doc. A/RES/69/38 of 11 Dec. 2014. Vienna: United Nations. http://www.un.org/en/ga/search/view_doc.asp?symbol=A/RES/69/38.

[59] United Nations General Assembly. Resolution adopted by the General Assembly—International cooperation in the peaceful uses of outer space—Sixty-Ninth Session. UN Doc. A/RES/69/85 of 16 Dec. 2014. Vienna: United Nations. http://www.un.org/en/ga/search/view_doc.asp?symbol=A/RES/69/85.

an arms race in outer space'.[60] In addition to welcoming the draft treaty on the prevention of the placement of weapons in outer space and of the threat or use of force against outer space objects introduced by China and the Russian Federation at the CD in 2008, and updated in 2014, it noted the importance of the political statements made by several States (i.e. Argentina, Armenia, Belarus, Brazil, Cuba, Indonesia, Kazakhstan, Kyrgyzstan, the Russian Federation, Sri Lanka and Tajikistan) that they would not be the first to place weapons in outer space. In addition to urging an early start of substantive work based on the updated draft treaty, it encouraged all States, especially space-faring nations, to consider the possibility of making a political commitment not to be the first to place weapons in outer space, and decided to include this item in the provisional agenda of its next session.[61]

Next, Resolution A/RES/69/38 on 'Transparency and confidence-building measures in outer space activities' built on the previous 2013 Resolution A/RES/68/50. It reaffirmed the right of free exploration and use of outer space by all states in accordance with international law, and also reaffirmed that the prevention of an arms race in outer space was in the interest of maintaining international peace and security, in addition to fostering international cooperation in the exploration and use of outer space for peaceful purposes.[62] To that end, the resolution encouraged Member States to continue to review and implement, to the greatest extent practicable, the proposed transparency and confidence building measures through relevant national mechanisms. It also decided to convene a joint ad hoc meeting of the Disarmament and International Security Committee (First Committee) and the Special Political and Decolonization Committee (Fourth Committee) to address possible challenges to space security and sustainability, and to include the item as a sub-item within the topic "General and complete disarmament" at the next session.

Resolution A/RES/69/85 on 'International cooperation in the peaceful uses of outer space' updated the similar Resolution from 2013. This year, the devastating effects of infectious diseases (e.g. Ebola virus disease) to the detriment of human life, society and development caused deep concern. Thus the UNGA urged scientific and academic institutions, and the rest of the international community, to undertake studies on the role of tele-epidemiology in monitoring, preparedness and response activities. It also noted the importance of information-sharing in

---

[60] United Nations General Assembly. Report of the First Committee—Prevention of an arms race in outer space—Sixty-Ninth Session. UN Doc. A/69/438 of 12 Nov. 2014. Vienna: United Nations. http://www.un.org/ga/search/view_doc.asp?symbol=A/69/438.

[61] United Nations General Assembly. Resolution adopted by the General Assembly—No first placement of weapons in outer space—Sixty-Ninth Session. UN Doc. A/RES/69/32 of 11 Dec. 2014. Vienna: United Nations. http://www.un.org/en/ga/search/view_doc.asp?symbol=A/RES/69/32.

[62] United Nations General Assembly. Resolution adopted by the General Assembly—Transparency and confidence-building measures in outer space activities—Sixty-Ninth Session. UN Doc. A/RES/69/38 of 11 Dec. 2014. Vienna: United Nations. http://www.un.org/en/ga/search/view_doc.asp?symbol=A/RES/69/38.

discovering, monitoring and physically characterizing potentially hazardous near-Earth objects to ensure that all countries, in particular developing countries with limited capacity in predicting and mitigating a near-Earth object impact, are aware of potential threats; and recognised the central role of the Office for Outer Space Affairs in fostering capacity-building in the use of space science and technology and their applications for the benefit of all countries.[63]

1.2.1.2 UNGA Committees

UNGA has a number of committees that address space policy and associated matters. The main ones are discussed below.

The Disarmament and International Security Committee
The Disarmament and International Security Committee works in close cooperation with the United Nations Disarmament Commission and the Geneva-based Conference on Disarmament to deal with disarmament, global challenges and threats to peace that affect the international community, and seeks solutions to the challenges in the international security regime, including the prevention of an arms race in outer space.[64] On 4 November 2014, the Disarmament and International Security Committee forwarded 57 draft resolutions and 6 draft decisions to UNGA for adoption, with priority attention put on nuclear weapons, although the deliberations on nuclear disarmament continued to be characterised as stalled due to the divergence in views on how to move forward.[65] Yet even with divergent positions, all of the forwarded drafts were ultimately adopted by the General Assembly in December 2014.[66]

The Committee on the Peaceful Uses of Outer Space (COPUOS)
The UN COPUOS activities included the Committee's plenary sessions on 11-20 June 2014, and the annual sessions of its scientific and technical and its legal subcommittees on 10-21 February 2014 and 24 March-4 April 2014, respectively. At the plenary session, the Committee considered ways and means to maintain outer space for peaceful purposes, emphasizing that international, regional and

---

[63] United Nations General Assembly. Resolution adopted by the General Assembly—International cooperation in the peaceful uses of outer space—Sixty-Ninth Session. UN Doc. A/RES/69/85 of 16 Dec. 2014. Vienna: United Nations. http://www.un.org/en/ga/search/view_doc.asp?symbol=A/RES/69/85.

[64] "Disarmament and International Security." General Assembly of the United Nation 3 Apr. 2015. http://www.un.org/en/ga/first/index.shtml.

[65] "Record-Breaking Participation in Debates, Passage of 63 Drafts — Largest Number in Decade — Notes First Committee Chair as Session Draws to Close." 4 Nov. 2014. UN 4 Apr. 2015. http://www.un.org/press/en/2014/gadis3517.doc.htm.

[66] "General Assembly Adopts 63 Drafts on First Committee's Recommendation with Nuclear Disarmament at Core of Several Recorded Votes." 2 Dec. 2014. UN 4 Apr. 2015. http://www.un.org/press/en/2014/ga11593.doc.htm.

interregional cooperation and coordination in the field of space activities were essential for strengthening the peaceful uses of outer space and assisting states in the context of the post-2015 development agenda and in the establishment of new sustainable development goals. It also endorsed the findings and recommendations of its subcommittees from their preceding sessions, and considered space and sustainable development; the spin-off benefits of space technology (review of current status); space and water, and space and climate change; the use of space technology in the United Nations system; the future role of the committee; and other related matters.[67]

At the 2014 scientific and technical subcommittee meeting, discussions and presentations built on a continuing range of space related issues, such as space technology for socioeconomic development in the context of the United Nations Conference on Sustainable Development and the post-2015 development agenda; remote sensing of the Earth by satellite (i.e. applications for developing countries and monitoring of the Earth's environment); space debris and the mitigation measures already implemented by some states; the further development of the space-system-based disaster management support programme (i.e. UN-SPIDER); recent developments in global navigation satellite systems; space weather for the purpose of understanding near-Earth space weather and to communicate that knowledge to interested parties; near-Earth objects; the use of nuclear power sources in outer space; and the long-term sustainability of outer space activities. The subcommittee also examined the physical nature and technical attributes of the geostationary orbit and its utilization and applications, including the field of space communications with particular note of the needs and interests of developing countries. The subcommittee received and considered information provided by the Member States on their activities in all these fields.[68]

The proceedings of the legal subcommittee focussed on exchanging information on the activities of international governmental and non-governmental organizations related to space law, in addition to the status and application of the five United Nations treaties on outer space.

Ongoing issues on the definition and delimitation of outer space were considered particularly with reference to the sometimes conflicting interests of states that desired legal certainty while also insisting on a demonstrable need and practical basis prior to developing a conclusion on the delimitation/definition of outer space. The subcommittee also discussed items on the character and utilization of the geostationary orbit, including consideration of ways and means to ensure the rational and equitable use of the geostationary orbit without prejudice to the role

[67] United Nations General Assembly. Report of the Committee on the Peaceful Uses of Outer Space on its Fifty-seventh session, Held in Vienna from 11-20 June 2014. UN Doc. A/69/20 of 1 July 2014. New York: United Nations. http://www.unoosa.org/pdf/gadocs/A_69_20E.pdf.

[68] United Nations General Assembly. Report of the Scientific and Technical Subcommittee on its fifty-first session, held in Vienna from 10 to 21 February 2014. UN Doc. A/AC.105/1065 of 4 March 2014. Vienna: United Nations. http://www.unoosa.org/pdf/reports/ac105/AC105_1065E.pdf.

of the International Telecommunication Union. National legislation relevant to the peaceful exploration and use of outer space was another agenda item; in addition to capacity-building in space law, considering a review and possible revision of the Principles Relevant to the Use of Nuclear Power Sources in Outer Space, and information and views on legal mechanisms relating to space debris mitigation measures.[69]

1.2.1.3 Other UN Bodies and Organs Monitoring Outer Space Activities

Beyond the UN General Assembly and its Committees, there are other UN bodies, programmes and organs related to space activities. The following discusses the ITU (a specialised agency of the UN), UN-SPIDER, the UN Programme on Space Applications, the International Committee on Global Navigation Satellite Systems (ICG), the United Nations Spatial Data Infrastructure (UNSDI), the Conference on Disarmament (CD) and UNIDIR.

International Telecommunication Union (ITU)

The International Telecommunication Union held its 2014 Council meeting at its headquarters in Geneva from 6 to 15 May 2014 and its 2014 World Radiocommunication Seminar (WRS) biennial conference in Geneva on 8-12 December 2014. During those meetings, a significant number of space related issues were addressed, including radio spectrum allocation and orbital slots procedures, emergency communications standardisation, and the coordination of satellite networks. The WRS focused on the regulatory aspects of the use of the radio-frequency spectrum and satellite orbits, in particular the application of the provisions of the ITU Radio Regulations.[70] Another ongoing issue is the topic of specific frequency band allocations to member states to facilitate the development of terrestrial mobile broadband applications, which in turn could potentially impact mobile satellite service operators that had planned on using the same frequency for emergency and disaster relief radio communication. These meetings provided a forum to exchange views on the associated technical, procedural and operational aspects of satellite communications and, particularly, technical issues related to technological advancements. The next World Radiocommunication Conference (WRC) is to be held in Geneva from 2 to 27 November 2015.[71]

---

[69] United Nations General Assembly. Report of the Legal Subcommittee on its fifty-third session, held in Vienna from 24 March to 4 April 2014. UN Doc. A/AC.105/1067 of 15 April 2014. Vienna: United Nations. http://www.unoosa.org/pdf/reports/ac105/AC105_1067E.pdf.

[70] "ITU World Radiocommunication Seminar focuses on regulatory aspects of the use of the radio-frequency spectrum and satellite orbits." 8 Dec. 2014. ITU 6 Apr. 2015. http://www.itu.int/net/pressoffice/press_releases/2014/CM11.aspx.

[71] Aubineau, Philippe. "WRC-15 Preparation." 7 Dec. 2012. ITU 2 Apr. 2014. http://www.itu.int/en/ITU-R/seminars/rrs/Documents/Tutorial/WRC-15.pdf.

UN-SPIDER

The United Nations Platform for Space-based Information for Disaster Management and Emergency Response (UN-SPIDER) organized several workshops and regional meetings in 2014. Set up by UNGA in 2006, the platform aims to provide universal access to all types of space-based information and services relevant to disaster management support. In this context, the International Charter on Space and Major Disasters was activated many times by the UN Office of Outer Space Affairs (UN OOSA) at the request of other UN entities and member states. Key geopolitical events that activated the Charter included floods in the UK, Zimbabwe, Burundi, Bolivia, the Solomon Islands, Afghanistan, Serbia, Argentina, Brazil, Sudan, Panama, India, Bangladesh, Pakistan, and France; hurricanes, tropical storms and torrential rains in the Réunion islands, Peru, Brazil, the Solomon Islands, Australia, China, Viet Nam, India, Bermuda, and the Philippines; landslides in Tajikistan, Sri Lanka, and Indonesia; wildfires in India, Chile, and the Dominican Republic; along with volcanic activities in Indonesia, an earthquake in China, a tsunami warning in Chile, and heavy snowfall in South Korea. The international charter was also activated in monitoring the Ebola epidemic in Western Africa, and to assist in the search for the missing Malaysia Airlines Flight MH370 aircraft, which helped to augment the crowd sourcing efforts initiated by the international community.[72]

UN Programme on Space Applications (SAP)

The UN Programme on Space Applications (SAP) is concerned with cooperation in space science and technology. In 2014, several activities were carried out under SAP's auspices dealing with topics including workshops on space technology for socio-economic benefits, on space law, and on the use of global navigation satellite systems for scientific applications. In addition, symposiums were arranged on space science and the United Nations, and on basic space technology: "Making Space Technology Accessible and Affordable". Moreover, an expert meeting was held on the International Space Station benefits for health, and lastly, an international conference on the use of space technology for water management.[73]

International Committee on Global Navigation Satellite Systems (ICG)

The aim of the International Committee on Global Navigation Satellite Systems (ICG) is to promote cooperation in matters of satellite navigation, through encouraging compatibility in global navigation satellite systems (GNSS)—to be used separately or together without interference among systems; interoperability, using satellite navigation services together to provide better capabilities than can be achieved by individual systems alone; and integration of GNSS services into

[72] "UN-SPIDER Monthly Updates." 2015. UNOOSA 7 Apr. 2015. http://www.un-spider.org/news-and-events/updates.

[73] "United Nations Programme on Space Applications Activities Schedule: 2014." United Nations Office for Outer Space Affairs 7 Apr. 2015. http://www.unoosa.org/oosa/en/SAP/sched/2014.html.

national infrastructures, particularly in developing nations. The UN OOSA serves as the Executive Secretariat of the ICG and the associated Providers' Forum. The ninth meeting of the ICG was organized by the European Commission and the European GNSS Agency (GSA) on behalf of the European Union, and took place in Prague, Czech Republic, on 10–14 November 2014. The meeting drew participants from industry, government, non-governmental officials and academia to review and discuss developments in global navigation systems.[74]

Throughout the sessions of its Plenary and working group meetings, presentations were made to update ICG Members, Associate Members, and Observers on the emergence of innovative applications in various domains. Moreover, the Office for Outer Space Affairs also contributed its "Programme on GNSS applications" presentation, which described the regional workshops and the work carried out through the regional centres for space science and technology education, affiliated to the United Nations, which also act as the information centres for ICG.

A keynote presentation was given by Bradford Parkinson (widely known as the founder of GPS) on the topic of "Assuring Positioning, Navigation and Timing (PNT): A PTA (protecting, toughening and augmenting) programme and recommendations of the United States PNT Advisory Board". Various new topics were presented including: "GNSS application market opportunities"; "High-precision agriculture and machine control with the Global Navigation Satellite System (GLONASS), GPS and other GNSS"; "Progress of BeiDou navigation satellite system applications"; "GNSS for train control and management systems: challenges and opportunities for global services"; and "Satellite services for personal air transportation".[75]

The four working groups also met during the Plenary, covering compatibility and interoperability; enhancement of performance of GNSS services; information dissemination and capacity building; and reference frames, timing and applications.[76] Within the Plenary, the ICG endorsed the decisions and recommendations of the working groups. Also, the Asia-Pacific Space Cooperation Organization (APSCO) was granted observer status during the Plenary.[77] The tenth ICG meeting will take place in Boulder, Colorado, USA from 2 to 6 November 2015.

---

[74] "ICG Meetings." United Nations Office for Outer Space Affairs 7 Apr. 2015. http://www.oosa.unvienna.org/oosa/en/SAP/gnss/icg/meetings.html.

[75] United Nations General Assembly. Ninth Meeting of the International Committee on Global Navigation Satellite Systems—Note by the Secretariat. UN Doc. A/AC.105/1083 of 26 Nov. 2014. New York: United Nations. http://www.unoosa.org/pdf/reports/ac105/AC105_1083E.pdf.

[76] ICG. Agenda—Ninth Meeting of the International Committee on Global Navigation Satellite Systems. UNOOSA 7 Apr. 2015. http://www.unoosa.org/pdf/icg/2014/icg-9/icg9-agenda.pdf.

[77] United Nations General Assembly. Ninth Meeting of the International Committee on Global Navigation Satellite Systems—Note by the Secretariat. UN Doc. A/AC.105/1083 of 26 Nov. 2014. New York: United Nations. http://www.unoosa.org/pdf/reports/ac105/AC105_1083E.pdf.

United Nations Spatial Data Infrastructure (UNSDI)
The United Nations Geographic Information Working Group (UNGIWG) held its fourteenth annual Plenary Meeting in New York, USA on 14—1 May 2014. The meeting was co-chaired by representatives from the United Nations Department of Safety and Security (UNDSS) and the United Nations Office for Outer Space Affairs (UNOOSA). The UNGIWG functions as a network of UN cartography and geospatial information management science professionals that lay the foundations for the United Nations Spatial Data Infrastructure (UNSDI). The UNSDI contributes substantively to the mission of the United Nations by functioning as a comprehensive, decentralised geospatial information network to facilitate decision-making mechanisms between member states, regional organizations and partners that advance social, economic development, environmental and humanitarian interests.[78]

Conference on Disarmament (CD)
The Conference on Disarmament (CD) is the key multilateral disarmament and arms control negotiating forum within the international community. While a deadlock continues regarding space security with several delegations reaffirming their respective positions in lengthy discussion on the issue of prevention of an arms race in outer space, in June 2014 the Permanent Representatives of the Russian Federation and of China to the CD transmitted the updated Russian and Chinese texts of the draft Treaty on Prevention of the Placement of Weapons in Outer Space and of the Threat or Use of Force Against Outer Space Objects (PPWT). The U.S. provided its analysis of the 2014 Russian-Chinese draft treaty in a note verbale submitted in September. The 2014 session occurred in three parts: the first part on 19 January–27 March, the second on 25 May–10 July, and the third on 3 August–18 September 2014, with twenty-nine informal plenary meetings, including three of the informal working group meetings, in February and in May through September 2014. As in previous years, some of the topics that were addressed by the CD included the cessation of the nuclear arms race and nuclear disarmament, prevention of nuclear war including all related matters, prevention of an arms race in outer space, effective international arrangements to assure non-nuclear-weapon states against the use or threat of use of nuclear weapons, and the prohibition of the development and manufacture of new types of weapons of mass destruction and new systems of such weapons—e.g. radiological weapons, transparency in armaments.[79]

United Nations Institute for Disarmament Research (UNIDIR)
The United Nations Institute for Disarmament Research (UNIDIR) conducts some projects that cover, both directly and indirectly, issues of space security. It seeks to review former proposals and to propose new options for breaking the deadlock in space

[78] "14th UNGIWG Plenary Meeting." UNGIWG 7 Apr. 2015. http://www.ungiwg.org/meetings/14th-ungiwg-plenary-meeting.

[79] Conference on Disarmament. Report of the Conference on Disarmament to the General Assembly of the United Nations. UN Doc. CD/2004 of 10 Sept. 2014. Geneva: United Nations. http://daccess-dds-ny.un.org/doc/UNDOC/GEN/G14/164/98/PDF/G1416498.pdf.

weaponisation matters at the CD. In July 2014, the UNGA was informed of the activities of the UNIDIR for the period of August 2013 to July 2014. In the field of space security, the UNIDIR with the support of the EU was working to build international understanding of, and support for, a code of conduct for outer space activities through a series of regional meetings, background papers, expert consultations and support activities. The project was extended through August 2014, with plans under consideration for a follow-on project. Moreover, the UNIDIR was selected as an expert consultant for the Group of Governmental Experts on Developments in the Field of Information and Telecommunications in the Context of International Security, with the first meeting held from 21 to 25 July 2014; the Group will remain active through 2015.[80] At the annual UNIDIR space security conference on the topic "Implementation and compliance", held on 19–20 April 2014, the UNIDIR aimed to provide an overview of current initiatives, an update on implementation of and adherence to existing instruments.[81]

1.2.2 The Group on Earth Observation

The Group on Earth Observation (GEO), with its system-of-systems cooperation platform (GEOSS), is a voluntary partnership of governments and international organisations that facilitate the coordination of current and future Earth observing systems, while also providing structured and sustained data worldwide as a decision support tool for its users. In 2014, GEO was very active in workshops and symposia, with Earth observation topics covering environmental monitoring in terms of agriculture, crops and rangeland assessments, inland and coastal water quality, and urban development. Certain workshops were dedicated to definite areas, such as the GEOGLAM initiative that is tasked to "coordinate satellite monitoring observation systems in different regions of the world in order to enhance crop production projections and weather forecasting data."[82] Moreover, other diversified and notable activity included conferences such as The Fifth Asia/Oceania Meteorological Satellite Users' Conference; the 10th International Conference of the African Association of Remote Sensing of the Environment (AARSE 2014); Information for Meeting Africa's Agricultural Transformation and Food Security Goals (IMAAFS); the 2nd Integrated Research on Disaster Risk Conference "Integrated Disaster Risk Science: A tool for sustainability"; and the UN/Morocco International Conference on the Use of Space Technology for Water Management.[83]

---

[80] United Nations General Assembly. United Nations Institute for Disarmament Research—Note by the Secretary General, on its Sixty-ninth session. UN Doc. A/69/176 of 23 July 2014.. http://www.unidir.org/files/files/GA_report_2014_Dir_E_A-69-176.pdf.

[81] "Space Security 2014: Implementation and Compliance." UNIDIR 8 Apr. 2015. http://www.unidir.org/programmes/emerging-security-threats/space-security-2014-implementation-and-compliance.

[82] "About the GEOGLAM Initiative." GEOGLAM 9 Apr. 2015. http://geoglam-crop-monitor.org/.

[83] "Meetings." GEO 9 Apr. 2015. http://www.earthobservations.org/meetings.php.

1.2.3 Europe

1.2.3.1 European Space Agency

The European Space Agency (ESA) Ministerial Council meeting was held on 2 December 2014. At the meeting, ESA's member governments agreed to develop the next generation Ariane 6 launcher, to upgrade its small-satellite Vega launcher, and to commit funding for the international space station (ISS) through 2017 by supplying the propulsion module for NASA's Orion capsule in lieu of the completed ATV cargo supply missions.[84] And instead of a mid-life upgrade of the Ariane 5 ECA, additional funding will be provided for the use of its current configuration, and continued used of the current Vega launcher. The next ESA Ministerial Council meeting is scheduled for 2016 in Switzerland.

The outcome of the 2014 Ministerial Council meeting resulted in ESA's total programme commitments from its 20 member states reaching 15 billion euros ($18.75 billion) over the next 3 years, with 74 % of the funding coming from four members: France (28 %), Germany (25 %), Italy (13 %), and the UK (8 %). Funding for the development of the 4 billion euros ($5 billion) Ariane 6, expected to launch in 2020, will come mainly from France which has a 52 % stake, in addition to Germany's 23 % share,[85] followed by Italy and the other five ESA member states (Belgium, Spain, the Netherlands, Sweden, and Switzerland) that are currently funding the optional Ariane 6 launcher programme.[86]

In contrast to the Ariane 5 launcher, which received state funding to offset periodic losses in times of low commercial launch demand, responsibility for the commercial exploitation of the Ariane 6 and Vega C will rest with industry. Hence, to balance the commercial risk, industry will also become the design authority. Airbus Defence & Space and Safran will lead the development of Europe's next generation launch systems through a joint venture called 'Airbus Safran Launchers' which was to start operations on 1 January 2015, with a work force of 450 people.[87]

At the 2014 Ministerial Council meeting, ESA was given 800 million euros ($990.5 million) by ten of its participating member governments (Belgium, Denmark, France, Germany, Italy, the Netherlands, Norway, Spain, Sweden and

[84] De Selding, Peter B. "ESA Members Agree To Build Ariane 6, Fund Station Through 2017." 2 Dec. 2014. SpaceNews 22 June 2015. http://spacenews.com/42699esa-members-agree-to-build-ariane-6-fund-station-through-2017/.

[85] De Selding, Peter B. "ESA Ministerial Produces a Few Surprises." 12 Dec. 2014. SpaceNews 22 June 2015. http://spacenews.com/esa-ministerial-produces-a-few-surprises/.

[86] "Ariane 6 | Les Chiffres Clés." 2 Dec. 2014. CNES 12 Aug. 2015. https://cnes.fr/sites/default/files/drupal/201506/default/is_ariane6-chiffres.pdf.

[87] Henry, Caleb. "ESA Hands Reins to Industry on New Launchers." 3 Dec. 2014. Satellite Today 22 June 2015. http://www.satellitetoday.com/launch/2014/12/03/esa-hands-reins-to-industry-on-new-launchers/.

Switzerland) to cover ISS costs for the 3 year period between 2015 and 2017.[88] Following the ESA ministerial, Germany remains the lead contributor to the ISS programme, followed by France, and Italy. Moreover, the UK's share in the ISS has increased to 4.6 % with its commitment of 40 million euros at the meeting.[89]

ESA's contribution to the development of public private partnerships (PPPs) in European space programmes progressed further in 2014, with several launches to be conducted in 2015. The European Data Relay System (EDRS), developed between ESA and Airbus Defence & Space, will relay data from satellites in non-geostationary orbit, to dramatically decrease latency time in signal transmission and provide near-real-time services on a global scale. The first EDRS payload (EDRS-A) will be hosted on a Eutelsat-9B commercial telecommunications satellite that will be launched by mid-2015; however, a demonstration EDRS terminal was also launched in April 2014 on the Sentinel-1A to test the EDRS business case by sending imagery to a laser terminal aboard Inmarsat's Alphasat satellite in geostationary orbit.[90] Another PPP exists between ESA and an industrial team lead by OHB System AG, for the development of the SmallGEO, intended as a general-purpose small geostationary satellite platform to buttress European industry's position in the commercial telecom market. The SmallGEO is expected to launch in the second half of 2015.[91]

ESA's budget decreased by nearly 200 million euros to 4.102 billion euros ($5.7 billion) in 2014. This 4.2 % decline mainly came from a reduction in funding from the EU, whereas fluctuations in the level of investment from ESA's member states tended to be modest by comparison. Despite reduced funding by Italy, the UK, and Spain, overall funding by ESA member states grew, with substantial increases by the Netherlands, Switzerland, and Sweden.[92] Following the 2014 Ministerial meeting, ESA's budget for 2015 is to grow by 8 % to be $4.433 billion ($5.222 billion).[93]

Shortly after the 2014 Ministerial Council meeting, it was announced that Germany's Johann-Dietrich Woerner was elected as Director General of ESA, to take effect at the beginning of July 2015.[94] Woerner had been the Chairman of the

---

[88] De Selding, Peter B. "Italian Space Increase Falls Short, Clouding Outlook for ESA Ministerial." 1 Dec. 2014. SpaceNews 22 June 2015. http://spacenews.com/42638italian-space-increase-falls-short-clouding-outlook-for-esa-ministerial/.

[89] De Selding, Peter B. "ESA Ministerial Produces a Few Surprises." 12 Dec. 2014. SpaceNews 25 June 2015. http://spacenews.com/esa-ministerial-produces-a-few-surprises/.

[90] De Selding, Peter B. "ESA Members Defer Funding Guarantee for Airbus-backed Data-relay System." 5 Sept. 2014. SpaceNews 25 June 2015. http://spacenews.com/41767esa-members-defer-funding-guarantee-for-airbus-backed-data-relay-system/.

[91] "No 48-2014: Press Breakfast With ESA's Director General." 18 Dec. 2014. SpaceNews 26 June 2015. http://www.esa.int/For_Media/Press_Releases/Press_Breakfast_with_ESA_s_Director_General.

[92] "ESA Budget for 2014." ESA 26 June 2015. http://www.esa.int/For_Media/Highlights/ESA_budget_2014.

[93] "ESA Budget for 2015." ESA 26 June 2015. http://www.esa.int/For_Media/Highlights/ESA_budget_2015.

[94] De Selding, Peter B. "How ESA's Next Director-General Got the Job." 23 Dec. 2014. SpaceNews 22 June 2015. http://spacenews.com/dlrs-woerner-to-head-esa/.

Executive Board of the German Space Agency (DLR) since 1 March 2007.[95] He will succeed ESA's current Director General, Jean-Jacques Dordain, who has been in that role for the past 12 years.

Despite geopolitical tensions between Russia and the U.S. and Europe related to Russia's actions in Ukraine, it seems unlikely that the European-Russian ExoMars programme and other joint space ventures (such as the ISS) will be impacted by the current situation. However, European funding for the ExoMars programme remained an issue leading up to the 2014 Ministerial Council meeting. Prior to the meeting, the 1.2 billion euros mission planned for January 2016 and May 2018 still lacked around 185 million euros ($240 million). While funds for the 2016 portion of the mission were secure, funding for the 2018 portion was expected to last only until mid-2015.[96] Fortunately, the ESA member governments at the 2014 Ministerial Council meeting secured the missing funds.[97]

As regards ESA's rapprochement with the EU, decisions on the direction of ESA's evolution (e.g. improved cooperation under the "status quo"; bringing ESA as an intergovernmental organisation under the authority of the European Union (following, to a certain extent, the model of the European Defence Agency); transforming ESA into an EU agency (following the model of existing regulatory agencies); etc.) were delayed pending further studies of how changes would affect Europe's space industrial base.[98]

1.2.3.2 European Union

In 2008, the European Union initiated the procedure to develop a draft International Space Code of Conduct for Outer Space Activities (ICoC) in response to UN General Assembly Resolutions 61/75 of 2006 and 62/43 of 2007, which had invited all members to submit concrete proposals on international outer space TCBMs in the interest of maintaining international peace and security and promoting international cooperation and the prevention of an arms race in outer space. Initial draft versions of ICoC had received lukewarm receptions due to the fact that the code had been developed within the EU system and lacked a parallel diplomatic process

[95] "Johann-Dietrich Wörner—Chairman of the Executive Board." 30 Oct. 2013. DLR 22 June 2015. http://www.dlr.de/dlr/en/desktopdefault.aspx/tabid-10329/510_read-204.

[96] De Selding, Peter B. "65th International Astronautical Congress | ExoMars Funding Commitment Needed in December, Thales Alenia Says." 1 Oct. 2014. SpaceNews 25 June 2015. http://spacenews.com/4206165th-international-astronautical-congress-exomars-funding-commitment-needed/.

[97] De Selding, Peter B. "ESA Members Agree To Build Ariane 6, Fund Station Through 2017." 2 Dec. 2014. SpaceNews 25 June 2015. http://spacenews.com/42699esa-members-agree-to-build-ariane-6-fund-station-through-2017/.

[98] De Selding, Peter B. "European Union Delays Decision on ESA Changes." 26 May 2014. SpaceNews 25 June 2015. http://spacenews.com/40685european-union-delays-decision-on-esa-changes/.

involving the considerations of the international community. However, following several revisions and three open-ended multilateral consultations (held in Kiev, Ukraine in May 2013, in Bangkok, Thailand in November 2013, and Luxembourg in May 2014), the fifth draft of ICoC, issued on 31 March 2014, has gained wider acceptance.[99]

Intended to be a non-binding set of principles and guidelines agreed to on a voluntary basis, the ICoC is based on three broad main principles: (1) freedom for all to use outer space for peaceful purposes, (2) preservation of the security and integrity of space objects in orbit, and (3) due consideration for the legitimate security and defence needs of states. Following the completion of the consultative phase, the EU aims to begin a multilateral "negotiation process" by the middle of 2015, after which the ICoC will be opened for signature by individual states. In parallel to the development of the non-binding, soft law of the ICoC—Russia and China are promoting a binding draft Treaty on Prevention of the Placement of Weapons in Outer Space and of the Threat or Use of Force Against Outer Space Objects (PPWT), which was introduced to the UN Conference on Disarmament (CD) in 2008, and subsequently updated in June 2014 following the last open-ended multilateral consultation on the ICoC in Luxembourg. The PPWT focuses mainly on security issues regarding the placement of weapons in outer space and does not address issues of ASAT missile testing and space debris.

The EU Copernicus flagship programme (formerly known as the Global Monitoring for Environment and Security, or GMES) made some headway in 2014, following the successful launch of the Sentinel-1A satellite on 3 April 2014. However, in previous years funding had become uncertain, e.g. during the Commission's development of the 7-year Multiannual Financial Framework (MFF) for 2014–2020, the 5.8 billion euros requested for Copernicus was taken out of the intended budget; instead, it was expected to be funded on an à la carte basis by individual participating states. However, following scrutiny by European industry officials, ESA, and the European Parliament, the Commission re-introduced the Copernicus programme in the new MFF budget, albeit with a 2 billion euros reduction, thus amounting to 3.8 billion euros (in 2011 prices).[100] The 34.4 % budget cut was approved by the European Parliament, meaning that replacement satellites likely will not be built in time to take over from Sentinel satellites, if they are to have 7-year life spans.[101] In coordinating the evolution of the Copernicus Space Component, ESA has prepared a long-term plan for the content and

[99] *See further* "Code of Conduct for Outer Space Activities." European Union External Action 24 June 2015. http://eeas.europa.eu/non-proliferation-and-disarmament/outer-space-activities/index_en.htm.

[100] "For Europe's Embattled GMES, Good and Bad News." 7 Dec. 2012. SpaceNews 25 Apr. 2014. http://spacenews.com/article/civil-space/32717for-europe%E2%80 %99 s-embattled-gmes-good-and-bad-news.

[101] Cf. De Selding, Peter B. "Copernicus Backers Worry EU Cuts Will Discourage Investment." 8 Feb. 2013. SpaceNews 30 May 2014. http://www.spacenews.com/article/civil-space/33564copernicus-backers-worry-eu-cuts-will-discourage-investment.

associated funding needs, covering the operation of the Sentinels up to 2020, and the procurement of recurrent Sentinel satellites and instruments and access to data available from contributing missions up to 2028.[102] The second Sentinel-1B satellite is expected to launch in late 2015.[103]

In the development of the EU's Horizon 2020 programme, other space programmes were also detrimentally affected. By February 2013, the Galileo flagship GNSS programme faced a 10 % reduction in funding, in addition to a scaled-back space research package in Horizon 2020 that faced up to 25 % in budget reductions. Where previously, Galileo's system had been allocated 7 billion euros over the 7-year period, the Commission allocated 6.3 billion euros to the programme, or 900 million euros each year.[104] On 22 November 2013, the European Parliament in its plenary approved the Galileo's programme's 7-year 6.3 billion euros budget at 2011 prices.[105]

The EU Galileo flagship GNSS programme experienced a significant setback with the 22 August 2014 launch of the first pair of Galileo's fully operational satellite models that placed the satellites into useless orbits due to a malfunction in the upper stage of the Soyuz Fregat launcher.[106] While ESA had begun taking steps to correct the orbits of both satellites by the end of 2014, lingering doubts about the Soyuz Fregat motivated Europe to reschedule the December launch of two additional Galileo satellites to 2015.[107] In other matters, Switzerland joined the EU's Galileo programme at the beginning of 2014, following an agreement that requires Switzerland to make catch-up payments to the EU totalling 80.05 million euros ($108.05 million), for the period 2008 to 2013, along with annual fees of 27 million euros ($36.44 million) in return for access to the Galileo's restricted Public Regulated Service (PRS) signals that will start in 2014.[108] Moreover, the EU is considering whether to make the PRS signals mandatory for the 28 member states, in addition to whether to make it available to Non-EU ESA member states, Norway

[102] "Green Light for GMES Copernicus." 4 July 2013. ESA 28 Apr. 2014. http://www.esa.int/Our_Activities/Observing_the_Earth/Copernicus/Green_light_for_GMES_Copernicus.

[103] "European Commissioner for Industry presented Sentinel—1A." 25 Nov. 2013. New Europe 25 Apr. 2014. http://www.neurope.eu/article/copernicus-satellite-will-launch-april-or-may.

[104] "Space Programs Facing Cuts in Seven-year EU Budget." 1 Feb. 2013. SpaceNews 25 Apr. 2014. http://spacenews.com/article/space-programs-facing-cuts-in-seven-year-eu-budget.

[105] "Galileo Funding: A 'Small' Difference of €700 Million." 22 Nov. 2013. The European Sting 25 Apr. 2014. http://europeansting.com/2013/11/22/galileo-funding-a-small-difference-of-e700-million/.

[106] De Selding, Peter B. "Galileo Launch, Initially Hailed as Success, Is a Failure." 23 Aug. 2014. SpaceNews 19 May 2015. http://spacenews.com/41650galileo-launch-initially-hailed-as-success-is-a-failure/.

[107] De Selding, Peter B. "Lingering Doubts Drove Europe To Sideline Galileo Launches until Next Year." 22 Oct. 2014. SpaceNews 19 May 2015. http://spacenews.com/42270lingering-doubts-drove-europe-to-sideline-galileo-launches-until-next-year/.

[108] De Selding, Peter B. "Switzerland To Join Galileo Program under Agreement with EU." 4 Feb. 2014. SpaceNews 19 May 2015. http://spacenews.com/39388switzerland-to-join-galileo-program-under-agreement-with-eu/.

and Canada, and whether to grant PRS signal access to the U.S. and Chinese governments.[109]

In February 2013, funding for Horizon 2020, the EU's Research and Innovation programme which includes a large variety of space research efforts, was estimated to be at 1.4 billion euros over the 7-year period, i.e. 200 million euros per year.[110] Nevertheless, by 10 December 2013 the actual amount allocated to space research under Horizon 2020 had fallen short of that estimate for at least 2014 and 2015. Under the Horizon 2020 Work Programme 2014–2015, the 2014 space research budget was 165.75 million euros, while the 2015 indicative budget was slightly higher at 181.9 million euros.[111]

Following the EU's new space competence under Article 189 of the Treaty on the Functioning of the European Union (TFEU), a number of communications have been exchanged between the European Commission, the Council, and the European Parliament in developing a space strategy, and establishing appropriate relations with ESA. While earlier proposals sought to transform ESA into a European Union agency, the idea was met with opposition from several major ESA players including Germany and the UK, which prefer ESA's geographic return rule over the EU's stricter value-for-money basis under its competition rules. Ultimately, a 6 February 2014 progress report by the Commission determined that transforming ESA into an EU agency would require political consensus that may be difficult to reach in the foreseeable future.[112] On 26 May 2014, a meeting in the EU Competitiveness Council '*Towards a shared EU-ESA vision for space fostering competitiveness*' removed the earlier proposals from consideration, instead suggesting to further study how ESA work for the Commission could be subject to different rules within ESA through the creation of an "EU pillar"; or alternatively, whether the current ESA-EU convention could be modified slightly in the short term, while in the long term ESA would progressively align its accounting, internal control and audit procedures with the corresponding EU rules.[113] On 5 December 2015 the Council conclusions of the Competitiveness Council meeting, entitled '*Underpinning the*

[109] De Selding, Peter B. "Late or Not, Galileo Positioned for a Global Embrace." 13 June 2014. SpaceNews 19 May 2015. http://spacenews.com/40893late-or-not-galileo-positioned-for-a-global-embrace/.

[110] "Space Programs Facing Cuts in Seven-year EU Budget." 1 Feb. 2013. SpaceNews 25 Apr. 2014. http://spacenews.com/article/space-programs-facing-cuts-in-seven-year-eu-budget.

[111] Commission of the European Communities. Horizon 2020—Work Programme 2014-2015. Annex 8. European Commission Decision C(2013)XXX of 10 December 2013. Brussels: European Union. http://ec.europa.eu/enterprise/policies/space/files/research/horizon-2020/h2020-leit-space-work-programme-2014-2015_en.pdf.

[112] European Commission. Report from the Commission—Progress report on establishing appropriate relations between the European Union and the European Space Agency (ESA). European Commission COM(2014) 56 final of 6 February 2014. Brussels: European Union. http://register.consilium.europa.eu/doc/srv?l=EN&f=ST%205978 %202014%20INIT.

[113] De Selding, Peter B. "European Union Delays Decision on ESA Changes." 26 May 2014. SpaceNews 25 June 2015. http://spacenews.com/40685european-union-delays-decision-on-esa-changes/.

European space renaissance: orientations and future challenges' invited both the Commission and ESA to review the European Space Policy and to identify to what extent it may require to be updated, with regard to the TFEU and ESA Convention, to appropriately meet the long-term needs of European users.[114]

1.2.3.3 EUMETSAT

The European Organisation for the Exploitation of Meteorological Satellites (Eumetsat) is an intergovernmental organisation that supplies weather and climate-related satellite data to the National Meteorological Services of its Member and Cooperating States in Europe, and other users worldwide. While support for Eumetsat's second-generation European Polar System (EPS-SG) had been hindered by the sovereign-debt crisis in previous years, the programme's approval process appeared to be progressing following its 82nd council meeting in Darmstadt, Germany, held on 26 November 2014. At the meeting, 86.86 % of the funding needed for the EPS-SG was secured; yet the programme will need to secure 95 % of its financial envelope before it can be started. The EPS-SG sets of satellites will succeed the current Metop three-satellite programme, and are expected to make observations from polar orbit in the 2021–2042 timeframe.[115]

Eumetsat's current Metop program consists of three identical satellites launched at 6-year intervals between 2006 and 2018, with operations running until 2023. The second polar-orbiting meteorological satellite, Metop-B, was successfully placed into orbit on 17 September 2012, where it began operating in tandem with Metop-A, which had launched in October 2006. The third identical Metop satellite, Metop-C, is expected to launch sometime between late 2016 and late 2018. The entire cost of the Metop program (including development of the three satellites, their launches and the related ground infrastructure) reached 3.2 billion euros (at 2011 prices), with Eumetsat covering 75 % of the cost, and ESA covering the rest of the amount.[116]

The Metop Second Generation (Metop-SG) is the space segment of Eumetsat's EPS. In terms of funding for the Metop-SG, the 20 member ESA will spend about 800 million euros ($1.1 billion) on the design and construction of the two Metop-SG satellites, while Eumetsat has budgeted the program at 3.323 billion euros over

[114] Council of the European Union. Underpinning the European space renaissance: orientations and future challenges. Competitiveness Council meeting of 5 December 2014. Brussels: European Union. http://italia2014.eu/media/4061/council-conclusions-on-underpinning-the-european-space-renaissance-orientations-and-future-challenges.pdf .

[115] "The 82nd meeting of the EUMETSAT Council took place in Darmstadt, Germany, on 26 November." 28 Nov. 2014. Eumetsat 24 June 2015. http://www.eumetsat.int/website/home/News/DAT_2430686.html?lang=EN&pState=1.

[116] "European Weather Satellite Launched after 4-Month Delay." 27 Sept. 2012. SpaceNews 28 Apr. 2014. http://spacenews.com/article/european-weather-satellite-launched-after-4-month-delay.

more than 20 years, including the construction of the four satellites under full Eumetsat responsibility.[117] The poor financial shape of many member governments had prompted Eumetsat's Council to reduce costs for Metop-SG by 5 % from the earlier estimate of 3.4 billion euros. On 7 April 2014, Airbus Defence and Space was selected to be the prime contractor for the Metop-SG.[118] The Metop-SG will feature 6 satellites, operating in 3 pairs, each carrying a different but complementary suite of instruments.[119] Airbus will also provide the Sentinel-5 spectrometer payload for the first Metop-SG satellite to be launched in 2021.[120]

The Meteosat Third Generation (MTG) system is a series of sounding and imaging satellites in geostationary orbit, being developed by France's Thales Alenia Space and Germany's OHB AG, and is aimed at providing services for 20 years starting in 2018. In July 2014, the Eumetsat Council approved contracts to develop the ground infrastructure for the MTG system, with 18.9 million euros ($25.7 million) going toward the Mission Data Acquisition Facility, including 5 years of operations; and 11.7 million euros ($15.9 million) for the telemetry, tracking and command system to monitor the satellites.[121]

The existing Meteosat programme consists of two generations of active satellites, i.e. the Meteosat First Generation (MFG) and the Meteosat Second Generation (MSG), operating in geostationary orbit over Europe and Africa. While only one satellite, Meteosat-7, remains operating until 2016 under the MFG programme; the MSG programme has three satellites in operation, Meteosat-8 to -10, which are expected to end service in 2019, 2021, and 2022 respectively. The primary role of the Meteosat satellites is to help detect and forecast rapidly developing high impact weather, including thunderstorms, volcanic ash, and fog, up to six hours ahead through continuous monitoring.[122]

By mid-2014, Eumetsat's membership had grown to 30 member states and 1 cooperating state (Serbia) from 27 member states in 2013. Lithuania became a

[117] De Selding, Peter B. "Formal Approval for Lower-cost Metop-SG Expected by Year's End." 10 July 2014. SpaceNews 8 Jan. 2015. http://spacenews.com/41212formal-approval-for-lower-cost-metop-sg-expected-by-years-end/.

[118] De Selding, Peter B. "Airbus Bests Thales Alenia-OHB Team for Billion-dollar ESA Weather Satellite Contract." 7 Apr. 2014. SpaceNews 8 Jan. 2015. http://spacenews.com/40109airbus-bests-thales-alenia-ohb-team-for-billion-dollar-esa-weather/.

[119] "MetOp-SG (MetOp-Second Generation Program)." eoPortal Directory 29 Apr. 2014. https://directory.eoportal.org/web/eoportal/satellite-missions/m/metop-sg.

[120] De Selding, Peter B. "ESA Orders Nearly $200 Million Weather Instrument from Airbus." 28 Mar. 2014. SpaceNews 7 Jan. 2015. http://spacenews.com/40016esa-orders-nearly-200-million-weather-instrument-from-airbus/.

[121] De Selding, Peter B. "Formal Approval for Lower-cost Metop-SG Expected by Year's End." 10 July 2014. SpaceNews 8 Jan. 2015. http://spacenews.com/41212formal-approval-for-lower-cost-metop-sg-expected-by-years-end/.

[122] "There are two generations of active Meteosat satellites, Meteosat First Generation (MFG) and Meteosat Second Generation (MSG), providing images of the full Earth disc, and data for weather...." Eumetsat 30 May 2014. http://www.eumetsat.int/website/home/Satellites/CurrentSatellites/Meteosat/index.html.

full member on 1 January, while Iceland joined on 22 January, and Bulgaria became a member as of 30 April 2014.[123] Their membership allows them to be fully involved in the strategic decisions of Eumetsat's Council, and allows their respective industries to bid for contracts. While their contributions are not expected to be substantial, the inclusion of these new members has symbolic significance for the organization as it moves forward with its flagship programmes.[124]

1.2.3.4 National Governments

France

In 2014, France returned to the top position in terms of ESA spending, having grown by 7.8 % to 854.4 million euros to meet its commitments for the next generation launcher. However, its spending was offset by a decrease in France's non-ESA-related space activities; thus overall spending at CNES increased by only 1.3 % to 1.43 billion euros for 2014.[125]

In April 2014, the French government proposed its Law on Space Operations requiring French launch providers to direct the upper stages of their launchers onto courses that would cause them to drop into open waters or disintegrate immediately after launch in Earth's atmosphere. The new national space law is meant to reduce orbital debris and prevent liability from upper stages re-entering the atmosphere and causing injury or property damage. The first of its kind, this new law will gradually take effect in the coming years, and will be applicable to all European launch systems. While this law is meant to contribute to the space environment, the fact that France is the only launching state to have this requirement is likely to have a negative impact when it comes to the commercial launch environment, as competitors in the U.S., Russia, China, and India are not yet operating under equivalent constraints.[126] Even so, other debris mitigation guidelines do exist, which have been deliberated on in the UNCOPOUS,[127] and have led to a voluntary international standard for debris mitigation (ISO 24113), in addition to several

[123] See further "EUMETSAT is a user-governed operational organisation, formed in 1986, which serves the needs of its Member States." Eumetsat 24 June 2015. http://www.eumetsat.int/website/home/AboutUs/WhoWeAre/MemberStates/index.html.

[124] "Eumetsat Welcoming New Members as Investments Loom." 30 Aug. 2013. SpaceNews 28 Apr. 2014 http://spacenews.com/article/civil-space/37012eumetsat-welcoming-new-members-as-investments-loom.

[125] "France Favors CNES with Better-than-inflation Budget Bump." 26 Sept. 2013. SpaceNews 28 Apr. 2014. http://spacenews.com/article/civil-space/37395france-favors-cnes-with-better-than-inflation-budget-bump.

[126] De Selding, Peter B. "French Debris-mitigation Law Could Pose Issue for Arianespace." 10 Apr. 2014. SpaceNews 25 June 2015. http://spacenews.com/40171french-debris-mitigation-law-could-pose-issue-for-arianespace/.

[127] "Space Debris Mitigation Guidelines of the Committee on the Peaceful Uses of Outer Space." 2010 United Nations Office for Outer Space Affairs 13 Aug. 2015. http://orbitaldebris.jsc.nasa.gov/library/Space%20Debris%20Mitigation%20Guidelines_COPUOS.pdf.

standards and technical reports; yet unlike the French law, they are not enforceable.[128]

Germany

On 20 May 2014, Germany's BAE System Intelligence and Security, and Airbus Defence & Space announced that they would market data from German radar satellite data to the U.S. government and certain commercial and international customers. More specifically, this is radar data from the TerraSAR-X, launched in June 2007, and TanDEM-X, launched in June 2010, while BAE systems will provide software integration and other services. The commercial sale of TerraSAR-X and TanDEM-X data is exclusively done by Airbus, which was granted the commercial license by DLR.[129]

In May 2014, Germany informed ESA that German spending on launch vehicles would remain flat for the upcoming decade, complicating ESA's push to secure funding and consensus for the future development of the Ariane launcher. Germany's position on launcher development softened at the December 2014 Ministerial Council meeting, where it agreed to take on a 22 % stake in the Ariane 6 launcher, and forego the development of the mid-life evolution to the Ariane 5 launcher.[130] Barring an increase of launcher development funding by any other member ESA, Germany's decision means that ESA's annual launcher budget will be 850 million euros per year from 2015 to 2024.[131]

Another outcome of the 2014 Ministerial Council meeting came from the German government's agreement to take a 7 % stake in the Vega launcher, amounting to 15 million euros ($18.75 million), in addition to a 23 % stake in the P120 launcher stage programme, amounting to around 147 million euros ($182.8 million); putting Germany on an equal footing with Italy as a programme contributor. The P120 will be the enhanced first stage for the upgraded Vega-C launcher, and will also serve as strap-on boosters for the Ariane 6 launcher. Two boosters will be used for the Ariane 62 configuration, while four will be used on the Ariane 64.[132]

Italy

In Italy, following the resignation of former ASI president Enrico Saggese on 7 February 2014, the Italian government appointed Aldo Sandulli, a law professor

---

[128] Finkleman, Dave. "Commentary | Rethinking Space Debris Mitigation." 25 Aug. 2014. SpaceNews 25 June 2015. http://spacenews.com/41659rethinking-space-debris-mitigation/.

[129] De Selding, Peter B. "BAE To Market German Radar Satellite Data to U.S. Government." 21 May 2014. SpaceNews 25 June 2015. http://spacenews.com/40644bae-to-market-german-radar-satellite-data-to-us-government/.

[130] De Selding, Peter B. "ESA Ministerial Produces a Few Surprises." 12 Dec. 2014. SpaceNews 25 June 2015. http://spacenews.com/esa-ministerial-produces-a-few-surprises/.

[131] De Selding, Peter B. "Germany's Budget Straitjacket Complicates Europe's Ariane Funding Outlook." 22 May 2014. SpaceNews 25 June 2015. http://spacenews.com/40655germanys-budget-straitjacket-complicates-europes-ariane-funding-outlook/.

[132] De Selding, Peter B. "ESA Ministerial Produces a Few Surprises." 12 Dec. 2014. SpaceNews 25 June 2015. http://spacenews.com/esa-ministerial-produces-a-few-surprises/.

at a Naples university, to oversee ASI's affairs until a new agency management team was in place.[133] On 17 May 2014, ASI's new president Roberto Battiston, a former advanced physics professor at Trento University, took over the post expressing Italy's intent to push for upgrades to the Italian-led Vega small-satellite launcher at the Ministerial Council meeting in December 2014.[134]

Following its sharp reduction of ISS funding in 2012 to 10 % due to the crippling effect of the financial crisis, and further to around 8 % by mid-2014,[135] on the insistence of Germany's delegation at the 2014 Ministerial Council, it agreed to return to its previous contribution level of 19 % in total contributions over the next 6 years, instead of the 3 year commitment that was requested by ESA. The reason for the 6-year commitment was due to the fact that Italy's contribution will be lower than the 19 % share level during 2015–2017; but as the funding costs for the ISS are expected to decline during 2018–2020, Italy's contribution will be above the 19 % share level for the latter 3 years. In return, Germany redirected to the Italian-led ExoMars project 15 million euros in funding it had designated to go to ISS.[136]

Prior to the meeting, budget constraints had already been visible, as funding for the Italian Space Agency (ASI) had fallen short of expectations. Italy was able to raise its share of contribution to the ISS thanks to a 200 million euros ($250 million) funding boost over 2015–2017 provided by the Italian government. The infusion is also intended to help finance continued work on the second generation of Italy's Cosmo-SkyMed radar Earth Observation system, which could otherwise face production delays.[137] The first satellite of the second generation system is to be launched in 2016, with a second satellite to be launched the following year.

The United Kingdom

During the sixth high-level Conference on EU Space Policy, 'What direction for Europe in space between now and 2020?', held on 28–29 January 2014 in Brussels, Belgium, it was revealed that the UK government had decided to block a European Commission effort to set common European regulations on the sale of high-resolution imagery outside of Europe. Despite the fact that the European industry

[133] De Selding, Peter B. "Italy Taps Law Professor To Lead ASI after Saggese Resigns amid Corruption Probe." 14 Feb. 2014. SpaceNews 30 May 2014. http://www.spacenews.com/article/civil-space/39491italy-taps-law-professor-to-lead-asi-after-saggese-resigns-amid-corruption.

[134] De Selding, Peter B. "Italy's New Space Chief Seeks To Close the Chapter on Corruption." 26 May 2014. SpaceNews 30 May 2014. http://www.spacenews.com/article/civil-space/40683italy%E2%80 %99 s-new-space-chief-seeks-to-close-the-chapter-on-corruption.

[135] De Selding, Peter B. "Profile | Roberto Battiston, President, Italian Space Agency." 26 Nov. 2014. SpaceNews 25 June 2015. http://spacenews.com/42669profile-roberto-battiston-president-italian-space-agency/.

[136] De Selding, Peter B. "ESA Members Agree To Build Ariane 6, Fund Station Through 2017." 2 Dec. 2014. SpaceNews 22 June 2015. http://spacenews.com/42699esa-members-agree-to-build-ariane-6-fund-station-through-2017/.

[137] De Selding, Peter B. "Italian Space Increase Falls Short, Clouding Outlook for ESA Ministerial." 1 Dec. 2014. SpaceNews 22 June 2015. http://spacenews.com/42638italian-space-increase-falls-short-clouding-outlook-for-esa-ministerial/.

has become transnational, the UK considered the action to be overreaching into the domain of national satellite imagery-export policies. It concluded that a blanket EU regulation would undermine Europe's commercial high-resolution satellite industry, as the regulation may limit the industry's ability to compete with non-European companies that also offer high-resolution imagery.[138]

While France has taken steps to reduce orbital debris and prevent liability from upper stages re-entering the atmosphere and causing injury or property damage through its Law on Space Operations, the British government has taken a different approach on risk by focussing on reducing space insurance requirements, and easing small satellite licensing rules. On 30 April 2014, the UK government published its 'Government Response to the UK Space Innovation and Growth Strategy 2014–2030', where it seeks to reduce the amount of insurance needed by satellite companies before government guarantees on third-party liability take effect from £80 million ($134.5 million) to a ceiling of £60 million ($100.9 million) by modifying the UK's Outer Space Act.[139] It also is reassessing how cubesats are regulated in terms of obtaining an operating licence to improve the UK space sector's international competitiveness. Moreover, the UK Space Agency (UKSA) will soon issue initial conclusions from its National Space Flight Coordination Group on developing a UK spaceport and starting commercial space flight from the UK.[140]

On 30 April 2014, the UK government published its 'U.K. National Space Security Policy' stressing the need for a terrestrial alternative to space-based navigation, positioning, and timing systems in case of potential loss of access to GNSS signals, either through malice or due to space weather.[141] The document highlighted that many critical infrastructures, e.g. military, civil government, and commercial, have become dependent on GPS, and that some crucial services may already have lost the knowhow needed to survive a satellite outage. Hence the need for increased resiliency measures including added redundancy, encryption and anti-jamming measures.[142]

138 De Selding, Peter B. "U.K. Blocks Bid To Create Common European Imagery Policy." 30 Jan. 2014. SpaceNews 24 June 2015. http://spacenews.com/39328uk-blocks-bid-to-create-common-european-imagery-policy/.

139 "Government Response to the UK Space Innovation and Growth Strategy 2014—2030." 30 Apr. 2014. UKSA 29 June 2015. https://www.gov.uk/government/uploads/system/uploads/attachment_data/file/307347/Government_response_-_space_growth_action_plan.pdf.

140 De Selding, Peter B. "Britain To Reduce Space Insurance Requirements, May Ease Smallsat Licensing Rules." 30 Apr. 2014. SpaceNews 25 June 2015. http://spacenews.com/40417britain-to-reduce-space-insurance-requirements-may-ease-smallsat-licensing/.

141 "National Space Security Policy." 30 Apr. 2014. UKSA 29 June 2015. https://www.gov.uk/government/uploads/system/uploads/attachment_data/file/307346/National_Space_Security_Policy.pdf.

142 De Selding, Peter B. "U.K. Policy Stresses Terrestrial Backup for Space-based Navigation." 1 May 2014. SpaceNews 25 June 2015. http://spacenews.com/40430uk-policy-stresses-terrestrial-backup-for-space-based-navigation/.

With the upcoming 2015 launch of British astronaut Tim Peake to the ISS, the UK increased its funding to the ISS at the 2014 Ministerial Council meeting to around 40 million euros or 4.6 % of total ESA funding through 2017.[143] Whereas in previous years, the UK's involvement in the ISS was minimal, its investment will now provide a UK built communications terminal, and will allow its researchers new opportunities such as the deployment of cubesats as part of tech demonstrations and science experiments conducted from the ISS.

1.2.4 The United States

As ties between the U.S. and Russia continue to deteriorate, U.S. lawmakers are increasingly realising the impact that inconsistent funding support has had on the state of U.S. human spaceflight capabilities, especially in regard to its Commercial Crew Program (CCP) and the development of the Space Launch System (SLS). On 27 March 2014, at a budget hearing before members of the House Appropriations commerce, justice, science subcommittee, NASA Administrator Charles Bolden said that had NASA received the amounts the White House had requested each year for the CCP program, at least one of the three competing companies would have been ready before 2015. The program is at risk of slipping to 2017 if it continues to receive less funding than requested. However, funding requests for NASA's SLS rocket and deep space Orion capsule might overshadow the requests for CCP. The CCP has received about $2 billion in total since 2010, while the SLS receives nearly $3 billion annually. Nevertheless, to Bolden, commercial access to LEO orbit is a precursor to NASA's deep space exploration programme, as the ISS will allow NASA to test some elements of programmes such as its Asteroid Redirect Mission (ARM), which will also be used as a stepping-stone for reaching Mars.[144] While NASA has not formally committed to this Asteroid Redirect Mission (ARM)—especially following the lukewarm response from Congress—should the mission proceed, it would use a solar-powered robotic spacecraft to push a 10-m asteroid into a deep retrograde orbit around the Moon. While no formal cost estimate for ARM has been determined, estimates put the mission cost at approximately $2.6 billion. To be successful, it will need to fit within overall financial considerations, in addition to satisfying political interests.[145] By 8 May, the House Appropriations Committee had approved $17.9 billion in funding for NASA for the fiscal year 2015; however, its CCP programme received 7.5 % less

[143] De Selding, Peter B. "ESA Ministerial Produces a Few Surprises." 12 Dec. 2014. SpaceNews 22 June 2015. http://spacenews.com/esa-ministerial-produces-a-few-surprises/.

[144] Leone, Dan. "Bolden, Lawmakers Point Fingers about State of U.S. Human Spaceflight." 27 Mar. 2014. SpaceNews 22 July 2015. http://spacenews.com/40007bolden-lawmakers-point-fingers-about-state-of-us-human-spaceflight/.

[145] Leone, Dan. "Bolden: Capabilities-based Approach Is All NASA Can Afford." 22 Oct. 2013. SpaceNews 29 Apr. 2014. http://www.spacenews.com/article/civil-space/37808bolden-capabilities-based-approach-is-all-nasa-can-afford.

than its request for 2015, i.e. $785 million. Nevertheless, that amount exceeded its 2014 request by nearly 13 %.[146]

The U.S. National Defense Authorization Act for 2014 (NDAA 14) was signed into law on 26 December 2013. A notable provision of the bill related to U.S. Missile Defense Programs is its measure to bar the use of government funds to integrate Chinese-built missile defence systems into the missile defence systems of the U.S. That prohibition extends to U.S. developed missile defence architectures such as such as forward-based X-band radars in Japan, Israel, Turkey, and the area of responsibility of the Commander of the United States Central Command.[147] In effect, the bill is meant to restrict the integration of Chinese systems into the NATO architecture. The inclusion of the measure into the bill stemmed from Turkey's September 2013 decision to purchase a Chinese-built long-range missile and air defence system for its own territorial use. As a result for NATO member Turkey, its NATO missile defence architecture will not be interoperable with the Chinese system, leading to some reduced capability.[148]

On 19 December 2014, the U.S. National Defense Authorization Act for 2015 (NDAA-15) was signed into law. With tensions increasing between the U.S. and Russia, the NDAA-15 draws a line on the U.S. continued use of Russian made rocket engines for its launchers. The act prohibits U.S. companies from contracting with Russian suppliers of rocket engines or renewing current contracts for space launch activities.[149] As the United Launch Alliance (ULA) Atlas 5 launcher uses the Russian built RD-180 engine for its main stage, this restriction essentially prohibits the use of the RD-180 engine after 2019.[150] However, the act exempts ULA's use of the nearly 2-year supply of RD-180 engines already in its possession; it also exempts a batch of 29 RD-180 engines that were ordered while the NDAA-15 was still pending.[151] Moreover, the NDAA-15 calls for a demonstration of a U.S. made replacement engine by 2019.[152]

---

[146] Leone, Dan. "NASA Budget Bill Would Boost Commercial Crew, Keep SOFIA Flying, Kill Hosted Climate Instrument." 7 May 2014. SpaceNews 22 July 2015. http://spacenews.com/40494nasa-budget-bill-would-boost-commercial-crew-keep-sofia-flying-kill-hosted/.

[147] Section 233. National Defense Authorization Act for Fiscal Year 2014, Pub. L. no 113-66 (2013). http://www.gpo.gov/fdsys/pkg/CPRT-113HPRT86280/pdf/CPRT-113HPRT86280.pdf.

[148] Gruss, Mike. "U.S. Defense Bill Could Undercut Turkey's Missile Defense Deal with China." 10 Jan. 2014. SpaceNews 26 May 2015. http://spacenews.com/39038 us-defense-bill-could-undercut-turkeys-missile-defense-deal-with-china/.

[149] Section 1608. Carl Leven and Howard P. "Buck" McKeon National Defense Authorization Act for Fiscal Year 2015, Pub. L. no 113-291 (2014). http://www.gpo.gov/fdsys/pkg/CPRT-113HPRT92738/pdf/CPRT-113HPRT92738.pdf.

[150] Gruss, Mike. "Compromise Appropriations Bill Includes $220 Million for RD-180 Replacement." 11 Dec. 2014. SpaceNews 22 July 2015. http://spacenews.com/compromise-appropriations-bill-includes-220-million-for-new-rocket-engine/.

[151] Gruss, Mike. "House-Senate Conference Measure To End Pentagon Use of RD-180." 5 Dec. 2014. SpaceNews 22 July 2015. http://spacenews.com/42701house-senate-conference-measure-to-end-pentagon-use-of-rd-180/.

[152] Gruss, Mike. "Compromise Appropriations Bill Includes $220 Million for RD-180 Replacement." 11 Dec. 2014. SpaceNews 22 July 2015. http://spacenews.com/compromise-appropriations-bill-includes-220-million-for-new-rocket-engine/.

1.2.5 Canada

On 7 February 2014, the Canadian government released Canada's Space Policy Framework, which outlines broad national goals for its space programme. The framework was developed in response to a major study ordered by the Canadian government on the country's aerospace and space sectors, which was released on 29 November 2012, and had called on the government to recognize the importance of space to national security and economic prosperity. Its core principles are based on the following: Canadian Interests First, Positioning the Private Sector at the Forefront of Space Activities, Progress through Partnerships, Excellence in Key Capabilities, and Inspiring Canadians.[153] The new policy framework calls for the continuation of Canada's astronaut programme, further investments in the development of advanced systems and scientific instruments as part of major international missions, and an increase in support for technology development, especially in areas of proven strength among domestic firms.[154] Nevertheless, the space budget for the Canadian Space Agency (CSA) was expected to drop by 13 % to CDN$260 million ($235 million) by 2014–2015.

At the beginning of 2014, Canada's Department of National Defence (DND) and the U.S. Strategic Command signed an updated accord permitting the exchange of advanced space situational awareness data. The agreement streamlines the request process to gather SSA data for satellite manoeuvre planning, collision avoidance, and anomaly mitigation from the U.S. Strategic Command's Joint Space Operations Centre located at Vandenberg Air Force Base in California.[155]As of September 2014, the U.S. government has signed nearly 50 data-sharing agreements with other governments, e.g. the UK, Japan, Australia, Italy, France, and the Republic of Korea, as well as private sector entities.[156] Canada has also signed a memorandum of understanding with the U.S., Australia, and the UK, to work more closely together on SSA activities, allowing for more effective and coordinated use of capabilities.[157]

---

[153] "Canada's Space Policy Framework | Launching the Next Generation." 7 Feb. 2014. CSA 29 June 2015. http://www.asc-csa.gc.ca/pdf/eng/publications/space-policy/canadas-space-policy-framework.pdf.

[154] Pugliese, David. "Canadian Policy Outlines Broad National Goals for Space Program." 7 Feb. 2014. SpaceNews 29 June 2015. http://spacenews.com/39412canadian-policy-outlines-broad-national-goals-for-space-program/.

[155] Ferster, Warren. "U.S., Canadian Governments Sign SSA Data Sharing Accord." 9 Jan. 2014. SpaceNews 25 June 2015. http://spacenews.com/39003 U.S.-canadian-governments-sign-ssa-data-sharing-accord/.

[156] Gruss, Mike. "UK Joins List of U.S. Allies Agreeing To Strengthen Space Surveillance Sharing." 25 Sept. 2014. SpaceNews 25 June 2015. http://spacenews.com/41995uk-joins-list-of-U.S.-allies-agreeing-to-strengthen-space-surveillance/.

[157] Gruss, Mike. "News from the 30th Space Symposium | U.S., Three Allies Sign Space Situational Awareness Accord." 22 May 2014. SpaceNews 25 June 2015. http://spacenews.com/40651news-from-the-30th-space-symposium-us-three-allies-sign-space/.

1.2.6 Russia

Following a July 2013 review by Russia's Audit Chamber, which found Russia's Federal Space Program for 2006–2015 to be ineffective, due largely to the poor management of space activities and budget funds allocated for space projects, Russia was in the midst of revising its space programme in 2014.[158] Following a December 2013 decree by Russia's President Vladimir Putin ordering the creation of the United Rocket and Space Corporation (URSC), Russia's government began taking steps to reorganize its space industry within the umbrella of the state-owned URSC.[159] By the end of 2014, Russia's government decided to merge its Federal Space Agency, Roscosmos, with the URSC instead of arranging an arms-length customer and contractor relationship between the organizations, in hopes of improving the sector's efficiency.[160] The consolidation stage is to be completed in mid-2015.

Moreover, Russia plans to spend an estimated 2.1 trillion roubles (~$63 billion) including extra-budgetary sources, to boost the development of its national space activities from 2013 to 2020. In addition to enabling effective participation in forward-looking projects, such as the ISS, the study of the Moon, Mars and other celestial bodies in the solar system, the programme is designed to maintain Russia's leading position as a global space power, while also supporting its defence capability, and boosting economic and social development.[161] The government's decision to increase Russia's space budget will enable Russia to surpass China and reach spending parity with Europe. With only one-fifth of its domestic demand for geospatial imagery currently able to be met by Russia's own satellites and with Russian meteorological satellites being below international standards, this funding could help Russia to redress its past underinvestment in Earth observation and meteorology satellites.[162]

---

[158] "Russia's Space Program Is Ineffective—Audit Chamber." 4 July 2013. RIA NOVOSTI 30 Apr. 2014. http://en.ria.ru/russia/20130704/182063035.html.

[159] "Putin Signs Decree on Creation of United Rocket and Space Corporation." 2 Dec. 2013. ITAR TASS 30 Apr. 2014. http://en.itar-tass.com/russia/709849.

[160] "Duma proposes to merge Roscosmos with United Rocket Space Corporation." 20 Dec. 2014. TASS.ru 21 July 2015. http://tass.ru/en/non-political/768257.

[161] "Russia Launches $70 Bln Space Program for 2013–2020." 27 Dec. 2012. RIA NOVOSTI 30 Apr. 2014. http://en.ria.ru/science/20121227/178432916/russia_launches.html.

[162] "Russia Boosting Space Budget To Surpass China, Equal Europe." 5 June 2013. SpaceNews 30 Apr. 2014. http://www.spacenews.com/article/civil-space/35638russia-boosting-space-budget-to-surpass-china-equal-europe.

1.2.7 Japan

On 20 August 2014, Japan's government released its policy report "A Comprehensive Space Strategy to Implement Japan's National Strategy" as an update to Japan's 5-year Basic Plan of 2012, and to function as a blueprint to realign Japan's space activity and space spending with security and economic priorities. The report recommends the creation of a national security-focused space agency in addition to building closer collaboration with the U.S. in areas such as navigation, maritime surveillance, space situational awareness and missile warning.[163] The document recommends an expansion of Japan's QZSS navigation system to a full seven satellite constellation, in addition to expanding the number of Japan's optical and radar information gathering satellites (IGS) from the current four to ten satellites, including data-relay communication satellites.[164] Yet there remains some contention relative to Japan's Ministry of Education, Sports, Science and Technology (MEXT), which governs Japan's civilian space agency JAXA, and views the creation of another agency as affecting its own role and budget.

The new focus of Japan on space and security has been in development since 2008, when the Japanese government passed its 5-year Basic Plan. In June 2012, Japan's government passed legislation implementing the Basic Plan, enabling the Prime Minister's Cabinet Office to take control of the planning and budgeting of Japan's government space program, and providing JAXA with the ability to pursue military space programs.[165] On 25 January 2013, Japan's Space Strategy Headquarters released the latest version of its Basic Plan that lays out the priorities for most of Japan's space development for the 5 years that started in April 2013. The new Basic Plan has three main targets: national security and disaster management, development of industries, and space science.[166]

[163] Kallender-Umezu, Paul. "Profile I Hiroshi Imazu, former Chairman, Space Policy Committee, Liberal Democratic Party of Japan." 27 Oct. 2014. SpaceNews 30 June 2015. http://spacenews.com/42331profile-hiroshi-imazu-former-chairman-space-policy-committee-liberal/.

[164] "Recommendation for a Comprehensive Space Strategy to Implement Japan's National Strategy." 26 Aug. 2014. Hiroshi-i.net 30 June 2015. http://hiroshi-i.net/10/wp-content/uploads/b785de34e6cd85a6423eb90542522d2d.pdf.

[165] Kallender-Umezu, Paul. "Japan Passes Overhaul of Space Management Structure." 2 July 2012. SpaceNews 30 Apr. 2014. http://spacenews.com/article/japan-passes-overhaul-space-management-structure.

[166] "Profile I Naoki Okumura, President, Japan Aerospace Exploration Agency." 9 Dec. 2013. SpaceNews 30 Apr. 2014. http://spacenews.com/search?search_api_views_fulltext=Japan&sort_by=created&sort_order=DESC&page=2

1.2.8 China

Chinese space policy revolves around the country's 5-year economic development plans. The country's space programme is therefore meant to support its overall development objectives, while maintaining a comprehensive set of objectives for space activities.

In 17 October 2014, China completed the construction of its fourth and largest spaceport, the Wenchang Satellite Launch Centre on the north-eastern coast of the southern Hainan Island.[167] At an estimated total cost of around $810 million this new spaceport is meant to lift large space stations and manned missions to Moon and Mars.[168] The facility will accommodate launches of both Long March 5 and Long March 7; both of which are expected to fly before the end of 2015 as part of China's 12th 5-year plan for space. The Long March 5 is closely tied to the development of China's space station, in addition to its developing Moon exploration programme, in which it will be mainly used for the lofting of China's manned space station, with the capacity to carry an 18 metric tons payload to NEO. The Long March 7 will be able to launch 12 metric tons into NEO, and will be capable of launching a cargo spacecraft to the country's manned space station. That said, a new Tiangong 2 space laboratory is planned for launch into orbit by 2015, followed by an experimental core module of the larger space station, and by 2022, China's manned space station should be completed. Both the Long March 5 and Long March 7 have been improved by using non-toxic and pollution-free propellant.[169]

While the Long March 5 to Long March 7 should be able to undertake the country's space activities planned for the coming 10 years, their capabilities will not be enough for China's long-term space programmes which require a super-heavy rocket for deep-space exploration projects. Hence, China has begun planning a Long March 9 Super-Heavy launcher which is intended to be more powerful than the Saturn V launcher of the Apollo Moon missions, and match the lift capacity of NASA's planned SLS Block 2 launcher. Still on the drawing board, the Long March 9 is expected to have a maximum launch capacity of 130 metric tons to LEO, and is expected to conduct its first launch in 2028.[170]

In the field of satellite development, the new Chinese space policy describes a comprehensive programme embracing all fields of satellite and spacecraft use. It calls for the development of improved weather and communications satellites, as well as of

[167] Wan, Adrian. "China's fourth spaceport completed in boost for space programme." 17 Oct. 2014. South China Morning Post 20 July 2015. http://www.scmp.com/news/china/article/1618422/construction-mainlands-fourth-spaceport-finished.

[168] "Chinese Rocket Engine Test a Big Step for Space Station Project." 15 July 2013. Space.com 2 May 2014. http://www.space.com/21957-china-rocket-engine-test-space-station.html.

[169] "Long March 7 Launch Vehicle Completed Booster Engine Firing Test." 26 Nov. 2013. China Manned Space Engineering 2 May 2014. http://en.cmse.gov.cn/show.php?contentid=1355.

[170] Lei, Zhao. "New Long March launcher on the drawing board." 9 Dec. 2014. SpaceDaily 16 July 2015. http://www.spacedaily.com/reports/New_Long_March_launcher_on_the_drawing_board_999.html.

an entirely new Earth observation and electromagnetic monitoring satellite series. The key plank of this programme will be the development of a satellite fleet capable of all-weather 24-hour operations worldwide, which would imply making significant advancements in space borne SAR and high resolution optical instrument technologies. By the end of 2014, China had launched three Earth observation satellites, including the Gaofen 2, CBERS (a joint collaboration with Brazil), and Fengyun 2G, in addition to 11 Yaogan series remote sensing satellites (Yaogan 20A, 20B, 20C, 21, 22, 23, 24, 25A, 25B, 25C, and Yaogan 26) for China's defence programme.[171]

1.2.9 India

India's government approved its 12th 5-year plan (2012–2017) in October 2012, which puts emphasis on speeding up the implementation of infrastructure projects, and aims to achieve an annual average economic growth rate of 8.2 %.[172] However, India's space programmes are driven by a decade profile and directions for 2025. Over the next decade, India's space programme aims to strengthen its operational services in communications and navigation; enhance its remote sensing imaging capability; continue space science missions for better understanding of the solar system and the universe; and to build on its experience in planetary exploratory missions. For instance, ISRO's newly launched Mars Orbiter Mission (MOM), which lifted into space on 5 November 2013, reached the planet on 24 September 2014.[173] On reaching Mars, the spacecraft—carrying five indigenous scientific payloads consisting of a camera, two spectrometers, a radiometer, and a photometer—began observing the Mars surface, atmosphere and exosphere extending up to 80,000 km to gain a better understanding of the planet's evolution.[174] Also, in September 2014, NASA and ISRO officials signed an agreement to establish a working group to explore potential coordinated observations and science analysis of Mars.[175]

[171] Federal Aviation Administration. The Annual Compendium of Commercial Space Transportation: 2014. Washington DC: FAA, Feb. 2015: 190.

[172] "Government Approves 12th Five Year Plan." 4 Oct. 2012. The Times of India 3 May 2014. http://timesofindia.indiatimes.com/business/india-business/Government-approves-12th-five-year-plan/articleshow/16672927.cms.

[173] "Mars Mission: India creates history as Mangalyaan successfully enters Mars orbit in first attempt." 24 Sept. 2014. The Economic Times 22 Mar. 2015. http://economictimes.indiatimes.com/news/science/mars-mission-india-creates-history-as-mangalyaan-successfully-enters-mars-orbit-in-first-attempt/articleshow/43299562.cms.

[174] "Indian Space Research Organisation | Mars Orbiter Mission." Indian Space Research Organisation 31 Mar. 2014. http://www.isro.org/mars/home.aspx.

[175] Laxman, Srinivas. "Mars Orbiter Mission activates all science instruments as NASA, ISRO form joint Mars working group." 1 Oct. 2014. Planetary.org 22 Mar. 2015. http://www.planetary.org/blogs/guest-blogs/2014/10010914-mars-orbiter-mission.html.

India's space programme also aims to build on its development of heavy lift reusable launch vehicles, and to develop its own human space flight programme. While a human spaceflight mission will not be commenced before the year 2017, there are funds in the 12th five-year plan to continue with pre-project studies and to develop critical technologies associated with the proposed mission. Following the successful launch of India's Geosynchronous Satellite Launch Vehicle (GSLV)-Mark II on 5 January 2014, India now has a launcher capable of lifting a two-member crew to LEO orbit.[176] Furthermore, its GSLV-Mark III, which had its first sub-orbital test launch on 18 December 2014, will have additional mass left for conducting scientific experiments in addition to hosting a two-member crew.[177] The GSLV-Mark III launcher will begin commercial launches starting in 2017.[178]

Lastly, ISRO's Chairman Koppillil Radhakrishnan retired on 31 December 2014, following the success of ISRO's MOM orbiter, and the development of GSLV, and a list of achievements since his appointment as Chair in October 2009.[179] His replacement was to be announced at the beginning of 2015.

1.2.10 Brazil

Despite long-standing obstacles for Brazil's space and R&D development, and the lack of expertise in its workforce, Brazil is becoming an important space player of the future due to its geographic position close to the equator, and its growing economy.

In terms of partnerships with space powers around the globe, Brazil in collaboration with China, launched the CBERS-4 Earth observation satellite on 7 December 2014. The China-Brazil Earth Resources Satellite (CBERS) project has been an outlet for the two nations to have an ongoing global strategic partnership since 1988.[180] The first CBERS-1 launched in 1999, followed by CBERS-2 in 2003, CBERS-2B in 2007, and the failed launch of the CBERS-3 satellite on 9 December 2013 due to a malfunction in the third stage of China's Long March 4B rocket.[181]

---

[176] "GSLV." ISRO 19 Apr. 2013. http://www.isro.org/launchvehicles/GSLV/gslv.aspx.

[177] "India Not to Undertake Human Space Flight Before 2017: ISRO." 17 Sept. 2012. The Economic Times 3 May 2014. http://articles.economictimes.indiatimes.com/2012-09-17/news/33902713_1_cryogenic-engine-gslv-mk-iii-radhakrishnan-today.

[178] De Selding, Peter B. "65th International Astronautical Congress | India Poised To Expand Presence in Global Launch Market." 1 Oct. 2014. SpaceNews 14 July 2015. http://spacenews.com/4205565th-international-astronautical-congress-india-poised-to-expand-presence/.

[179] Jayaraman, K.S. "Nayak Named Interim Chief at ISRO." 6 Jan. 2015. SpaceNews 19 Aug. 2015. http://spacenews.com/nayak-named-interim-chief-at-isro/.

[180] Ibid.

[181] "China Blames Long March Failure on Third-stage Malfunction." 10 Dec. 2013. SpaceNews 3 May 2014. http://www.spacenews.com/article/launch-report/38585china-blames-long-march-failure-on-third-stage-malfunction.

Brazil is also undertaking cooperative programs with Ukraine to build six Cyclone-4 rockets that could launch heavy satellites into low Earth orbit (LEO) and small communications satellites to geostationary transfer orbit as part of Brazil's goal to have an autonomous operational launch capability. Brazil and Ukraine created the binational company "Alcântara Cyclone Space" (ACS), with financing divided equally, for commercial launches using the Ukrainian vehicle Cyclone-4 from the Alcântara Launch Center (CLA) whose proximity to the equator gives the launch vehicle more capacity to orbit than can be offered from Russian spaceports.[182] The launcher development program is estimated to cost 1.58 billion reals ($802.5 million) over the 10-year period. The Cyclone-4 program, listed as a separate budget item, is allocated an additional 459.8 million reals ($233.8 million) during the 10-year period. Moreover, around 1.9 billion reals ($970 million) has been allocated for space infrastructure improvements, mostly going toward development of the ALC.[183] In 2014, Brazil's launcher development programme continued to receive the lion's-share of funding within the space programme, with 53 % of the total space budget.[184] Brazil hopes to launch the Cyclone-4 rocket from the ALC sometime in 2015.

Other launch vehicles under development in Brazil include its Satellite Launch Vehicle (VLS-1) project and its Microsat Launch Vehicle (VLM) project. The VLS-1 project was paused in 2003, following an explosion on the launch pad killing 21 people, but development picked up again late in 2013, albeit with a shortage of funding and trained personnel. Under a previous agreement between Roscosmos and the Brazilian Space Agency (AEB), Russia will provide launcher technology to help complete Brazil's VLS-1 project.[185] Moreover, Brazil has partnered with the Germany DLR to develop its VLM, which was originally intended to be for suborbital mission purposes, but has evolved to have small satellite launch capabilities.[186]

And in December 2014, it appeared that Russia might increase its ties with Brazil, as it is considering moving its Sea Launch mobile platform and command ship from the coast of California, USA, to the coast of Brazil. However, for Brazil involvement with the Russian-Ukrainian owned Sea Launch AG has some drawbacks, including the use of foreign produced Zenit launchers, possible incompatibility with future plans for a follow-on Cyclone 5 rocket, and uncertainty regarding

---

[182] "News from the 64th International Astronautical Congress | Backers Insist Brazilian Spaceport Is Nearing Launch Readiness." 4 Oct. 2013. SpaceNews 3 May 2014. http://www.spacenews.com/article/launch-report/37550news-from-the-64th-international-astronautical-congress-backers-insist.

[183] "Brazil Scales Back Launch Vehicle Plans." 10 Feb. 2013. Parabolic Arc 3 May 2014. http://www.parabolicarc.com/2013/02/10/brazil-scales-back-launch-vehicle-plans/.

[184] Euroconsult 2015. Profiles of Government Space Programs.

[185] "Russia Offers Brazil New Joint Space Projects." 20 Oct. 2013. RIA NOVOSTI 3 May 2014. http://en.ria.ru/russia/20131020/184250734.html.

[186] "Brazil Scales Back Launch Vehicle Plans." 10 Feb. 2013. Parabolic Arc 3 May 2014. http://www.parabolicarc.com/2013/02/10/brazil-scales-back-launch-vehicle-plans/.

Sea Launch's reliability which has been plagued by failures, and a gap in planned launches.[187]

1.2.11 Emerging Space Powers

In line with the development of South Korea's upgraded KSLV-2 launcher, the government intends on sending the KSLV-2 to the Moon, carrying a probe and lunar rover to search for rare minerals on the surface by 2020.[188] In November 2014, the South Korean government reached the budget deliberation stage after the project had passed the feasibility study phase. South Korea's current Park Geun-hye government has made it a priority to develop its lunar orbiter and lunar module by 2020; 5 years earlier than planned by the previous Roh Moo-hyun administration's Space Development Roadmap and the Lee Myung-bak administration's Basic Promotion Plan for Space Development. The government anticipates the economic value of the mission to be approximately 3.8 trillion won ($3.469 billion), or 5.17 times the budget input, and it expects to create 324.6 billion won ($296 million) of added value along with 4800 jobs. To accelerate development, South Korea's government signed a joint research agreement with NASA in July 2014, wherein KARI benefits from the core orbiter and probe technology transferred, while the U.S. is able to do more economically attractive space research.[189]

In the Middle East region, Iran continued to establish itself as a regional space actor through the implementation of a comprehensive space programme. On 10 April 2014, Iran was said to have signed a protocol on cooperation with Russia's Roscosmos on wide cooperation in space exploration, ranging from training Iranian cosmonauts in Russia to potential technology transfers in developing remote sensing and telecommunication satellites. Iran also intends to build domestic communication stations capable of receiving information from Russia's Resurs-DK and Resurs-P satellites, and it may contract with Russia for the construction and launch of a reconnaissance satellite and a telecommunications satellite, which might concern the U.S. and parts of the Middle East due to its potential civilian and military applications.[190]

[187] Messier, Doug. "Will Sea Launch End Up in Brazil?" 24 Dec. 2014. Parabolic Arc 20 July 2015. http://www.parabolicarc.com/2014/12/24/sea-launch-brazil/.

[188] "South Korea Reveals Moon-lander plans." 13 Nov. 2013. Nature.com 4 May 2014. http://www.nature.com/news/south-korea-reveals-moon-lander-plans-1.14159.

[189] "Korea Ready to Become 7th Country to Reach the Moon." 13 Nov. 2014. Business Korea 20 July 2015. http://www.businesskorea.co.kr/article/7278/lunar-exploration-project-korea-ready-become-7th-country-reach-moon.

[190] "Russia to train Iranian cosmonauts, build recon sats—report." 7 May 2014. RT.com 20 July 2015. http://www.rt.com/news/157496-russia-iran-space-satellite/.

1.3 Worldwide Space Budgets and Revenues

Total government space expenditure in 2014 increased substantially, reaching \$79.17 billion from \$74.10 billion in 2013.[191] Total government expenditure for civil space programs amounted to \$43.84 billion in 2014. In 2014, the compound annual growth rate (CAGR) of the entire space industry, including commercial revenues and government expenditure was 9.1 %, increasing significantly from the 5.9 % growth experienced in 2013.[192] The following section provides a more detailed analysis of institutional budgets.

The 2015 Space Report indicates that the 2014 total revenue of commercial satellite services grew slightly, by 0.5 %, reaching \$123.18 billion including telecommunications, Earth observation and positioning services. However, revenue from space-related commercial infrastructure including manufacturing of spacecraft and in-space platforms, launch services as well as ground equipment is estimated to have grown by 8.6 % reaching \$127.65 billion. And overall, total commercial space revenue reached \$250.83 billion in 2014.[193]

1.3.1 Overview of Institutional Space Budgets

According to the Space Report 2015, total institutional spending on space programs in 2014, including that of intergovernmental organisations, increased by 6.8 % to \$79.17 billion, with a notable increase of 12.9 % by non-U.S. government space actors which are investing in new capabilities or expanding existing ones.[194] Global 2014 space spending was comprised of \$43.84 billion in civil expenditure (55.4 % of the total share), and \$35.33 billion in defence expenditure (44.6 %), showing a slight shift toward increased defence space spending from 2013.[195] In contrast, Euroconsult listed 2014 civil space expenditure to be \$42.03 billion and government expenditure for defence space programmes to be \$24.45 billion; changing the civil-to-defence ratio in the opposite direction, i.e. 63.2 % civil expenditure, 36.8 % defence.[196]

In defence spending, the Space Report 2015 estimates that worldwide defence related expenditure reached \$35.33 billion in 2014, with 63.6 % of this amount (\$22.483 billion) spent by the United States for space security programmes under its Department of Defence (DoD). The U.S. DoD's space budget funds its military space programmes, in addition to organisations such as the National

[191] The Space Report 2015. Colorado Springs: The Space Foundation, 2015: 14.

[192] Ibid. at 1.

[193] Ibid.

[194] Ibid. at 2.

[195] Ibid. at 23–24.

[196] Euroconsult 2015. Profiles of Government Space Programs: 16.

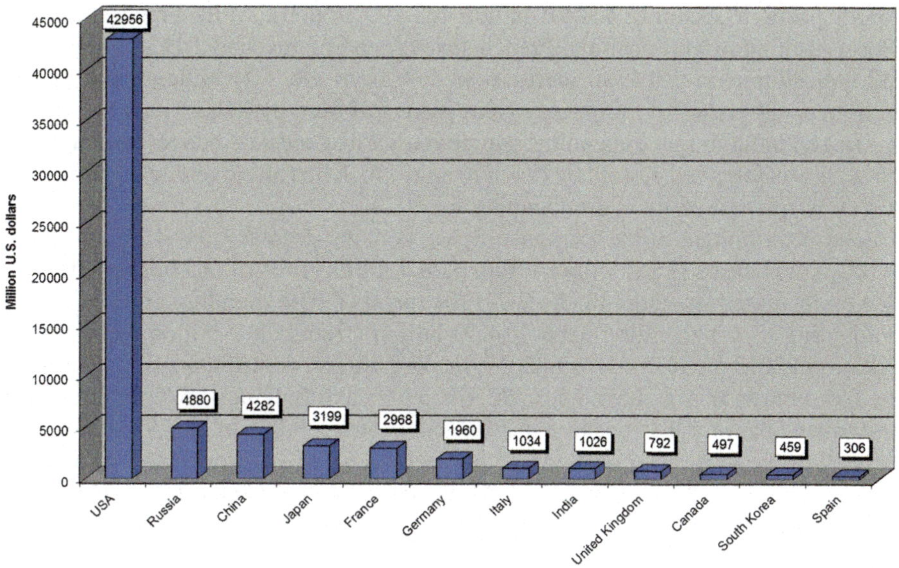

Fig. 1.1 Public space budgets of major space powers in 2014 (Based on Euroconsult and the Space Report 2015 data).

Reconnaissance Office (NRO) and the National Geospatial-Intelligence Agency (NGA). Moreover, total defence spending by non-U.S. government space actors has continued to increase, reaching a 36.4 % share ($12.84 billion), from 31.7 % ($10.33 billion) in 2013.[197] However, it should be noted that a degree of uncertainty exists regarding expenditures on defence space activities as not all relevant funding is made public.

The expenditure hierarchy among states saw significant changes in 2014 (Fig. 1.1).[198] The United States had the largest space budget, placing $20.473 billion toward civil purposes, and $22.483 billion toward defence.[199] On the other hand, prior to the reorganisation of Russia's space industry, the Roscosmos space budget for 2014 amounted to approximately 165.814 billion roubles ($4.88 billion).[200] China remained in third position, spending an estimated 26.564 billion Yuan ($4.282 billion) in space activity; this estimate comes from the average space spending as a percentage of GDP that is used in the Space Report 2015, i.e. 0.041 %, multiplied by China's 2014 GDP of 64,452.88 billion Yuan.[201] Japan remained in

[197] The Space Report 2015. Colorado Springs: The Space Foundation, 2015: 39.

[198] N.B.: Figures in this section are based on the Space Report 2015 data (U.S.A, Russia, Japan, China, and France), while all other values in Fig. 1.1 comes from the Euroconsult Report 2015—Profiles of Government Space Programs.

[199] The Space Report 2015. Colorado Springs: The Space Foundation, 2015: 23.

[200] Ibid. at 36.

[201] Ibid. at 27.

fourth position spending ¥324.0 billion ($3.199 billion), while France followed closely behind in fifth position. French space spending reached 2.153 billion euros ($2.968 billion) in 2014, an increase of 5 % from the 2.04 billion euros ($2.65 billion) spent in 2013.[202] Germany came next, followed by Italy, which had barely overtaken India in space spending. Moreover, while Canada's budget decreased by 26 %, it was followed closely by South Korea, which had also overtaken Spain, with a 46.6 % increase in its space spending for 2014.

The 2014 budget of the European Space Agency decreased by 4.2 % reaching 4.102 billion euros ($5.65 billion) from 4.282 billion euros ($5.65 billion) in 2013, due to decreased spending by the EU. ESA member state spending grew by 7.4 %, amounting to 3.339 billion euros ($4.60 billion) from 3.110 billion euros ($4.11 billion) in 2013.[203] In 2014, as in 2013, the five biggest contributors were Germany 18.7 %, France 18.4 %, Italy 8.5 %, the UK 6.6 %, and Belgium 4.6 %. Belgium had overtaken Spain for fifth position in 2013, contributing 6.0 % of the 2013 budget.[204] Among the countries assessed within this section, 13.9 % of world concentration of space expenditure was generated by Asian space powers, China, Japan, India, and South Korea, whereas the remaining 86.1 % came from the U.S., Europe (including ESA contributions), and Russia in 2014.

1.3.2 Overview of Commercial Space Markets

In 2013 (latest year for which figures were available), global industry revenues, including revenue from satellite services, satellite manufacturing, launch industry, and ground equipment continued to grow as shown in both Satellite Industry Association (SIA) and Space Report 2014 assessments. The SIA reported global satellite revenues at $195.2 billion in 2013, from $189.5 billion in 2012,[205] whereas the Space Report lists total commercial revenues to have grown to $240.07 billion in 2013, from $225.87 billion in 2012.[206] However, it must be clarified that these authorities appear to use different methods of assessment. As indicated in previous assessments, there is a discrepancy in the findings of SIA and the Space Report, amounting to a difference of $44.87 billion between the 2013 figures, and $36.37 billion in 2012 figures; a likely consequence of the different reporting methodologies. The following section presents key figures and data on commercial space

[202] Ibid. at 29–34.

[203] "ESA Budget for 2014." ESA 26 June 2015. http://www.esa.int/For_Media/Highlights/ESA_budget_2014.

[204] "ESA Budget 2013." 24 Jan. 2013. ESA 2 Feb. 2014. http://www.esa.int/About_Us/Welcome_to_ESA/Budget_as_presented_during_DG_press_conference_24_January_2013.

[205] State of the Satellite Industry Report." May. 2014. Satellite Industry Association and The Tauri Group 24 May 2014. http://www.sia.org/wp-content/uploads/2014/05/SIA_2014_SSIR.pdf .

[206] C.f. The Space Report 2014, 24, and The Space Report 2013, 26.

activities divided by field of activity, based primarily on available SIA figures generated by the Tauri Group, in addition to previous Futron reports.

Satellite Services

A sustained expansion of satellite capacity and corporate revenue was fuelled by continued demand from emerging economies and developing regions. Worldwide satellite capacity jumped by 64 % from 900 36 MHZ transponder equivalents in 2011 to exceed 1400 transponder equivalents in 2012.[207] However, in the following year, budget crunches in the U.S. likely resulted in the reduction of transponder equivalents to around 800 for the year 2013.[208] This notwithstanding, satellite services have maintained a consistent share of overall space related revenues: 60.8 % in 2011, 59.9 % in 2012, and 60.8 % in 2013. Satellite services earned $118.6 billion in 2013, from $113.5 billion in 2012, due mostly to the increase in direct to home (DTH) satellite services. However, the rate of revenue growth for satellite services has continued to decrease, declining from 16 % growth in 2008, to rest at 5 % for both 2012 and 2013.[209]

Satellite services can further be deconstructed into their component parts, including consumer services (satellite radio (DARS), and consumer satellite broadband), fixed satellite services (e.g. transponder agreements, and managed services), as well as mobile services (voice and data), and remote sensing. The following is a breakdown of the industry's key developments and trends, according to the nature of the services provided.

Consumer Services

As mentioned above, consumer services are made up of satellite television, radio, and broadband services. Direct Broadcast Services (DBS) also include direct-to-home satellite television. This section of the industry showed considerable development in 2013, fuelled by a quantitative expansion in emerging markets and a qualitative increase in new technologies and services in developed ones. While demand revenue has increased in recent years, its rate of growth slipped from 9.5 % in 2011 to 5.3 % in 2012 and 4.8 % in 2013. Moreover, DBS revenue increased by just 4.7 % to $88.4 billion in 2012, and by another 4.7 % to $92.6 billion in 2013. This growth was mainly attributed to the number of HDTV channels in recent years, this growth amounted to 3853 channels or a 42 % increase by 2011, 4768 channels or a 24 % increase by 2012, and around 6246 channels or a 31 % increase by 2013.[210] In 2012 and 2013, consumer satellite television services accounted for about 78 % of total satellite services revenue, with the share of available HDTV channels servicing Europe and Asia increasing to 41 %, while the share of available HDTV channels in the Americas reduced from 70 % in 2011 to around 59 % by 2013. In fact, while the annual growth rate of DBS was expected to exceed that of the rest of the satellite services' sector with DBS replacing more traditional services

[207] Satellite Telecommunications Report—2012 Year-End Summary. Futron: 1.

[208] Satellite Telecommunications Report—2013 Year-End Summary. Futron: 1.

[209] 2014 State of the Satellite Industry Report: 11.

[210] Ibid.

such as video distribution, the growth rate of satellite radio and satellite broadband advanced in both 2012 and 2013.[211] In 2013, satellite radio revenue grew from $3.0 billion in 2011 to $3.4 billion in 2012, a 13.3 % increase, and by another 11.8 % to reach $3.8 billion in 2013. Similarly, satellite broadband revenue grew by 25 %, from $1.2 billion in 2011 to $1.5 billion in 2012, and by 13.3 % to $1.7 billion in 2013.[212] Additionally, the majority of worldwide broadband satellite revenue continued to be generated in the U.S., earning around 70 % of the revenue of 2013. All major satellite operators have increased their investments in new technologies and products in developed markets, while new services entered into operation by the end of 2013.

Fixed Satellite Services

Fixed Satellite Services (FSS) refers to the use of spacecraft that utilise land terminals in fixed positions to broadcast. Whereas Consumer Services covers satellite broadband Internet, communications and network television and radio broadcasts, FSS relates to commercial signal agreements, such as transponder agreements and managed network services. From 2011 to 2013 the FSS outlook remained positive, as operators continued to profit from previous investments in new capacity.

While some reports had anticipated a decrease in new satellite investment leading up to 2012-2013, the effect of the current boom in FSS is expected to continue throughout the decade. Industry-wide FSS revenue climbed by 4.7 % to $15.7 billion in 2011, by another 4.5 % to $16.4 billion in 2012, and remained unchanged at $16.4 billion in 2013. The growth is explained by the continued demand for video and broadband, mainly from the Americas, but with additional growth in Europe and Asia. In 2014, the revenue ranking order of the top FSS operators maintained their position; however, the sharp rise of the U.S. dollar against most major currencies at the end of the year had the effect of showing revenue declines, whereas actual revenue growth had been reported in local currencies.[213]

To handle the increase in demand, coming especially from HDTV consumers, commercial operators invested in technological upgrades that let them meet the needs of consumers while still making significant profits. Transponder agreement revenue continued to expand in 2011 and 2013, with growth evident across multiple regions. Similarly, while the number of HDTV channels has jumped in successive years from around 1500 in 2009 to over 6000 in 2013, with nearly 59 % serving the Americas, growth continues in Europe and Asia.[214]

---

[211] Ibid.

[212] Ibid.

[213] De Selding, Peter B. "The List | Top Fixed Satellite Service Operators." 13 July 2015. Space News 18 Aug. 2015. http://spacenews.com/the-list-2014-top-fixed-satellite-service-operators/.

[214] State of the Satellite Industry Report." Oct. 2013. Satellite Industry Association and The Tauri Group 24 May 2014: 11–14. http://www.sia.org/wp-content/uploads/2014/05/SIA_2014_SSIR.pdf.

Mobile Satellite Services
Mobile satellite services offer both mobile data service and mobile voice service (including satellite phones). Over the course of 2011 and 2012, mobile satellite services earned revenue of $2.4 billion each year, thereafter increasing by 8.3 % to $2.6 billion in 2013. Within the Mobile Satellite Services segments, revenue earned by mobile voice services has remained relatively unchanged since 2009, staying at $700 million in 2012, and increasing to $800 in 2013; whereas mobile data services revenue has increased by $100 million increments since 2009, reaching $1.8 billion in 2012, and staying at that level in 2013. The latter segment comprises nearly 70 % of all mobile satellite services revenue.[215]

Remote Sensing
Remote sensing refers to commercial companies that provide optical and radar images to the open market; however, they are mostly used by government entities that have been increasingly outsourcing such capabilities over the past few years. While commercial remote sensing revenue increased by 42.9 % in 2009, it stayed at the $1 billion benchmark throughout 2010, growing by 10 % to $1.1 billion in 2011, by 18.2 % to reach $1.3 billion in 2012, and by 15.4 % to $1.5 billion in 2013.[216] U.S. government demand in addition to good industry performance were the dominant factors boosting remote sensing services providers' revenue over the course of 2011 to 2013.

In 2013, around 16 civilian remote sensing satellites were launched, in addition to 11 military surveillance satellites. While the total number of civilian and military remote sensing satellites in orbit increased in 2012 and 2013, the civil industry continues to outpace military spacecraft by a significant margin, indicating a shift in where government funding is directed, and suggesting that states are looking more toward PPPs in this field.

Satellite Manufacturing
The total revenue of satellite manufacturers that built satellites both for governmental and commercial launches in 2011 amounted to $11.9 billion; an increase of 11 % from 2010. Satellite manufacturing revenue increased by another 23 % in 2012, earning $14.6 billion and increased by a further 7.5 % to reach $15.7 billion in 2013.[217] As displayed in Fig. 1.2, manufacturing revenue exceeded the growth trend forecasted by Futron in 2010, with actual revenue surpassing the $15 billion revenue expected in 2013.[218] It should be noted that the U.S. earned 56.2 % of 2012 manufacturing revenue, amounting to $8.2 billion, with 61 % of those earnings coming from U.S. government contracts.[219] In 2013, the U.S. earned 69.2 % of

[215] Ibid. at 11.

[216] 2014 State of the Satellite Industry Report, 11.

[217] State of the Satellite Industry Report, 16.

[218] 2010 Futron Forecast of Global Satellite Services Demand: Executive Summary.

[219] State of the Satellite Industry Report, 16.

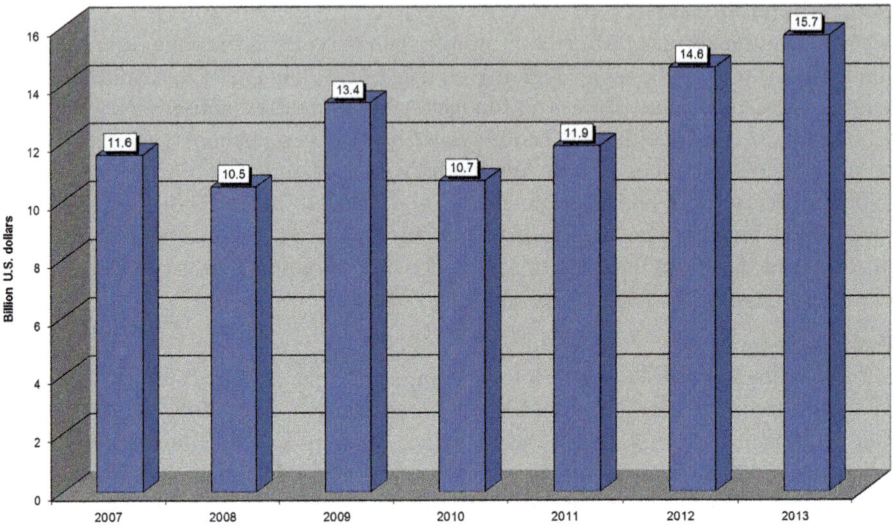

Fig. 1.2 World satellite manufacturing revenue (Source: SIA/Space Foundation)

2013 manufacturing revenue, amounting to $10.9 billion, with 75 % of those earnings came from U.S. government contracts.[220]

Commercial Space Launch

A total of 23 commercial launches were conducted in 2014; there was one launch failure of the Antares 120 cargo capsule, which carried 29 cube satellites intended to be released from the ISS. The 22 successful commercial launches carried 53 commercial services payloads into orbit (in addition to the 56 commercial cube satellites released from the ISS). Commercial launches in 2014 accounted for 25 % of the total 92 launches conducted for the year; and amounted to 36 % of the 295 payloads launched. Of the 295 payloads that were launched in 2014, 69 were cube satellites launched directly into orbit (not counting the failure of IFT 1 (Yui) and KickSat 1, and the launch of the 4M (Manfred Memorial Moon Mission) cube satellite to the Moon), while another 96 cube satellites were intended to be released into orbit from the ISS (the 29 cube satellites mentioned earlier had been destroyed, and 6 cube satellites were returned back to Earth from the ISS). When not considering cube satellites, the percentage of commercial payloads launched amounted to 30.7 %, or 39 commercial payloads out of a total of 127 - non-cube satellite payloads.

In 2014, U.S. launch providers conducted 11 commercial launches, while an additional 12 were non-commercial; its commercial launches amounted to 47.8 % of the total commercial launches for the year. Russia had the most launches for 2014, although only 4 out of its 28 launches were conducted for commercial

[220] 2014 State of the Satellite Industry Report, 18.

purposes; those launches held a 17.4 % share of total commercial launches for the year—substantially lower than its 52.2 % share in 2013. China did not have any commercial launches for the year, but did conduct 16 non-commercial launches. On the other hand, 6 out of the 11 European launches in 2014 were commercial; increasing its share of total commercial launches to 26.1 % for the year (not counting the actual number of payloads launched). And finally, both India and the multinational Sea Launch AG each had a single commercial launch for the year, amounting to a 4.3 % share for each launch provider.

The revenue from the 23 launches amounted to an estimated $2.36 billion, an increase of 23.6 % from $1.91 billion earned in 2013. In 2014, the U.S. earned its highest amount of commercial launch revenue since 1998, amounting to $1.107 billion—more than three times the amount it earned in 2013. Europe was a close second, earning $920 million for the year—an increase of 29.6 % from the $710 million earned in 2013). Russia followed in third position, earning $218 million (a 71.3 % reduction from the $759 million earned in the previous year). Next, multinational revenue took in $95 million—similar to its $100 million earned in 2013. And India earned $15 million in its single commercial launch for 2014, highlighting its competitive advantage as a low-cost launch provider.[221]

In 2014, the Ariane 5 had four commercial and two non-commercial launches, which lifted 7 commercial telecommunications satellites, 1 military communications satellite, two civil government communications satellites into GEO orbit, and the final ATV-5 to the ISS. The Europeanized Soyuz had two commercial launches, which placed 8 O3b commercial communications satellites to medium Earth orbit (MEO); and two non-commercial launches which lifted the Sentinel 1A for ESA and two Galileo navigation satellites for the European Commission. And the Vega launcher had one launch from French Guiana (Figs. 1.3 and 1.4).

Ground Equipment

Ground equipment revenue includes infrastructure elements, such as mobile terminals, gateways and control stations, and consumer equipment, such as very small aperture terminals (VSAT), ultra small aperture terminals (USAT), DTH broadcast dishes, satellite phones and digital audio radio satellite (DARS) equipment. Portable Navigation Devices (PND) form one of the sub-segments of end-user electronics that incorporate GPS chip sets.

In 2013, PND market revenue was $31.2 billion, having been in decline since 2008, and experiencing only marginal growth in subsequent years. This low growth can be attributed to the migration from standalone devices to embedded chipsets for devices such as smart phones.[222] However satellite navigation equipment represented a 56.2 % share of overall ground equipment revenue in 2013. Overall ground equipment revenues continued their slow expansion albeit with lowered growth rates since 2008; thus in 2013 ground equipment revenue grew by only

---

[221] Federal Aviation Administration. The Annual Compendium of Commercial Space Transportation: 2014. Washington DC: FAA, Feb. 2015: 12.

[222] State of Satellite Industry Report 2014, 27.

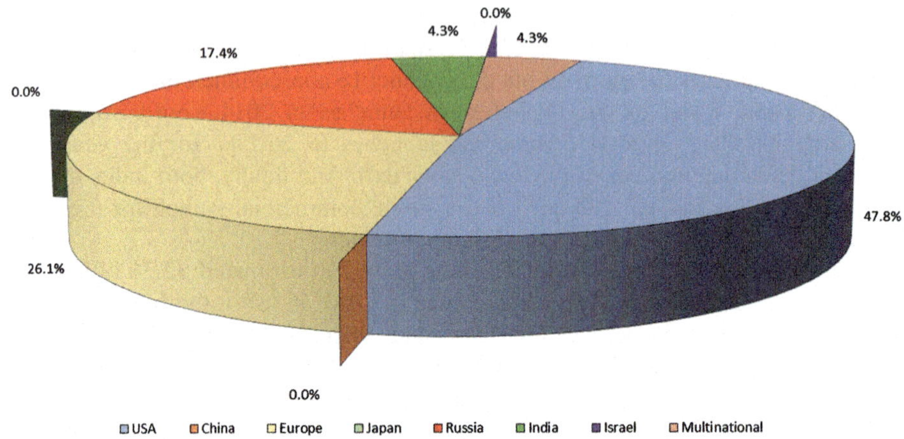

Fig. 1.3 Commercial Launch Activity by Country in 2014 (Source: FAA)

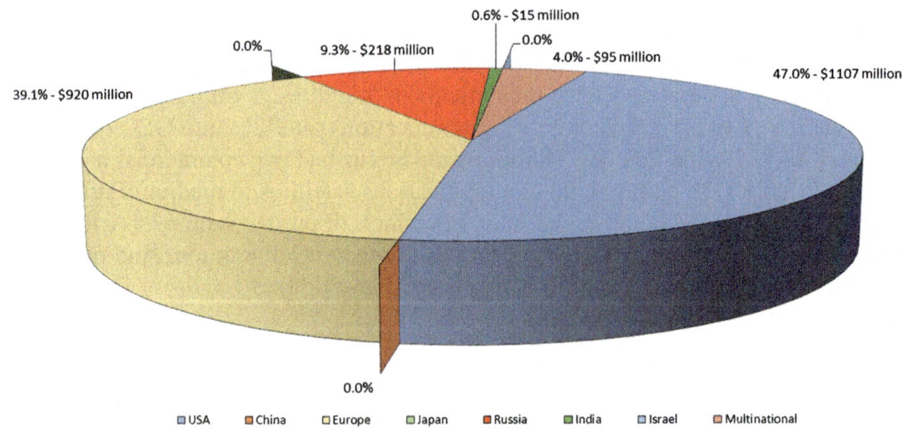

Fig. 1.4 Commercial Launch Revenues by Country in 2014 (Source: FAA)

1.1 % to $55.5 billion, representing a 28.4 % share of the $195.2 billion in world satellite industry revenue in 2013.

The revenue profiles for Garmin and TomTom, the two companies leading the PND market, diverged in 2014, with Garmin's revenue showing signs of growth, while TomTom's revenue continued to decrease. Following 3 years of reductions in revenue, in 2014 Garmin's total revenue increased by 9.1 % to $2.871 billion, following a drop to $2.632 billion in 2013.[223] However, its growth came from increased revenue in aviation, marine, fitness and outdoor products. On the other hand, TomTom's total revenue amounted to 950.292 million euros ($1.155 billion)

[223] "Garmin Ltd. Form 10-K for Period Ending December 27, 2014." Garmin 2 Apr. 2015: 41. http://www8.garmin.com/aboutGarmin/invRelations/reports/10-K_2014.pdf.

in 2014, a continued decrease of 1.3 % from the 963.454 million euros ($1.326 billion) earned in 2013 (Table 1.1).[224]

Insurance Sector

While in previous years insurance costs were in decline as the space industry continued to demonstrate increased hardware reliability, low accident rates and promising growth, the year 2013 was the first money losing year since 2007, with more than $800 million in expected claims.[225] The year 2014 did not look any better for insurance providers, with claims made by Orbital Sciences, ABS, Hispasat, the Russian Satellite Communications Company, and others appearing likely to result in another losing year, which could be expected to mean that insurance premiums will increase in coming years. Following the failure of the Orbital Sciences Antares 120 Commercial Resupply Services (CRS) launch on 28 October, having succeeded in attaining the milestone goal of ignition and lift-off (but not the milestone goal of launch success), Orbital Sciences was entitled to partial payment from NASA. Additionally, Orbital Sciences had taken out a $48 million insurance policy for the 'mission success' milestone, which will be paid by insurance providers and is expected to cover the entire final milestone payment.[226] Following the launch of its ABS-2 satellite on 6 February 2014, the Asia Broadcast Satellite (ABS) commercial satellite operator experienced an anomaly of a key Russia-directed beam by mid-summer, which has resulted in an insurance claim of $214 million.[227] Another claim is expected for Hispasat's Amazonas 4A, launched on 22 March 2014, which had a problem with a power subsystem that could halve its broadcasting capability; the Amazonas 4A is insured for 145 million euros ($199 million).[228] And the insurance claim for Russia's Express-AM4R satellite that failed due to the launch failure of the Proton M is expected to be 7.8 billion rubles ($225.4 million).[229] The failure of the Europeanized-Soyuz to place the two Galileo satellites into orbit on

[224] "TomTom Annual Report 2014." 21 Feb. 2015. TomTom 2 Apr. 2015: 47. http://annualreport2014.tomtom.com/xmlpages/resources/TXP/tomtom_ar_2014/pdf/TomTom_Annual_Report_2014.pdf.

[225] SpaceNews Staff. "SpaceNews 2014 Year in Review." 26 Dec. 2014. SpaceNews 13 July 2015. http://spacenews.com/spacenews-2014-year-in-review/.

[226] De Selding, Peter B. "Orbital Sciences Entitled To Partial NASA Payment for Antares Failure." 25 Nov. 2014. SpaceNews 13 July 2015. http://spacenews.com/42658orbital-sciences-entitled-to-partial-nasa-payment-for-antares-failure/.

[227] De Selding, Peter B. "ABS Files $214 Million Insurance Claim for Bad Satellite Beam." 17 Oct. 2014. SpaceNews 13 July 2015. http://spacenews.com/42218abs-files-214-million-insurance-claim-for-bad-satellite-beam/.

[228] Forrester, Chris. "Hispasat's $100 m claim on Amazonas 4A." 25 Apr. 2014. Advanced Television 13 July 2015. http://advanced-television.com/2014/04/25/hispasats-100m-claim-on-amazonas-4a/.

[229] "Russia: Express-AM4R's failed launch covered by insurance." 20 May 2014. Business Insurance 13 July 2015. http://www.businessinsurance.com/article/20140520/NEWS09/140529999.

Table 1.1 Understanding TomTom and Garmin variables

	Total revenue	2013	2014
TomTom		963.454 million euros ($1.326 billion)	950.292 million euros ($1.155 billion)
Garmin		$2.632 billion	$2.871 billion
	Related products		
TomTom	Automotive systems	110.932 million euros ($152.709 million)	109.409 million euros ($132.984 million)
Garmin	Automotive/ Mobile	$1.302 billion	$1.240 billion
	Geographical sales		
TomTom	Europe	710.101 million euros ($977.525 million)	718.767 million euros ($873.647 million)
	North America	177.725 million euros ($244.656 million)	163.461 million euros ($198.684 million)
	Rest of world	75.628 million euros ($104.110 million)	68.064 million euros ($82.730 million)
Garmin	Europe/Middle East/Africa	$955.900 million	$1.054 billion
	Americas	$1.433 billion	$1.538 billion
	Asia Pacific region	$243.056 million	$278.092 million

22 August 2014 did not lead to insurance claims, as the European Commission had chosen to self-insure, in line with common governmental practices.

Even with the launch and in-orbit failures in 2014, governments seem eager to stimulate the space sector by reducing the cost and associated red tape in launching a satellite. On 30 April 2014, the UK government agreed to adopt industry recommendations to reduce satellite insurance requirements for third party damage by 25 %, from £80 million to £60 million with the UK government covering the possible remaining losses beyond that figure, thus easing the way for satellite companies to obtain licenses for satellites (particularly small satellites) and to increase economic activity in the UK space sector.[230]

1.3.3 Developments in the Space Industry

1.3.3.1 Industrial Developments in Europe

The Arianespace commercial launch company reported revenue of 1.399 billion euros ($1.7 billion) in 2014; a 41.5 % increase from the 989 million euros ($1.32

[230] De Selding, Peter B. "Britain To Reduce Space Insurance Requirements, May Ease Smallsat Licensing Rules." 30 Apr. 2014. SpaceNews 25 June 2015. http://spacenews.com/40417britain-to-reduce-space-insurance-requirements-may-ease-smallsat-licensing/.

billion) in revenue for 2013. Its operating income for 2014 was at 3 million euros ($3.65 million), while its backlog of orders stood at 4.11 billion euros ($5.00 billion).[231] In 2013, Arianespace's revenue had dropped by 27.8 % from the 1.329 billion euros ($1.75 billion) it earned in 2012, due to late arriving satellites that necessitated launch in 2014.[232] This resulted in the postponement of two of the six planned Ariane 5 heavy-lift launches and two of the four planned Europeanized Russian Soyuz launchers from French Guiana in 2013.[233] In 2014, Arianespace won 14 additional launch contracts, including 9 contracts for commercial launches into GTO on the Ariane 5 ECA, in addition to two contracts to launch Galileo satellites on the Ariane 5 ES launcher, two contracts to launch four O3b satellites, and the Sentinel 1B on the Europeanized Soyuz launcher, and one Vega launch which will place the OPTSAT 3000 and Venμs spacecraft into Sun-synchronous orbit.

Eutelsat reported total revenue for the 12 months ending 30 June 2014 of 1.348 billion euros ($1.84 billion), a 5.0 % increase from the 1.284 billion euros ($1.67 billion) earned in the previous year.[234] The majority of revenue came from European regions (64.0 %), which diminished slightly from the previous year, whereas the Americas garnered the second highest share (15.6 %), followed by the Middle East at (12.6 %), Africa (5.5 %) and Asia (2.2 %).[235] In the 6 months ending 31 December 2014, Eutelsat reported revenue of 722.8 million euros ($878.55 million), up 11.6 % from the 647.4 million euros ($891.2 million) earned in the same 6-month period a year earlier.[236] These figures do not consider the impact of the acquisition of Satmex, which was finalized on 1 January 2014. The acquisition of Satmex, financed with a 930 million euros ($1.28 billion) 6-year bond issue on 13 December 2013, increased Eutelsat's presence in Latin America at a time when demand for satellite capacity in Latin America is expected to grow an average of 4.2 % per year over the 2013-2018 period.[237]

---

[231] 2014 Annual Report. Arianespace 23 July 2015. http://www.arianespace.com/about-us-corporate-information/Annual-Report-2014-EN.pdf.

[232] De Selding, Peter B. "Arianespace Expects To Notch at Least 11 Launches in 2015." 6 Jan. 2015. SpaceNews 9 Feb. 2015. http://spacenews.com/arianespace-expects-to-notch-at-least-11-launches-in-2015/.

[233] De Selding, Peter B. "Arianespace CEO Says Price Cuts Are Bearing Fruit." 25 June 2014. SpaceNews 29 Jan. 2015. http://spacenews.com/41023arianespace-ceo-says-price-cuts-are-bearing-fruit/.

[234] "Full Year 2013-2014 Results." 31 July 2014. Eutelsat 9 Feb. 2015. http://www.eutelsat.com/home/investors/financial-information/financial-press-releases/2013-2014/press-list-container/full-year-2013-2014-results.html.

[235] Reference Document 2013-2014. Eutelsat Communications 9 Feb. 2015: 137. http://www.eutelsat.com/files/contributed/investors/pdf/reference-document-2013-2014.pdf.

[236] Press Release. "First Half 2014-2015 Results." 12 Feb. 2015. Eutelsat Communications 18 Aug. 2015. http://www.eutelsat.com/files/contributed/news/press/en/2015/PR-H1-2014-15.pdf.

[237] Reference Document 2013-2014. Eutelsat Communications 9 Feb. 2015: 14. http://www.eutelsat.com/files/contributed/investors/pdf/reference-document-2013-2014.pdf.

By 3 June 2014, Abertis Telecom, the Spanish telecommunications infrastructure provider, had sold its remaining 5 % stake in Eutelsat to qualified investors for a total of 275 million euros.[238] Formerly Eutelsat's largest shareholder, Abertis Telecom began the divestment of its 32 % stake in January 2012, first selling a 16 % share of Eutelsat to investors in a transaction valued at about 1 billion euros,[239] followed by another 7 % stake in June 2012 for 385.2 million euros, and an additional 3 % in March 2013 for 182 million euros.[240]

Hispasat reported a nominal increase of 0.34 % in revenue generating 202.1 million euros ($245.6 million) by the end of 2014, from 201.4 million euros ($277.2 million) in 2013. Its EBITDA remained above 80 % in 2014 (lowering 1.3 points from 81.3 % in 2013), while its net profit reduced by 15.6 % to 45.6 million euros ($55.4 million), from 54.0 million euros ($74.3 million) in 2013. The company faced some challenges in 2014, such as the anomaly experienced on Hispasat's Amazonas 4A, which resulted in a reduced available capacity and a delayed release on the market, which had a negative impact on the revenue anticipated for the year. Moreover, unfavourable exchange rates and the downturn in demand in European markets also had a negative effect on returns. In 2014, 197.2 million euros ($239.7 million) of the total revenue came from space capacity rentals, whereof 60.6 % of the revenue came from clients in the Americas, where the operator has consolidated a strong presence and a significant market share, while the remaining 39.4 % came from clients in Europe and North Africa.[241]

Telenor Satellite Broadcasting of Norway reported a 1.20 % decrease in revenue for the year ending 2014, with the decline attributable to its divestment of its subsidiary Conax, which had been deconsolidated as of 25 March 2014. The Oslo-based satellite fleet operator reported revenue of 959 million kroner ($128.7 million) from 971 million kroner ($158.6 million) in 2013.[242] Telenor expects to enhance its commercial broadcasting capacity with the launch of the Thor 7 satellite in mid-2015.

Following the Airbus Group's recent consolidation of its Astrium space hardware and services division with the Cassidian and Airbus Military divisions to form

[238] "Abertis sells final 5 % Eutelsat stake for €275 million." 4 June 2014. DigitalTVEurope.net 9 Feb. 2015. http://www.digitaltveurope.net/188672/albertis-sells-final-5-eutelsat-stake-for-e275-million/.

[239] De Selding, Peter B. "Chinese Investment Firm Taking 7 Percent Stake in Satellite Fleet Operator Eutelsat." 22 June 2012. Space News 25 Feb. 2014. http://www.spacenews.com/article/chinese-investment-firm-taking-7-percent-stake-satellite-fleet-operator-eutelsat.

[240] "Abertis sells final 5 % Eutelsat stake for €275 million." 4 June 2014. DigitalTVEurope.net 9 Feb. 2015. http://www.digitaltveurope.net/188672/albertis-sells-final-5-eutelsat-stake-for-e275-million/.

[241] "Last year, HISPASAT generated a 9 % increase in its revenue from America." 19 Feb. 2015. Hispasat 2 Apr. 2015. http://www.hispasat.com/en/press-room/press-releases-2015/166/hispasat-aumenta-en-un-9-sus-ingresos-.

[242] Q4/2014—Interim report | January—December 2014. 10 Feb. 2015. Telenor Group 2 July. 2015. http://www.telenor.com/wp-content/uploads/2014/09/Telenor-Q4-report-2014-110215.pdf.

Airbus Defence and Space (Airbus D&S), Airbus D&S reported 13.025 billion euros ($15.832 million) in revenue for 2014, a near 1 % reduction from the restated revenue of 13.121 billion euros ($18.062 million) earned in 2013. Yet Airbus D&S' Space Systems accounted for 31 % of that 2014 amount, or 4.04 billion euros ($4.911 million). New orders for the company overall increased by 3.5 %, amounting to 12.225 billion euros ($14.859 million) from 11.808 billion euros ($16.255 million) in 2013.[243] While prior to the reorganisation, the Astrium division had been required to eliminate up to 2470 Astrium positions from its pool of 18,000 employees over the next 3 years, the four telecommunications satellites and 10 Earth observation satellite orders its satellite division won in 2014 has resulted in a lowering of planned cuts by 600 positions.[244]

The tight budgetary situation throughout Europe, and the increasing competition in the global market, has led Airbus to seek to divest its commercial satellite communications services business, which is mainly devoted to mobile satellite services, and focus on its core space, military aircraft, and missile divisions. Airbus D&S had purchased its commercial mobile satellite division, Vizada, in 2011 for $960 million, from which it earned around $660 million in revenue in 2011 and 2012. Despite divestment plans, Airbus D&S intends to keep the minor government portion of business services of Vizada.[245] The sale of Vizada is to be concluded by mid-2015. In November 2014, Inmarsat indicated its interest in purchasing Vizada back from Airbus D&S, as the sale would give Inmarsat a dominant share of the mobile satellite services reseller business in addition to its role as wholesale provider of satellite capacity.[246]

Thales Alenia Space posted total revenue of more than 2.1 billion euros ($2.55 billion) in 2014, of which 62 % came from Earth Observation, exploration and navigation satellites, while the remaining 38 % came from telecommunications satellites. The company also gained 2.2 billion euros ($2.67 billion) earned in new orders.[247] By mid-October, the company had begun the construction of a new building for the integration and testing of high-resolution optical observation instruments, as the last step in company's 2008 Odyssée programme to expand

[243] "FY Results 2014." 27 Feb. 2015. Airbus Group 1 Apr. 2015: 14. http://www.airbusgroup.com/dam/assets/airbusgroup/int/en/investor-relations/documents/2015/Closing/FY14/Airbus-Group-FY2014-Presentation/Airbus Group FY2014 Presentation.pdf.

[244] De Seldinh, Peter B. "Successes in 2014 Prompt Airbus To Reduce Space Layoffs." 20 Jan. 2015. SpaceNews 26 Aug. 2015. http://spacenews.com/successes-in-2014-prompt-airbus-to-reduce-planned-layoffs-in-space-business/.

[245] De Selding, Peter B. "Airbus To Divest Commercial Mobile Satellite Services Division." 16 Sept. 2014. SpaceNews 23 July 2015. http://spacenews.com/41881airbus-to-divest-commercial-mobile-satellite-services-division/.

[246] De Selding, Peter B. "Inmarsat Eyeing Airbus Mobile Satellite Services Division." 10 Nov. 2014. SpaceNews 23 July 2015. http://spacenews.com/42499inmarsat-eyeing-airbus-mobile-satellite-services-division/.

[247] De Selding, Peter B. "Thales Alenia Space Chief Touts Program Progress, Cost-cutting Efforts." 24 Mar. 2015. Thales Alenia Space 1 Apr. 2015. http://spacenews.com/thales-alenia-space-chief-touts-program-progress-cost-cutting-efforts/.

the company's Cannes site.[248] While marketing high-resolution optics presents a major economic and strategic challenge, it also brings into focus the limits that can be placed on high-resolution satellites sold for export, and whether the selling state can require "shutter-control". Yet, the company seeks to efficiently address the changing requirements of a booming global market.

OHB AG of Germany increased its total revenue to 728.15 million euros ($885.05 million) for the 2014 fiscal year; amounting to a 7.1 % increase from the 680.12 million euros ($936.25 million) earned in 2013. Its EBITDA had a small increase of 1.2 % reaching 53.42 million euros in 2014 from 52.80 million euros in 2013.[249] In January 2014, OHB's subsidiary CGS S.p.A was awarded a sub-contract worth 44 million euros ($59.9 million) from principal contractor Telespazio S.p.A for the Italian defence ministry's OPSAT (OPtical SATellite) 3000 programme, scheduled for launch sometime in 2016. In June 2014, OHB's Sweden branch opened its new production facility, while in the following month another facility was opened in Oberpfaffenhofen, near Munich, Germany. Also in July, OHB shareholders passed a resolution approving the conversion of OHB AG into a Societas Europae (SE) by mid-2015, which will allow for more Europeanization of space technology, as the change in corporate structure will make it easier for the group to establish new branches in other parts of the European Union, while also permitting a more uniform and clearer governance system.[250] At 31 December 2014, OHB's number of employees amounted to 2086 (a reduction of 326 from the previous year) which was due to the deconsolidation of its subsidiary Aerotech Peissenberg GmbH & Co.; by the end of the year, 78.9 % of OHB's employees were based in companies in Germany, while 15.9 % were in other parts of Western Europe, and the remaining 5.2 % located in Chile and French Guyana.[251]

RUAG Space is continuing toward an industry leadership role as Europe's largest independent supplier of space technology, recording a 4.9 % increase in revenue in 2013 earning CHF 299 million ($335.76 million),[252] followed by an additional 7.7 % increase in 2014 which brought CHF 322 million ($325.39 million).[253] Based in Switzerland, Sweden, and Austria, the company now employs 1163 personnel.[254] By the end of 2014, RUAG announced an agreement to purchase the space unit of Finnish company Patria. In the 17 December 2014

[248] "Thales Alenia Space to build new space optics facility—Invests 8 million euros to extend Cannes plant." 9 Dec. 2014. Thales Group 1 Apr. 2015. https://www.thalesgroup.com/en/worldwide/space/press-release/thales-alenia-space-build-new-space-optics-facility-invests-8-million.

[249] Annual Report 2014. 26 Mar. 2015. OHB 31 Mar. 2015: 2. http://www.ohb.de/tl_files/ohb/pdf/finanzberichte_hauptversammlung/2014/OHB_GB_14_E.pdf.

[250] Ibid.

[251] Ibid.

[252] RUAG Annual Report 2013. 18 Mar. 2015. RUAG 1 Apr. 2015: 98. http://www.ruag.com/fileadmin/ruag/Divisions/Aerostructures/Aerostructures_home/About_us/RUAG_GB_2013_EN.pdf.

[253] RUAG Annual Report 2014. 17 Mar. 2015. RUAG 1 Apr. 2015: 66. http://www.ruag.com/fileadmin/ruag/group/Annual_Report/GB/RUAG_2014_GB_EN.pdf.

[254] Ibid. at 18.

announcement, the transaction included Patria Space's business operations and assets, in addition to the transfer of 32 Patria employees to be employed in the newly founded company RUAG Space Finland.[255] As the key product areas of Patria's Space unit are in spacecraft control electronics, electrical power subsystems, electronic units and related test equipment, its acquisition will enhance RUAG's portfolio with the potential to broaden its technology base. On 18 December 2014, RUAG took a stake in INNOCAMPUS AG, the operating company of Swiss Innovation Park (SIP) Biel-Bienne, Switzerland, investing a total of CHF 132 million ($136.98 million).[256] RUAG's aim is to increase its expertise in additive manufacturing technology (i.e. 3D printing) for use in structures that will operate in space.

1.3.3.2 Industrial Developments in the United States

The Stratolaunch project, backed by billionaire Microsoft co-founder Paul Allen, is seeking to develop the largest air-launch system in the world, with the aim of launching commercial and government payloads and eventually conducting human missions. ATK Aerospace will build the first two stages of Stratolaunch's three-stage launcher in a joint venture with Orbital Sciences called Orbital ATK. The Orbital ATK joint venture is estimated to be worth $5 billion.[257] The first two stages will be fuelled by solid propellant, while the third stage was initially intended to be a restartable cryogenic engine—burning liquid hydrogen and liquid oxygen.[258] However, by November 2014, Stratolaunch had suspended work on the cryogenic upper stage, and will develop an all-solid vehicle for the time being.[259] Released from its carrier aircraft, operating from a large airport or spaceport, the air-launched rocket is planned to boost up to 6800 kg of payload to low Earth orbit, and smaller payloads to geostationary transfer orbit.[260]

---

[255] Media release. "RUAG acquires Patria's Space Business." 17 Dec. 2014. RUAG 1 Apr. 2015. http://www.ruag.com/space/media/media-releases/news/ruag-acquires-patrias-space-business/5faeadbc8094ec868a0d76b27f9b3a45/.

[256] Media release. "RUAG takes a stake in INNOCAMPUS AG." 18 Dec. 2014. RUAG 1 Apr. 2015. http://www.ruag.com/group/media/media-releases/news/ruag-takes-a-stake-in-innocampus-ag/fe80c95c94dc7c2230bb47dad3ba5dd8/.

[257] Boyle, Alan. "Orbital ATK Space Venture Created in $5 Billion Merger." 30 Apr. 2014. NBC News 17 July 2015. http://www.nbcnews.com/science/space/orbital-atk-space-venture-created-5-billion-merger-n93911.

[258] Leone, Dan. "ATK To Supply Stratolaunch Rocket Stages." 13 Aug. 2013. Space News 28 Feb. 2014. http://www.spacenews.com/article/launch-report/36764atk-to-supply-stratolaunch-rocket-stages.

[259] "Stratolaunch Drops Liquid Stage From Rocket System." 7 Nov. 2014. Aerospec International 17 July 2015. http://aerosmart.com/stratolaunch-drops-liquid-stage-from-rocket-system/.

[260] Leone, Dan. "Orbital Tapped To Build Stratolaunch Rocket." 5 June 2013. Space News 28 Feb. 2014. http://www.spacenews.com/article/launch-report/35647orbital-tapped-to-build-stratolaunch-rocket.

Following GeoEye's acquisition by DigitalGlobe in 2013 and following uncertainty regarding U.S. government commitments, it was business as usual for DigitalGlobe, with U.S. government revenue rising 53 % from the previous year.[261] For the year ending 31 December 2014, DigitalGlobe reported revenue of $654.6 million; a 6.8 % increase from the $613 million earned in 2013.[262] The image provider had initially predicted revenues of $699 million but unexpected competition in the emerging market business, and a 2.5 month delay in the launch of its WorldView-3 satellite resulted in reduced returns for the year.[263] By mid-year, business with Russia had nearly halted due to the Ukrainian crisis, yet, this loss of business did not have a substantial impact due to matched growth from India, the Middle East, oil and gas companies, and NGOs.[264]

Inmarsat reported a 1.9 % increase in revenue for the year ending 31 December 2014, generating $1.286 billion from the $1.262 billion earned in 2013. In 2014, its EBITDA grew by 8 % to $701.0 million from $648.8 million in 2013. In addition to the underlying growth stemming from Maritime, Enterprise, and Aviation, the increase also reflected the LightSquared Cooperation Agreement (which earned $75.4 million of the total revenue) and additional revenues following the acquisition of Globe Wireless. Yet, Inmarsat has seen a continued slowdown in its government business, which decreased by 21.7 % to $319.9 million from $408.3 million in 2013, and a reduction in revenue due to the disposal of the majority of the retail energy business to the company RigNet.[265]

SpaceX intends to develop a reusable first stage to reduce the cost of the Falcon 9 v1.1 launcher by up to 75 %.[266] On 18 April 2014, SpaceX demonstrated its first controlled soft water landing of the Falcon 9's first stage in the Atlantic Ocean during the successful launch of its third CRS mission to the ISS.[267] The booster was equipped with landing legs that extended as the booster gently approached the water

[261] De Selding, Peter B. "DigitalGlobe Revenue Up Sharply Despite U.S. Spending Slowdown." 1 Nov. 2013. Space News 3 Mar. 2014. http://www.spacenews.com/article/military-space/37960digitalglobe-revenue-up-sharply-despite-us-spending-slowdown.

[262] DigitalGlobe Annual Report 2014. DigitalGlobe 17 July 2015. https://www.last10k.com/sec-filings/DGI/0001558370-15-000193.htm.

[263] De Selding, Peter B. "Unexpected Competition, Launch Delay Are Drag on DigitalGlobe Revenue." 27 Feb. 2014. SpaceNews 19 May 2014. http://www.spacenews.com/article/financial-report/39648unexpected-competition-launch-delay-are-drag-on-digitalglobe-revenue.

[264] De Selding, Peter B. "DigitalGlobe Revenue up Despite Steep Drop in Russian Business." 1 Aug. 2014. SpaceNews 29 Jan. 2015. http://spacenews.com/41459digitalglobe-revenue-up-despite-steep-drop-in-russian-business/.

[265] "Inmarsat plc reports Preliminary Full Year Results 2014." 5 Mar. 2015. Inmarsat 18 July 2015. http://www.inmarsat.com/press-release/inmarsat-plc-reports-preliminary-full-year-results-2014/.

[266] Ferster, Warren. "Upgraded Falcon 9 Rocket Successfully Debuts from Vandenberg." 29 Sept. 2013. Space News 26 Mar. 2014. http://www.spacenews.com/article/launch-report/37450upgraded-falcon-9-rocket-successfully-debuts-from-vandenberg.

[267] Kremer, Ken. "SpaceX Makes Strides Towards 1st Stage Falcon Rocket Recovery during Space Station Launch." 19 Apr. 2014. Universe Today 21 July 2015. http://www.universetoday.com/111377/spacex-makes-strides-towards-1st-stage-falcon-rocket-recovery-during-space-station-launch/.

with zero spin before splashing down. By the end of 2014, SpaceX was constructing a floating landing platform for use in controlled landing tests planned for early 2015.[268]

SpaceX is also planning to use a new launch pad at Vandenberg U.S. Air Force Base to launch the standard Falcon 9 rocket and its upcoming heavy-lift variant. SpaceX currently launches its Falcon 9 only from Canaveral Air Force Station, whereas the new pad will be built to also accommodate the developmental Falcon 9 Heavy launch vehicle, now expected to debut in 2015.[269] The company will spend between $20-30 million to renovate the site, unused since 2005; the Titan 4 was the last rocket launched from there. SpaceX will also update its launch facility in Cape Canaveral, enabling the heavy-lift rocket to launch from both coasts. SpaceX's Florida base is undergoing general upgrades; additional hangars are being built to prepare the Falcon 9 rockets and customer payloads for launch. With an expected flight rate of 10 to 12 launches per year, the current facility (Space Launch Complex 40) will receive a 16,000 square-meter addition, including an unused Delta 2 processing building. SpaceX will receive $7.3 million from Space Florida, a state-funded agency, toward the upgrades. These upgrades include, *inter alia*, a clean room, a hazardous hypergolic fuelling facility and enough volumetric space to encapsulate a payload in a fairing in a vertical position.[270]

By the end of 2014, Virgin Galactic had begun rebuilding its SpaceShipTwo (SST), following the tragic destruction on 31 October that killed one of its pilots and injured another. Prior to the loss, the SST was in the final phases of obtaining a commercial launch licence for the SST, having successfully completed its third rocket-powered test flight on 10 January 2014, in addition to a "cold flow" test without ignition on 28 August after switching to a plastic-based fuel source.[271] A second SST is expected to be ready for testing in 2015, with commercial service beginning in 2016. Designed to be released from its carrier at an altitude of 15 km, its rocket engine will ignite to propel the craft to an apex altitude of up to 110 km. By 28 May 2014, more than 700 passengers had reserved a flight on the spacecraft, each at $250,000 per seat.[272] Among these customers are scientists along with their experiments and space tourists. Soon, other companies might also be able to

[268] Foust, Jeff. "Next Falcon 9 Launch Could See First-stage Platform Landing." 24 Oct. 2014. SpaceNews 21 July 2015. http://spacenews.com/42305next-falcon-9-launch-could-see-first-stage-platform-landing/.

[269] "SpaceX Breaks Ground on West Coast Launch Pad." Space News 25 July 2011: 8.

[270] Klotz, Irene. "SpaceX Expanding Florida Facilities to Meet Launch Demand." Space News 28 Nov. 2011: 5.

[271] Foust, Jeff. "Virgin Galactic Delays First Commercial Flights to 2015." 11 Sept. 2014. SpaceNews 9 Apr. 2015. http://spacenews.com/41837virgin-galactic-delays-first-commercial-flights-to-2015/.

[272] Klotz, Irene. "Virgin Galactic Hoping for SpaceShipTwo Altitude Boost with New Fuel." 28 May 2014. SpaceNews 9 Apr. 2015. http://spacenews.com/40719virgin-galactic-hoping-for-spaceshiptwo-altitude-boost-with-new-fuel/.

provide similar suborbital space travel products, including *inter alia* XCOR Aerospace, Blue Origin and Armadillo Aerospace.[273]

1.3.3.3 Industrial Developments in Russia

On 2 December 2013, Russia's President Vladimir Putin signed a decree ordering the creation of the United Rocket and Space Corporation (URSC), wherein several federal state-owned unitary space enterprises will be reorganized into open joint stock companies to be fully owned by the federal government, with each enterprise contributing 100 % of shares minus one-share to the new corporation's authorized capital.[274] The consolidation stage is to be completed in mid-2015. Initial steps taken in early 2014 were intended to reduce costs in the industry by consolidating the country's space industry into a single company to eliminate excess manufacturing capacity, in addition to streamlining the purchasing of foreign electronic components in order to negotiate volume-based discounts.[275] Initially, the reform was meant to split the Federal Space Agency, Roscosmos, between a demand and supply side, where Roscosmos would act as a customer, responsible for space policy, research and ground infrastructure,[276] while URSC would act as the contractor.[277] However, by the end of 2014, the Russian government proposed to unify Roscosmos as one agency with URSC, as the division of functions of customer and contractor had not considerably improved the sector's efficiency.[278]

The major catalyst for the reversal was serious shortcomings that arose in the construction of the Vostochny Cosmodrome. By mid-2014, there were signs that the new spaceport would miss its aim of being ready before the end of 2015, as construction was behind schedule by up to 3 months, and the 6000 workers at the site less than half of what was needed. The delay required a cash infusion of an additional 50 billion roubles ($1.3 billion) to finish construction on time.[279] Moreover, in October, the head of a company involved in building the spaceport was arrested and charged with

[273] Wall, Mike. "Virgin Galactic Aims for 1st Rocket-Powered Flight This Year." 28 Feb. 2012. SPACE.com 3 May 2012. http://www.space.com/14706-virgin-galactic-spaceshiptwo-powered-flight.html.

[274] "Putin Signs Decree on Creation of United Rocket and Space Corporation." 2 Dec. 2013. ITAR TASS 30 Apr. 2014. http://en.itar-tass.com/russia/709849.

[275] "Putin Signs Legal Decree Consolidating Russian Industry." 9 Dec. 2013. SpaceNews 30 Apr. 2014. http://www.spacenews.com/article/civil-space/38551putin-signs-legal-decree-consolidating-russian-industry.

[276] "Russia Bets on Sweeping Reform to Revive Ailing Space Industry." 26 Dec. 2013. Reuters 4 Mar. 2014. http://www.reuters.com/article/2013/12/26/us-russia-space-idUSBRE9BP02S20131226.

[277] "Overhaul pending in Russian space sector." 3 Dec. 201. ITAR-TASS News Agency 4 Mar. 2014. http://en.itar-tass.com/russia/763105.

[278] "Duma proposes to merge Roscosmos with United Rocket Space Corporation." 20 Dec. 2014. TASS.ru 21 July 2015. http://tass.ru/en/non-political/768257.

[279] Bodner, Matthew. "Cash Infusion Reaffirms Putin's Commitment to New Launch Complex." 3 Sept. 2014. SpaceNews 16 July 2015. http://spacenews.com/41747cash-infusion-reaffirms-putins-commitment-to-new-launch-complex/.

embezzling 1.8 billion roubles ($43 million) from the project.[280] The Vostochny Cosmodrome will give Russia an alternative launch port to the Baikonur Cosmodrome that has been leased from Kazakhstan since the Soviet Union's collapse in 1991.[281]

The Russian-Ukrainian-owned Sea Launch AG consortium has continued to experience setbacks stemming from launch failures and difficulties in winning launch contracts in 2014, requiring a series of cost-reduction measures to remain afloat before activities are expected to resume in mid-2015. Those measures will likely include laying-off staff, and reducing operating expenses by laying-up the Sea Launch Commander and Odyssey vessels. For the remainder of the year the Russian government is considering whether to move Sea Launch operations to Russia, which will give it access to Russian institutional launch contracts that are currently out of reach for the international company.[282] Other proposals have included moving launch operations to Vietnam, or Brazil, or selling Sea Launch to an Israeli company.[283] Sea Launch hopes to have 3 missions in 2015, followed by an even greater launch capacity in the years that follow.[284]

1.3.3.4 Industrial Developments in Japan

Mitsubishi Electric Co. (Melco) of Japan develops satellites within its Information and Communication Systems (ICS) business segment; however, as this segment does not separate satellite-related revenue from its telecommunication, information systems, and electronic systems business, it should only be seen as generating a small portion of the total revenue earned by this segment. In the year ending 31 March 2014, the ICS segment generated 11.8 % of Melco's total sales. The ICS segment increased net sales by 4.9 %, reaching ¥548.2 billion ($5.33 billion) in 2014, from ¥522.4 billion ($5.54 million) in 2013. Operating income began climbing to previous levels, increasing to ¥5.5 billion ($53.49 million) in 2014, from ¥1.6 billion ($16.98 million) in 2013, due partly to an increase in sales.[285]

[280] Sweet, Rod. "Russia tightens control of troubled cosmodrome project." 24 Nov. 2014. Global Construction Review 16 July 2015. http://www.globalconstructionreview.com/news/russia-tightens-control-trou7b7led-c0os6mo6dro4m4e/.

[281] Soldatkin, Vladimir. "Putin orders building hastened at new Russian spaceport." 27 Aug. 2014. Reuters 16 July 2015. http://www.reuters.com/article/2014/09/02/us-russia-space-idUSKBN0GX1AV20140902.

[282] Messier, Doug. "Sea Launch Cuts Back as Manifest Gap Looms." 23 Aug. 2014. Parabolic Arc 21 July 2015. http://www.parabolicarc.com/2014/08/23/sea-launch/.

[283] Messier, Doug. "Will Sea Launch End Up in Brazil?" 24 Dec. 2014. Parabolic Arc 20 July 2015. http://www.parabolicarc.com/2014/12/24/sea-launch-brazil/.

[284] Henry, Caleb. "A Heavyweight Battle: How Sea Launch Plans to Stay Afloat With SpaceX in its Waters." 13 Dec. 2013. ViaSatellite 6 Mar. 2014. http://www.satellitetoday.com/launch/2013/12/13/a-heavyweight-battle-how-sea-launch-plans-to-stay-afloat-with-spacex-in-its-waters/.

[285] 2014 Annual Report. Mitsubishi Electric 10 Feb. 2015: 12. http://www.mitsubishielectric.com/company/ir/library/annual_report/pdf/ar2014.pdf.

The revenue generated for the 2014 fiscal year stemmed from an increase in the communications infrastructure business, along with growth in the system integration business, and increased sales due to progress in orders already received for projects in the space system business.[286]

On 6 November 2014, the Japanese government launched its Advanced Satellite with New System Architecture for Observation, or ASNARO-1, on a Russian Dnepr rocket from the Russian Yasny launch base. The ASNARO-1 is the first of a new generation of low-cost Earth observation satellite platforms built by NEC Corp. Rather than develop a more powerful bus for Western customers, its focus is on emerging Asian and South American countries that do not yet require powerful bus systems. With over 40 years of experience in building around 7000 communication subsystems for more than 200 satellites (including 67 satellite development projects under its leadership),[287] NEC plans to offer its low cost satellites in package deals that include communications, ground support and, in the case of Earth observation satellites, geospatial information products and services to enable full use of the systems.[288]

1.3.3.5 Industrial Developments in China

AsiaSat of Hong Kong had a 8.9 % decrease in revenue and a 19.1 % decrease in operating profit for the year ended 31 December 2014, with sales reaching HK $1.365 billion ($175.95 million) from HK$1.499 billion ($193.3 million) in 2013, while operating profit was HK$726.3 million ($93.62 million) from HK$897.7 million ($115.76 million).[289] The lowered profitability during the first half of 2014 had been expected by AsiaSat, which was partly related to a reduction in turnover stemming from a major customer's significant contract extension at a lower rate, along with incremental depreciation on AsiaSat 7, and an increase in operating expenses.[290] AsiaSat now has six satellites in operation; its two most recent satellites, AsiaSat 8 and AsiaSat 6, were launched on 5 August 2014 and 7 September 2014, respectively.[291]

---

[286] "Investor Relations—Results by Business Segment." 28 Apr. 2014. Mitsubishi Electric 10 Feb. 2015. http://www.mitsubishielectric.com/company/ir/highlights/segment/index.html.

[287] "Satellite Systems | Heritage—Major Japanese Satellites Integrated by NEC." NEC 18 Aug. 2015. http://www.nec.com/en/global/solutions/space/satellite_systems/index.html.

[288] Kallender-Umezu, Paul. "Japan's NEC Looks To Expand Commercial Market Footprint." 24 Nov. 2014. SpaceNews 11 Feb. 2015. http://spacenews.com/42644japans-nec-looks-to-expand-commercial-market-footprint/.

[289] Reaching Further, Bringing You Closer—2014 Annual Report 2014. AsiaSat 17 July 2015: 54. http://www.asiasat.com/asiasat/EN/upload/doc/support_reports/ar2012_eng.pdf.

[290] Ibid. at 4.

[291] "Satellite Fleet." 2014. AsiaSat 10 Feb. 2015. http://www.asiasat.com/asiasat/contentView.php?section=3&lang=0.

1.3.3.6 Industrial Developments in India

Very high regulatory barriers in India make access to its satellite telecommunications market unusually difficult. Yet India is expected to remain a growing market for non-Indian providers for the long term, as India will not launch enough bandwidth on its own to meet demand.[292]

While India was developing a new satcom policy in 2014, with satellite operators, equipment vendors and service providers all anticipating revised regulatory restrictions,[293] discussions by India's Department of Space in consultations with the telecom department (DoT) did not specifically address easing foreign ownership rules.[294] In April 2012 the Indian government proposed a new tax that would impose a 10 % royalty fee on foreign satellite communications services that would be retroactive covering the past 36 years. The government is looking for ways to reduce its deficit without regard to the crippling effect this tax will have on broadcast and communications. Such a move would put India in breach of its international tax treaties, as compliance with the royalty tax would amount to double taxation. The matter is now on appeal before the Supreme Court of India. However, if the tax is enforced and foreign satellite providers are unable to obtain a tax credit in their own tax homes, they will be forced to pass on the royalty to their customers in India. The 130-member Cable and Satellite Broadcasting Association of Asia (Casbaa), including many of the world's biggest commercial satellite fleet operators, is also attempting to persuade the Indian government that retroactivity violates basic international rules and principles of fair play.[295]

1.3.3.7 World

The Canadian satellite component manufacturer Com Dev International reported a decrease in total revenue for the fiscal year-ended 31 October 2014, earning C $208.2 million ($185.98 million), down 3.4 % from C$215.5 million ($205.94 million) in 2013 due mainly to continued U.S. government spending constraints.[296]

---

[292] De Selding, Peter B. "Despite Barriers, SES Invests Heavily in Indian Satellite Market." Space News 6 June 2011: 6.

[293] Krishnan, Deepu. "Challenges and Opportunities in the Indian Satcom Market." 17 Feb. 2014. SpaceNews 18 May 2014. http://www.spacenews.com/article/opinion/39541challenges-and-opportunities-in-the-indian-satcom-market.

[294] Parbat, Kalyan. "Department of Space, DoT to overhaul satellite communications policy." 19 Apr. 2014. The Economic Times 17 July 2015. http://articles.economictimes.indiatimes.com/2014-04-19/news/49252215_1_satcom-orbital-policy.

[295] De Selding, Peter B. "Satellite Fleet Operators Protest Indian Tax Proposal." 5 Apr. 2012. Space News 9 May 2012. http://www.spacenews.com/policy/120405-sat-operators-protest-indian-tax.html.

[296] "COM DEV Announces Fourth Quarter and Year-End Fiscal 2014 Results." 15 Jan. 2015. COM DEV International 10 Feb. 2015. http://www.comdev.ca/images/financial-reports/CDV_Q4_14_Financial_Release2.pdf.

By the end of the fiscal year ending 31 October 2014, Com Dev's backlog had decreased by 5.8 % to C$155.1 million ($138.55 million) down from C$164.7 million ($157.39 million) in 2013. However, the company's full year revenue from its commercial segment showed 26.5 % growth in 2014, earning C$137.4 million ($122.74 million) from C$108.6 million ($103.78 million) in 2013.[297]

In 2014, Thaicom of Thailand generated revenue of 10.004 billion Thai Baht ($303.41 million), an increase of 26 % from the 7.896 billion Thai Baht ($239.86 million) earned in 2013. Thaicom also enjoyed substantial profits from its operations for its third consecutive year, generating 1.832 billion Thai Baht ($55.56 million) in 2014, an increase of 62 % from the 1.128 billion Thai Baht ($34.27 million) earned in 2013.[298] Satellite services amounted to 87.8 % of the revenue generated in 2014, a marginal change from the 88.2 % share of revenue generated in the previous year. And following the launch of Thaicom's Thaicom 6 satellite on 6 January 2014, along with Thaicom 7 on 7 September 2014, transponder lease revenue had a dramatic increase of 26.1 %, though the transponders available to the local and regional markets are not sufficient to meet the demand.[299]

1.3.4 Industrial Overview

1.3.4.1 Launch Sector

The launch sector is an enabler rather than as a primary economic activity. Yet, with the growth of low-cost launch services, the revenue it generates is becoming more significant.

The year 2014 experienced increased activity for the launch sector, with a total of 92 launches conducted by launch providers from Russia, the United States, Europe, China, Japan, India, Israel, and the multinational Sea Launch AG. There were some important events in 2014, such as the 2 non-commercial launch failures: a Proton M launch from Baikonur, Kazakhstan, carrying Russia's Express AM4R satellite in May; and a Europeanized Soyuz 2.1b launch from Kourou, French Guiana carrying the first two fully operational satellite models of Europe's Galileo navigation system in August—the failure of the launcher's upper stage and premature release of the satellites compromising the mission. There was also the commercial launch failure of the Antares-120 launcher, carrying a Cygnus commercial cargo capsule including 29 cube satellites to be released from the ISS, in October.

When looking into the launches of specific countries (Table 1.2), Russia continued as the world leader in the number of launches for 2014, accounting for 34.8 %

[297] Ibid.

[298] "Connecting The Future—Annual Report 2014." 6 Mar. 2015. Thaicom Public Company Limited 2 Apr. 2015: 150. http://thcom.listedcompany.com/misc/AR/20150327-thcom-ar2014-en.pdf.

[299] Ibid. at 19, 49–51.

Table 1.2 Worldwide launches in 2014 per country, number of launched systems, and commercial status (Source: FAA)

Launchers	Number of launch systems active in 2014	Total number of launches	Commercial launches	Non-commercial launches
Russia	9	32	4	28
United States	9	23	11	12
China	7	16	0	16
Europe	5	11	6	5
Japan	1	4	0	4
India	3	4	1	3
Israel	1	1	0	1
Multinational	1	1	1	0
Total	**36**	**92**	**23**	**69**

of the total launches. The United States had the second largest share, accounting for 25 % of the total. China was in third position in 2014, with a 22.2 % share, while Europe followed in fourth position with a 12.0 % share. Both Japan and India each had a 4.3 % share, while Israel and the Multinational Sea Launch AG each provided 1.1 % of the total launch figure (see Fig. 1.5).[300]

Russia launched 32 vehicles using nine different launch system configurations.[301] The U.S. used nine different launch system configurations[302] for a total of 23 launches, while seven different launch configurations[303] were used by China for its 16 launches. Europe used its Ariane 5 ECA and ES-ATV launchers, the Soyuz 2.1 launcher, and its Vega launcher for its 11 launches (5 Ariane 5 ECA, 1 Ariane 5 ES-ATV, 1 Soyuz 2.1a, 3 Soyuz 2.1b, and 1 Vega). Japan had 4 launches using its the H-11A 202 launch configuration; while India used three launcher configurations (i.e. PSLV CA, PSLV XL, and GSLV MK2) for its 4 launches. Israel used the Shavit launcher for its single launch, while Sea Launch AG used its Zenit 3SL for its launch.

The launch system utilization diminished in this period, moving from 40 active launch systems in 2013 to the use of 36 in 2014. While most launch activity was conducted by four countries i.e. Russia, the U.S., China, and Europe; Japan and India saw increased activity. However, it should be recalled that certain launchers have dual-launch capabilities, e.g. Europe's Ariane 5 launcher is able to lift two standard-size payloads to orbit. Hence the number of launches does not reflect the number of payloads brought to orbit.

[300] Federal Aviation Administration. The Annual Compendium of Commercial Space Transportation: 2014. Washington DC: FAA, Feb. 2015: 10.

[301] i.e. Angara A5, Dnepr, Proton M, Rockot, Soyuz 2.1a, Soyuz 2.1b, Soyuz FG, Soyuz U, and Strela.

[302] i.e. Antares-120, Antares-130, Atlas 5 401, Atlas 5 541, Delta 2 7320-10, Delta 4 Heavy, Delta 4 Medium +(4,2), Falcon 9, and Falcon 9 v1.1

[303] i.e. Kuaizhou, Long March 2C, Long March 2D, Long March 3A, Long March 3C, Long March 4B, and Long March 4C.

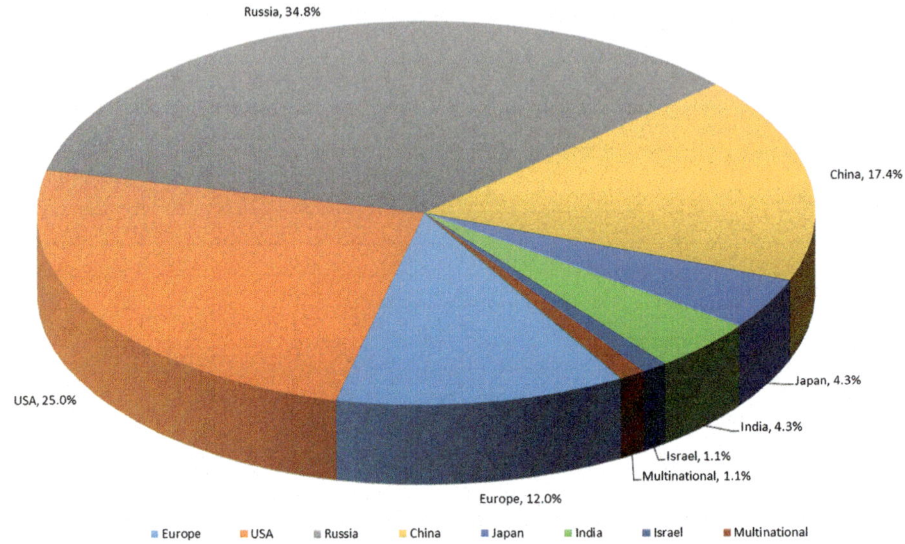

Fig. 1.5 Worldwide launches by country in 2014 (Source: FAA)

The U.S. was the leader in the number of commercial launches in 2014. The Ariane 5 dual payload capability paved the way for Europe to earn the second highest amount of commercial launch revenue. And while Russia had the same number of launches as the previous year, its four commercial launches placed it at a distant third position in terms of commercial revenue generated. When considering non-commercial launches, Russia had a substantial lead, conducting 40.6 % of the launches, China was in second position with 23.2 %, followed by the U.S. with a 17.4 % share. Europe, Japan, India, and Israel split the remaining 18.8 % share of non-commercial launches.

Whereas in the two previous years, it appeared that the number of commercial launches was increasing relative to non-commercial payload launches, the ratio returned to the 1-to-3 equilibrium experienced in earlier years; yet the number of payloads launched increased significantly in 2014, due mainly to the exponentially increasing number of cube satellites and microsatellites that were launched.

In terms of the global share of payloads launched in 2014 (Fig. 1.6), the U.S. surpassed Russia for the number of payloads launched, lifting 131 payloads, which represents 44.4 % of the total. Russia was in second place, launching 88 payloads (on par with the previous year) resulting in a 29.8 % share. In third position, China's 24 payloads amounted to an 8.1 % share, followed closely by Europe with 23 payloads (7.8 %), and Japan with 18 payloads (6.1 %). India launched 9 payloads (3.1 %), while Israel and Multinational each had 1 payload (0.3 %). However, when excluding the total 168 cubesat payloads from the assessment, the U.S. drops to second position with a 23.6 % share, while Russia's share increases to 34.6 % of the total. Moreover, Europe's share increases to 18.1 %, while China drops to fourth position with 15.7 %, with little change to the remaining launch country shares.

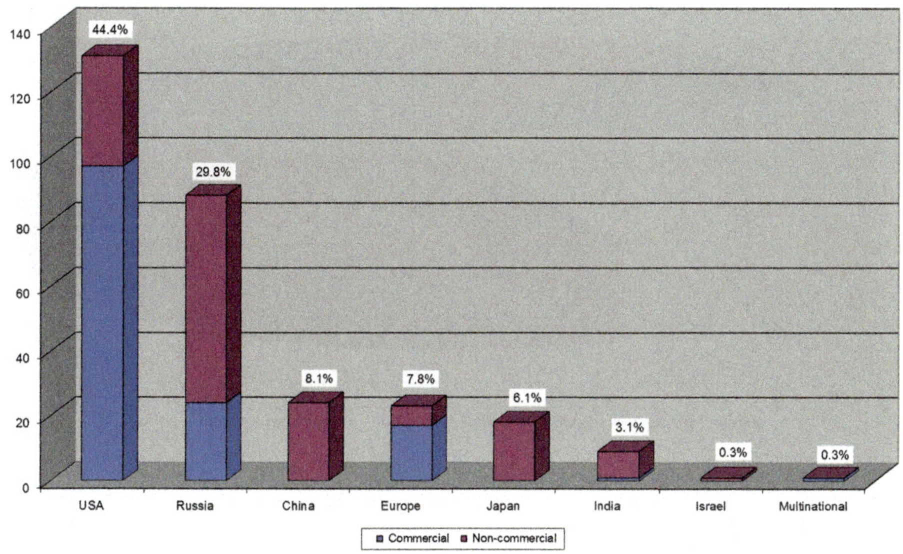

Fig. 1.6 Total payloads launched in 2014 by country, share and commercial status (Source: FAA)

Some notable changes in the distribution of payload sizes occurred in the 2014 reporting year (Fig. 1.7). With 168 payloads, "Micro" sized payloads accounted for 56.9 % of the total payloads launched, which is not surprising considering this increasing trend was visible when it reached 46.2 % of the total in 2013. The average mass of these cube satellite payloads was around 10.2 kg, with the sum of their mass reaching 1721.3 kg. While 96 cube satellites were to be released from the ISS, 27 were destroyed in the 28 October Antares 120 launch failure, in addition to the return of 6 cube satellites from the ISS. "Small" satellites were in second position with 41 payloads launched, accounting for 13.9 % (decreasing by 6.9 points from 2013). "Medium" sized payloads were in third position with 33 payloads at 11.2 % (up 0.8 points), 26 "Large" payloads at 8.8 % (down 5.8 points), 25 "Intermediate" payloads at 8.5 % (up 1.9 points), and 2 "Heavy" payloads at 0.7 % (down 0.7 %).[304]

In 2014, Arianespace and the International Launch Services (ILS) held the top position in terms of launches in GEO, each with 27.8 %, followed by United Launch

[304] Micro payloads have a mass of 91 kg or less, and are mainly science satellites, technological demonstrators or small communications satellites. Small payloads weigh between 92 and 907 kg and are very often Earth Observation satellites, similar to the Jason or the RapidEye series. Medium payloads weigh between 908 and 2268 kg, and feature the most diverse set of satellites, including small satcoms in geostationary orbit, Earth Observation satellites, and most of the Russian military satellites from the Kosmos series. Intermediate payloads, weighing between 2269 and 4536 kg, comprise medium satcoms and big scientific satellites. Large payloads, between 4537 and 9072 kg, refer to big satcoms, as well as to the Soyuz and Progress spacecraft flying to the ISS. Finally, Heavy payloads, exceeding 9072 kg, are linked to ISS activity, such as the cargo spacecraft, ATV, HTV, etc. *See* Commercial Space Transportation: 2011 Year in Review, 32.

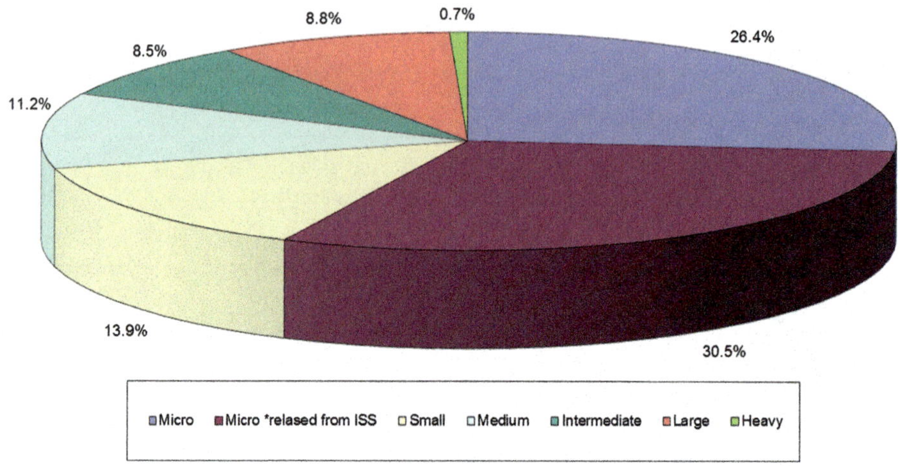

Fig. 1.7 Distribution of the payloads launched in 2014 by mass class (Source: FAA)

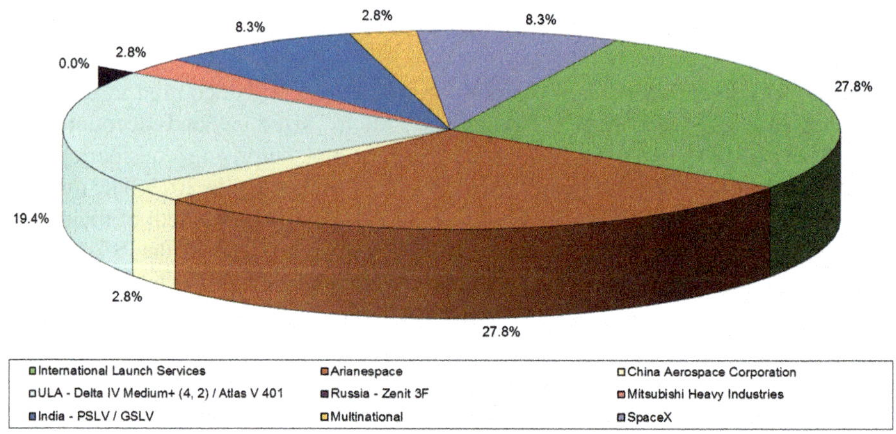

Fig. 1.8 Share of launch contracts for GEO satellites in 2014 by launch service provider

Alliance (ULA) with 19.4 % (Fig. 1.8). Arianespace placed 10 communications satellites into GEO orbit with the use of five Ariane 5 ECA launchers. The Russian ILS launched 10 GEO communication satellites on eight Proton M launchers; including a failed launch of the Express AM4R satellite. The U.S. ULA conducted 7 launches to place seven payloads into GEO orbit. India's PSLV launcher was used twice to launch the IRNSS 1B and IRNSS 1C navigation satellites to GEO orbit, while its GSLV launched the Gsat 14 communications satellite. And the SpaceX Falcon 9 was used three times to launch three commercial communications satellites (Asiasat 8, 6, and Thaicom 6) into GEO orbit. Lastly, Japan used its H-IIA 202 launcher to place its Himawari 8 Meteorological satellite into GEO orbit, while the Multinational Sea Launch placed the Eutelsat 3B commercial communications satellite into orbit.

1.3.4.2 Satellite Manufacturing Sector

In 2014, 295 payloads were launched (including an estimated 168 cube satellites, 14 crewed or cargo missions to the ISS, 1 test launch of the U.S. Orion EFT-1 capsule, and the launch of China's Chang'e 5 experimental lunar mission). The U.S. manufactured 47.5 % of the launched payloads (including 110 cubesats), while Russia accounted for 13.2 %, and Europe produced 11.5 %. China accounted for 8.5 % of the payloads launched, while Japan produced 8.1 % and India accounted for 1.4 %. The remaining 9.8 % of payloads were from various parts of Asia, the Middle East, and the Americas.[305]

Of the 279 satellites launched in 2014,[306] 143 were non-commercial. As in recent years, China's CAST led in manufacturing 16 non-commercial satellites, followed by the Reshetnev Company which produced 13 non-commercial satellites, while the regions of Asia, North America, Europe, and Russia also showed a strong presence in this market. Boeing produced only 5 non-commercial satellites, followed by Orbital Sciences with 4 non-commercial satellites, and 3 - non-commercial satellites developed by ISRO. Both Thales Alenia Space and Lockheed Martin each provided 2 non-commercial satellites, while Canada's Space Systems Loral (SS/L), Airbus Defence & Space, and Mitsubishi each developed 1 non-commercial satellite (Fig. 1.9).[307]

This year, Europe's Thales Alenia Space took the lead among the top commercial prime contractors with the launch of 9 of its commercial satellites. SS/L took second position manufacturing 6 commercial satellites, while Airbus Defense & Space was third in terms of its 5 commercial satellites. Orbital Sciences built 2 commercial satellites, while ISRO, Mitsubishi, and Ball Aerospace each developed 1 commercial satellite (Fig. 1.9).[308]

In 2014, 12.5 % of the 279 satellites launched were geostationary satellites (Fig. 1.10). In this field, 42.9 % of the 35 GEO satellites came from North America (i.e., 6 by SS/L, 5 by Orbital Sciences, 1 by Boeing, 1 by Lockheed Martin, and 2 additional satellites listed under Other—North America). Europe provided 17.1 % of the GEO satellites (i.e., 4 by Airbus Defence & Space and 2 by Thales Alenia Space). Russia's Reshetnev Company held a 20.0 % share of GEO satellites launched into orbit, launching a total of 7 for Russia. Japan's Mitsubishi provided 2 satellites, while India's ISRO launched 4, in addition to 1 other satellite by Other—Asia/ME. China's CAST did not launch any satellite to GEO orbit in 2014. However, when contrasting the 244 non-GEO orbiting satellites, North America maintained 47.0 % share (i.e., 4 by Boeing, 1 by SS/L, 1 by Orbital

305 Federal Aviation Administration. The Annual Compendium of Commercial Space Transportation: 2014. Washington DC: FAA, Feb. 2015: 9–10. Payloads are assigned to the nation that commissioned them, not according to the nationality of the manufacturer.

306 i.e. Not counting the 14 crewed or cargo missions to the ISS, the 1 test launch of the U.S. Orion EFT-1 capsule, and the launch of China's Chang'e 5 experimental lunar mission.

307 Federal Aviation Administration. The Annual Compendium of Commercial Space Transportation: 2014. Washington DC: FAA, Feb. 2015: 190–194.

308 Ibid.

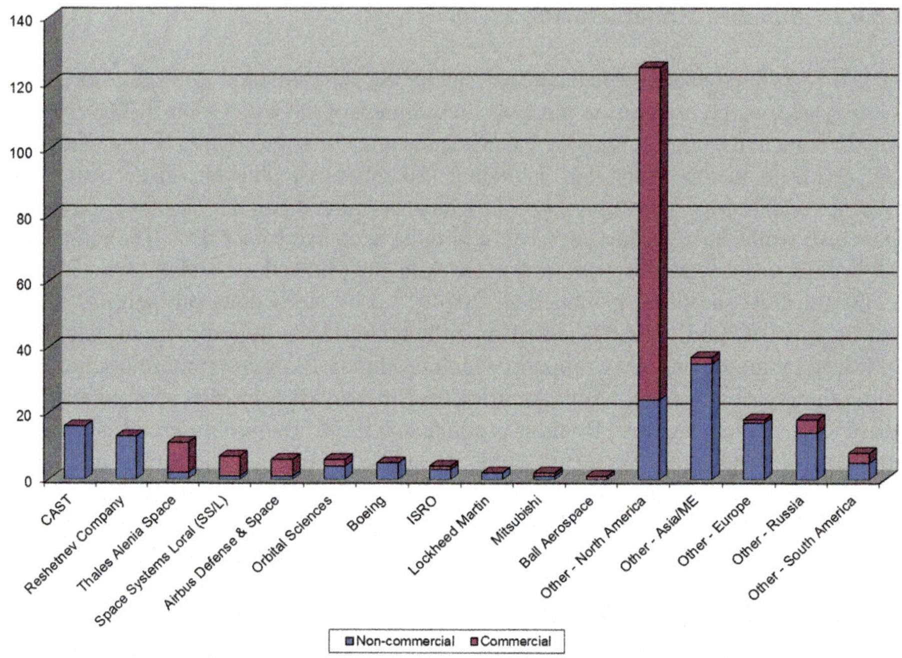

Fig. 1.9 Satellites launched in 2014 by manufacturer and commercial status (Source: Futron)

Sciences, 1 by Lockheed Martin, 1 by Ball Aerospace, and another 123—mostly cube satellites—from other parts of North America).[309] Asian manufacturers held a 21 % share with 52 non-GEO satellites (i.e., 16 by CAST, while another 36 was developed by in Other—Asia/ME). Europe's share was 11.9 % with its 29 non-GEO satellites (i.e., 9 by Thales Alenia Space, 2 by Airbus Defense & Space, and the remainder from other European makers). Russia had a 9.8 % share, with its 24 - non-GEO satellites (i.e. 6 by the Reshetnev Company, and the remaining by Other Russian manufacturers). The remaining 8 non-GEO satellites were developed in South America, and accounted for a 3.3 % share (Fig. 1.11).[310]

In 2014, North American prime spacecraft manufacturers held a strong lead in orders for commercial GEO communications satellites, accounting for 48.1 % of the contracts awarded, European contracts grew to 33.3 % of the available awards. SS/L took the competitive lead in commercial orders, winning 9 of 27 contracts (including the Amazonas 5, BRIsat, BulgariaSat1, Echostar23, Hispasat1F, Intelsat 36, JCSat15, JCSat16, and PSN 6). Europe's Thales Alenia Space came next with 5 orders (i.e. EuropaSat/HellasSat3, Koreasat5A, Koreasat7, Telkom 3S, and

[309] Discounting the 101 U.S. cube satellites, the U.S. share was 21.0 % of the remaining 143 - non-GEO satellites that were launched.

[310] Federal Aviation Administration. The Annual Compendium of Commercial Space Transportation: 2014. Washington DC: FAA, Feb. 2015: 190–194.

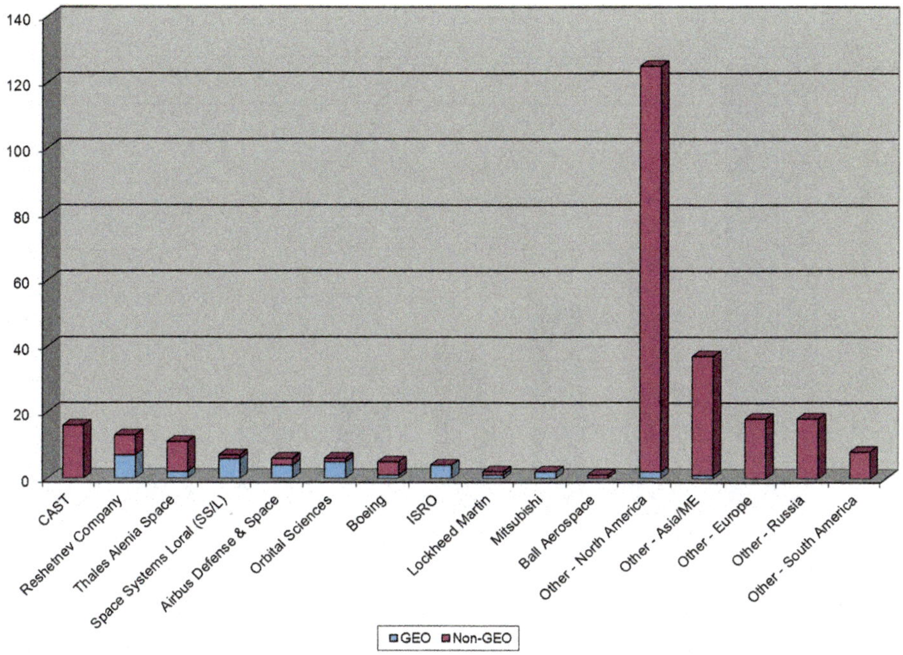

Fig. 1.10 Satellites launched in 2014 by manufacturer and orbit type (Source: Futron)

Yamal 601), followed by Airbus Defence & Space, which had 4 orders (i.e. Echostar105/SES11, Eutelsat172B, SES 10, and SES 12). Next U.S. prime contractors' Orbital Sciences had 3 contracts (i.e. Al Yah 3, Hylas 4, and Thaicom 8), while Boeing won 1 commercial contract for the year (i.e. Intelsat 35e). Russia and Japan each took a 7.4 % share by winning 2 contracts; Russia's Duaria Aerosapace will build the NexStar 1, and NexStar 2 satellites, and Japan's Mitsubishi will build the Es'hail2, and Superbird8 satellites. Lastly, Turkish Aerospace Industries (TAI) won a domestic contract to build the Türksat 6A, resulting in a 3.7 % share.[311]

1.3.4.3 Satellite Operators Sector

In 2014, the satellite operator business growth rate improved following the financial crisis, with growth partly attributable to the industry's long investment cycles. Following a slump in global orders for commercial GEO telecommunications satellites in the years following the crisis, orders in 2014 increased to 27 satellites. The ranking of the four largest satellite fleet operators: Intelsat (Luxemburg,

[311] "Satellite Orders Report—2014 Year-End Summary." Futron 13 July 2015. http://www.futron.com/upload/wysiwyg/Resources/FoF/2015/FutronSM2014-EOY.pdf.

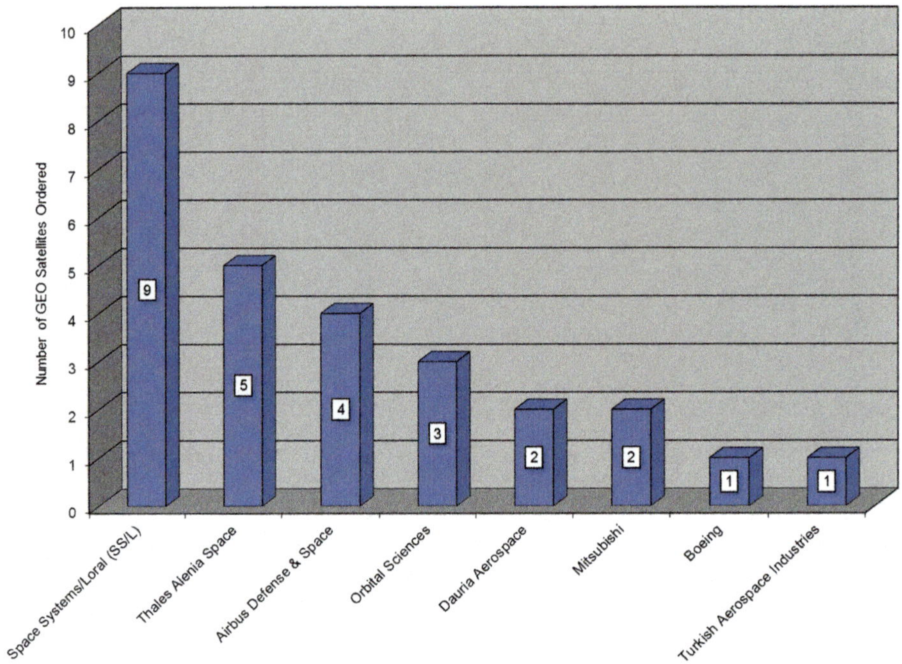

Fig. 1.11 GEO satellite orders in 2014 by manufacturer

Washington), SES (Luxemburg), Eutelsat (Paris) and Telesat (Canada) remained unchanged for 2014 in terms of revenue earned. By the end of 2014 Intelsat, as the top FSS operator earned $2.47 billion in revenue, and had 55 satellites in orbit and 7 satellites on order. Next, SES with $2.33 billion in revenue, had 53 satellites in orbit and 7 satellites on order, while Eutelsat whose 2014–2015 revenue is estimated to be $1.65 billion, had 34 satellites in orbit and 7 satellites on order. And Telesat with $793.7 million in revenue had 10 satellites and 1 satellite on order in 2014.[312]

1.4 The Security Dimension

1.4.1 The Global Space Military Context

The Space Report 2015 listed space-related military spending in 2014 as having reached $35.325 billion, an increase from $31.9 billion in 2013 following renewed defence spending in the U.S., and by non-U.S. space actors.[313] On the other hand,

[312] De Selding, Peter B. "The List | Top Fixed Satellite Service Operators." 13 July 2015. Space News 18 Aug. 2015. http://spacenews.com/the-list-2014-top-fixed-satellite-service-operators/.

[313] See: The Space Report 2014. Colorado Springs: The Space Foundation, 2014: 40.

Euroconsult noted a reduction in defence spending, listing world government expenditures for defence space programs as $24.45 billion in 2014, down from $27.97 billion in 2013; the reduction came mainly from an anticipated reduction in U.S. defence spending.[314] While there is a significant discrepancy between the two authorities, it is likely due to the availability of information at their times of publication as well as the different programmes considered. As is typical with the nature of dual-use technology in space activity, there is a risk that certain military activities have been already included in larger budgets, which can result in double counting. Moreover, while missions, often listed as civil programmes, may also serve dual-purpose military objectives, their expenditure is not included in this section.

The Space Report 2015 estimated the U.S. to account for about 63.6 % of global military space spending in 2014, less than the 68.3 % it estimated for 2013. In contrast, non-U.S. global military spending grew to 36.4 % in 2014 from 31.7 % in 2013.[315] Interestingly, the Euroconsult report presented the same defence expenditure ratio of 63.6 % of global military space spending attributed to U.S. defence expenditure, while non-U.S. military spending accounted for the remaining share; however, U.S. spending was estimated to be $15.56 billion, while the rest of the world spent an estimated $8.90 billion.[316] Russia, China, and Japan were next in line in military space spending estimates, with $4.33 billion, $1.91 billion, and $842 million in 2014, wherein Russia and Japan's budget decreased by 5.5 % and 12.8 % respectively, while China's budget increased by 14.1 % from 2013.[317] However, due to fluctuating exchange rates, variations in purchasing power, and different employment costs, a direct comparison of the budgets of these countries in fixed dollar values does not present a clear picture of their relative space defence efforts.

1.4.2 Europe

EU Member States continued to dominate spending on European military space programmes. Military space programmes are undertaken by all major European space faring nations, with an even larger number of European countries participating on the basis of bilateral or multilateral agreements and arrangements. From Euroconsult estimates, total European space defence spending in 2014 appears to have increased by 15.0 % from 2013, with spending by France, the United Kingdom, Italy, and Germany amounting to $1.146 billion in 2014 from the $995 million estimated for 2013.[318] Moreover, Euroconsult's estimate for total European

[314] Euroconsult 2015. Profiles of Government Space Programs: 16.

[315] The Space Report 2015. Colorado Springs: The Space Foundation, 2015: 39.

[316] Euroconsult 2015. Profiles of Government Space Programs: 16.

[317] Ibid.

[318] Euroconsult 2015. Profiles of Government Space Programs: 16.

government expenditure in defence programmes is $1.23 billion in 2014 from $1.07 billion in 2013.[319] France's space defence budget increased by 15.6 % reaching $585.20 million (440 million euros) in 2014; on the other hand, the Space Report 2015 lists France's national security space budget at $486.74 million (355.57 million euros).[320] Euroconsult estimates that the UK spends $249.15 million (£151 million) annually, with nearly 93 % of expenditure directed toward its military satellite communications in 2014.[321] Italy's space defence budget is estimated to be $247.38 million (186 million euros) for the purchase of space capability from commercial operators for security, defence and commercial purposes, whereas in 2013 it was estimated at $190 million.[322] Lastly, Germany's space defence budget is estimated by Euroconsult to be $65.17 million (49 million euros), however that amount might be due to the nature of dual-use technology in space activity that is funded through larger budgets, as SpaceNews reported Germany's military space spending to be about $546.38 million in 2014 from $219 million in 2013.[323] The other military space budgets in Europe remained relatively unchanged from the 2013 military space budgets.[324]

On 1 October 2014, Germany became the 10th member of an EDA-led programme to pool and share purchases of commercial satellite bandwidth for military and government use in the EU. It joins Belgium, Finland, France, Greece, Italy, Luxemburg, Poland, Romania, and the UK in a pay-per-use model for services, rather than having to pay a regular fee. The EDA acts as the central purchasing body on behalf of the Contributing Members. Since May 2013 a total of 18 orders for satellite bandwidth have been placed for a total value of around 2 million euros ($2.6 million).[325] The EDA hopes that this programme will give EU member states a sample of the type of savings that can be attained by group purchases, as another step towards consolidation of military satellite communications efforts in Europe.[326]

On 30 October 2014, ESA and the U.S. Strategic Command signed a space surveillance data-sharing agreement to exchange space situational data as the space

---

[319] Ibid.

[320] The Space Report 2015. Colorado Springs: The Space Foundation, 2015: 30.

[321] Euroconsult 2015. Profiles of Government Space Programs. Paris: Euroconsult, 2015.

[322] C.f. Euroconsult 2015. Profiles of Government Space Programs. Paris: Euroconsult, 2015, and "Piano Triennale di Attività 2013-2015." 25 Jan. 2013. ASI 29 Mar, 2015: 8, 71. http://www.asi.it/files/PTA_2013-2015_INTEGRATO_DELIBERE_132-159.pdf.

[323] De Selding, Peter B. "Work Underway in Germany on Radar Satellite with 25-centimeter Resolution." 24 Sept. 2013. SpaceNews 29 Sept. 2015. http://spacenews.com/37369work-underway-in-germany-on-radar-satellite-with-25-centimeter/.

[324] Euroconsult 2015. Profiles of Government Space Programs: 16.

[325] "Germany joins the EU Satcom Market." 1 Oct. 2014. EDA 23 July 2015. https://www.eda.europa.eu/info-hub/news/article/2014/10/01/germany-joins-the-eu-satcom-market.

[326] De Selding, Peter B. "Germany Joins European Military Bandwidth-sharing Program." 1 Oct. 2014. SpaceNews 23 July 2015. http://spacenews.com/42048germany-joins-european-military-bandwidth-sharing-program/.

environment continues to become more congested. The U.S. has signed nearly 50 similar data-sharing agreements with other governments and private-sector entities, including France, Italy, the UK, Canada, Japan, Australia, and South Korea. The agreement provides ESA with more timely information about the space environment, in exchange for position, radio-frequency information and planned orbit manoeuvres of some European satellites.[327]

France and Italy's military and civil space authorities have shared the cost of the Athena-Fidus broadband satellite that launched on 6 February 2014. The Athena-Fidus carries an extremely high frequency/Ka-band payload for Italy and a Ka-band payload for France.[328] The Defence Ministries of both France and Italy will also have separate telecommunications payloads on the Sicral 2 satellite, being constructed by Thales Alenia Space, which should now launch sometime in 2015.

1.4.3 The United States

While the U.S. DoD space budget slightly increased by 3.5 %, to $22.483 billion in 2014 from $21.717 in 2013, the budget for the U.S. Missile Defense Agency (U.S. MDA) continued to decrease in 2014, reflecting overall changes in U.S. military space spending to prioritize new capabilities. The U.S. MDA had a 1.9 % decrease in its budget, receiving $7.6 billion in 2014; moreover it has requested $7.5 billion for 2015. However, the new request allocates nearly $99.5 million to redesign its Exoatmospheric Kill Vehicle (EKV) that is launched upon the Ground-based Midcourse Defence (GMD) interceptor. The redesign stemmed from concerns about the system's reliability following three intercept failures that were attributable to issues with the EKV. This new Capability Enhancement-2 version of the EKV will have a modular open architecture, along with common interfaces and standards to allow for easier upgrades and a wider vendor and supply base. The 2015 budget request also includes $435 million to buy 30 Standard Missile 3 Block 1B interceptors, which will complement the U.S. Pentagon's purchase of 44 of the interceptors announced in January and March 2014.[329]

Throughout 2014, SpaceX aggressively vied to attain certification to launch military payloads, in addition to bringing injunction proceedings to U.S. federal courts to challenge the USAF's $11 billion 36 rocket core block-buy arrangement

---

[327] Gruss, Mike. "U.S. Strategic Command, ESA Sign Space Surveillance Data-sharing Agreement." 31 Oct. 2014. SpaceNews 23 July 2015. http://spacenews.com/42384 us-strategic-command-esa-sign-space-surveillance-data-sharing-agreement/.

[328] De Selding, Peter B. "Italian Military Buys $100 M Spy Satellite from Israel in Exchange Deal." 6 July 2012. SpaceNews 6 May 2014. http://spacenews.com/article/italian-military-buys-100 m-spy-satellite-israel-exchange-deal.

[329] Gruss, Mike. "$7.5B MDA Budget Request Includes $100 M for Kill Vehicle Redesign." 10 Mar. 2014. SpaceNews 23 July 2015. http://spacenews.com/3978975b-mda-budget-request-includes-100 m-for-kill-vehicle-redesign/.

with the DoD's traditional prime contractor United Launch Alliance (ULA).[330] The matter arose from the U.S. DoD's November 2012 decision to authorize the USAF to purchase up to 50 rocket cores during the next 5 years under its Evolved Expendable Launch Vehicle (EELV) programme. While 14 cores were to be procured competitively, with the first of these competitive awards expected in 2015 for a mission launching in 2017, the other 36 will be procured from ULA on a sole-source basis.[331] Another matter complicating the situation is the USAF's recent disaggregation approach to funding, as it now intends to halve the number of competitively procured space launches to 7 rocket cores, limiting the available contracts on which SpaceX may competitively bid.[332]

1.4.4 Russia

Despite a long tradition of military space activity, yearly reviews are made in the current day context. As in previous years, Russian military spending appears to be dwarfed by U.S. military expenditure, the ratio of Russian to U.S. military space spending being estimated to be about 1:4 in 2013. Euroconsult estimates the total amount Russia puts towards its military program to have been at least $4.33 billion (i.e. 137.36 billion roubles) in 2014 (including dual-use programmes), however that value might overlook purchase power parities in that spending, and is based on fluctuating exchange rates.[333] Overall by 2014, Russia reached the midpoint of a 10-year $700 billion weapons modernisation programme with a focus on strategic nuclear weapons, fighter aircraft, ships and submarines, air defences, communications, and intelligence.[334]

Russia formed its Aerospace Defence Forces (ADF) on 1 December 2011, to monitor its airspace and track space objects above its territory. ADF completed development of its first military space surveillance complex, in Altai, West Siberia, in December 2014, wherein it began tests of its laser-optical and radio engineering complexes that will provide the capability to track small objects and debris currently in orbit. Russia also conducted tests of the modernized Okno-M optoelectronic space

[330] Gruss, Mike. "SpaceX Challenge to ULA Block Buy Could Hinge on Questions of Timing." 9 May 2014. SpaceNews 23 July 2015. http://spacenews.com/40524spacex-challenge-to-ula-block-buy-could-hinge-on-questions-of-timing/.

[331] Leone, Dan. "Pentagon Approves EELV Block Buy, with Competitive Twist." 4 Dec. 2012. SpaceNews 15 May 2014. http://www.spacenews.com/article/military-space/32657pentagon-approves-eelv-block-buy-with-competitive-twist.

[332] Gruss, Mike. "U.S. Air Force Halves Size of Competitive EELV Procurement." 7 Mar. 2015. SpaceNews 23 July 2015. http://spacenews.com/39772 us-air-force-halves-size-of-competitive-eelv-procurement/.

[333] Euroconsult 2015. Profiles of Government Space Programs. Paris: Euroconsult, 2015.

[334] Masters, Jonathan. "How Powerful Is Russia's Military?" 14 Nov. 2014. Defense One 19 June 2015. http://www.defenseone.com/threats/2014/11/how-powerful-russias-military/99062/.

surveillance station, located in Tajikistan. Russia plans to build around ten specialized space control complexes, to track satellites and other space objects, by 2018.[335]

In November 2014, a Russian space object began exhibiting unusual behaviour. The object had been assumed to be space debris stemming from the launch of three Rodnik military communication satellites in May 2014. Following launch, the object known as 2014-28E changed its orbit, and then manoeuvred back to a similar orbit to the Rokot launcher's spent upper stage, with the two objects rendezvousing shortly after.[336] Whether for civilian or military purposes, the object has piqued the interest of observers tracking the object.[337] One speculation is whether the object, and similar manoeuvres performed by China's Shijian 15 satellite in May 2014, is related to Russia and China's push for a binding treaty on the Prevention of the Placement of Weapons in Outer Space and of the Threat or Use of Force Against Outer Space Objects (PPWT).

1.4.5 Japan

Japan's space budget for the fiscal year 2014, running from 1 April 2014 through 31 March 2015, increased by 18.8 %, totalling $3.78 billion (¥382.7 billion) for space programs across the government of Japan, with $737 million (¥74.6 billion) budgeted for defence spending.[338] On the other hand, Euroconsult estimates Japan's space spending (including dual-use systems), to have decreased by 3.5 % to $2.60 billion (¥274.7 billion), with $840 million (¥89 billion) allocated for defence purposes. By its estimate, in 2014, 77 % of Japan's defence space expenditure went toward Earth observation, and about 23 % toward satellite communications; spending in defence areas is expected to continue to drive increased spending through 2015.[339]

Geopolitical tensions in the region, particularly with neighbours such as North Korea, and recent natural disasters have motivated the country to refocus its efforts from its traditional multilateral "peaceful-use-only" position in space activities to the space security and defence areas. The need for closer integration of space into the nation's defence strategy is becoming more visible, which is likely to continue with closer cooperation with the U.S. military, and increased communications, remote sensing, and navigation capabilities. For instance, the 20 August 2014

---

[335] Staff Writers. "Russia Tests First Military Space Surveillance Complex." 3 Dec. 2014. SpaceDaily.com 19 May 2015. http://www.spacedaily.com/reports/Russia_Tests_First_Military_Space_Surveillance_Complex_999.html.

[336] Boyle, Alan. "Russian Space Object 2014-28E Sparks Worries About 'Satellite Killer'." 18 Nov. 2014. NBC News 19 June 2015. http://www.nbcnews.com/science/space/russian-space-object-2014-28e-sparks-worries-about-satellite-killer-n251111.

[337] Jones, Sam. "Object 2014-28E—Space junk or Russian satellite killer?" 17 Nov. 2014. Financial Times 19 June 2015. http://www.ft.com/cms/s/2/cdd0bdb6-6c27-11e4-990f-00144feabdc0.html.

[338] The Space Report 2015. Colorado Springs: The Space Foundation, 2015: 34.

[339] Euroconsult 2015. Profiles of Government Space Programs. Paris: Euroconsult, 2015.

update to Japan's 5-year Basic Plan of 2012 recommends that the QZSS be increased from one satellite to four, and that a full seven satellite constellation should be developed as soon as possible, as it would enable stable management of the stand-alone regional navigation system.[340]

1.4.6 China

In recent years China's space defence budget is estimated to have seen considerable increases on a year-to-year basis, having increased by 17.6 % to $1.41 billion in 2012, followed by 18.6 % to $1.67 billion in 2013.[341] China's overall defence budgets have also seen significant increases in recent years, growing by 12.2 % from $126 billion in 2013 to $132 billion in 2014. It is expected that this figure could reach about $145 billion in 2015, while previous U.S. DoD estimates suggest actual defence spending in 2015 could be greater than $175 billion.[342]

The manner in which China develops its technical capabilities and the depth of its military interest in the space program merits analysis. It seems to be clear that Chinese space efforts are intimately connected to the Chinese army, principally because many space activities are under the direct control of the People's Liberation Army (PLA). Moreover, all Chinese space operation facilities are entirely manned and operated by the PLA. In the case of manned space activities, all development and plans are directly under the control of the Chinese military and political bureau. This does not mean that all space activities conducted by the Chinese military are military per se, but it shows the central role of the Chinese military in the Chinese space architecture.

China's space defence perspective continues to be difficult to identify due to a dearth of unclassified sources. In April 2014, the U.S. DoD presented its Annual Report to Congress on 'Military and Security Developments Involving the People's Republic of China 2014', which highlighted the activities conducted by the country in 2013, and brought to light increasing concern regarding counterspace capability as central to enabling modern, "informationized" warfare. Among other developments and trends mentioned, the report says some PLA writings emphasize the necessity of using anti-satellite weapons to 'blind and deafen the enemy', and that destroying or capturing satellites and other sensors will deprive an opponent of

[340] GPS World Staff. "QZSS May Expand to Meet Japan's Surveillance Needs." 22 Aug. 2014. GPS World 15 June 2015. http://gpsworld.com/qzss-may-expand-to-meet-japans-surveillance-needs/.

[341] Profiles of Government Space Programs. Paris: Euroconsult, 2014: 14.

[342] Fisher (Jr.), Richard D. "China's Rising Defense Budgets Stoke Regional Concerns." 8 Jan. 2015. Aviation Week & Space Technology 12 June 2015. http://aviationweek.com/defense/china-s-rising-defense-budgets-stoke-regional-concerns.

initiative on the battlefield, making it difficult to bring their precision guided weapons into full play.[343] China conducted at least eight space launches in 2013 to expand its space-based intelligence, surveillance, reconnaissance, meteorological, and communications satellite constellations.

In May 2014, the Chinese satellite Shijian 15, in orbit since July 2013, began to converge toward another Chinese satellite Shijian 7, exhibiting atypical propulsion capabilities, followed by an interception with the possible use of a remote capture arm and close proximity operations. While such capability can have civil applications, such as to refuel or tow a satellite, military applications are also possible, such as the possibility of disabling an active payload belonging to a foreign nation without creating additional debris in space.[344]

Moreover, by 21 November 2014, China had launched its second Kuaizhou ("quick vessel") launcher, which is designed to launch a Kuaizhou small satellite quickly into a low-Earth orbit to support "natural disaster monitoring." The first Kuaizhou launch occurred on 25 September 2013. However, there are some indications that the Kuaizhou's Feitan Space Emergency Response System, FT-1, may be related to the Dong Neng 2, DN-2 ASAT system.[345]

Renewed concern over ASAT testing occurred throughout the reporting period, following a March 2014 paper by the Secure World Foundation which highlighted facts that strongly suggested that a May 2013 scientific launch mission by China may have also tested the rocket component of new direct ascent anti-satellite weapons system derived from a road-mobile ballistic missile.[346] Furthermore, on 23 July 2014, China is believed to have conducted a "non-destructive" test of an anti-satellite weapon, while China has responded that the launch was a ballistic missile intercept, similar to the tests conducted by the U.S. and other countries.[347]

[343] Office of the Secretary of State. "Annual Report to Congress—Military and Security Developments Involving the People's Republic of China 2014." 5 June 2014. Defense.gov 12 June 2015: 40. http://www.defense.gov/pubs/2014_DoD_China_Report.pdf.

[344] Jones, Sam. "Object 2014-28E—Space junk or Russian satellite killer?" 17 Nov. 2014. Financial Times 19 June 2015. http://www.ft.com/cms/s/2/cdd0bdb6-6c27-11e4-990f-00144feabdc0.html.

[345] Fisher (Jr.), Richard D. "China launches second Kuaizhou mobile SLV." 26 Nov. 2014. HIS Jane's 360 16 July 2015. http://www.janes.com/article/46360/china-launches-second-kuaizhou-mobile-slv.

[346] Gruss, Mike. "Report Rekindles Suspicions About Chinese Rocket Launch." 21 Mar. 2014. SpaceNews 12 May 2015. http://spacenews.com/39936report-rekindles-suspicions-about-chinese-rocket-launch/.

[347] Gruss, Mike. "U.S. State Department: China Tested Anti-satellite Weapon." 28 July 2014. SpaceNews 12 June 2015. http://spacenews.com/41413 us-state-department-china-tested-anti-satellite-weapon/.

1.4.7 India

India is developing its own space military programme; however, the majority of the activities of the Indian Space Research Organization (ISRO) still focus on civil applications.

On 27 April 2014, India's Defence Research Development Organization (DRDO) conducted the first test launch of its newly developed interceptor missile from launch pad-IV of the Integrated Test Range at Wheeler Island, about 100 km from Odisha. Expecting to connect with its target at an altitude of 120 km, the missile's infrared seeker successfully tracked the target missile throughout the test, however no direct contact was made as the distance between the interceptor and the target was more than what was expected.[348] Even so, the test was announced to be at least a partial success by DRDO officials, who said the main objective was to track the target missile and see the performance of the infrared (IR) seeker in the interceptor.[349] India's Prithvi Defence Vehicle (PDV), part of India's Ballistic Missile Defense (BMD) programme, has the potential to destroy incoming missiles with a strike range of around 2500 km outside the earth's atmosphere (at an altitude of over 150 km).[350]

India's PDV can also be seen as a further step toward developing its own anti-satellite capabilities. In this pursuit, the DRDO is looking at the feasibility of developing such an anti-satellite vehicle by integrating its Angi-3 missile with its PDV. If it succeeds, the anti-satellite missile would have an effective range of about 1400–1500 km, and would advance India's missile capabilities to be on a par with U.S. and China.[351]

[348] Rout, Hemant Kumar. "Prithvi Defence Vehicle Fails to Intercept." 15 May 2014. New Indian Express 12 June 2015. http://www.newindianexpress.com/states/odisha/Prithvi-Defence-Vehicle-Fails-to-Intercept/2014/05/15/article2225945.ece.

[349] "Interceptor spot on, though without blast: DRDO." 28 Apr. 2014. The Hindu 12 June 2015. http://www.thehindu.com/news/national/interceptor-spot-on-though-without-blast-drdo/article5953934.ece#comments.

[350] Rout, Hemant Kumar. "DRDO Planning to Test-fire High-altitude 'Killer' Missile in January." 30 Nov. 2013. The New Indian Express 12 May 2014. http://www.newindianexpress.com/nation/DRDO-Planning-to-Test-fire-High-altitude-Killer-Missile-in-January/2013/11/29/article1917837.ece.

[351] "India Contemplates Anti-Satellite Vehicle Integration with Agni-III Ballistic Missile." 15 Oct. 2013. Missile Threat 12 May 2014. http://missilethreat.com/india-contemplates-anti-satellite-vehicle-integration-with-agni-iii-ballistic-missile/.

Chapter 2
Developments in Space Policies, Programmes and Technologies Throughout the World and in Europe

Cenan Al-Ekabi

2.1 Space Policies and Programmes

The major space policy developments worldwide were presented in Chap. 1, above, to identify the principal space faring nations' strategies in 2014. In the section below, there will be a brief discussion of developments in technology related areas, including policies and access to space technologies. The aim of this section is to clarify how these strategies interact with and influence specific space programmes, and related research and development projects.

2.2 Space Transportation

2.2.1 *Europe*

With the conclusion of the 2 December 2014 ESA Ministerial Council meeting, ESA has decided to proceed with the Ariane 6 as the next generation launcher, to be developed by 2020. A major distinction of the Ariane 6 from its progenitors is in its governance arrangement, wherein industry partners Airbus and Safran, under a joint venture called Airbus Safran Launchers, will take over as the launcher's design authority, in addition to taking on the responsibility of the commercial exploitation of these launchers. Meanwhile, ESA member states have agreed to continue supporting the Ariane 5, Vega, and Europeanized-Soyuz launchers until the 2016 ESA Ministerial Council meeting. At the next meeting, ESA will decide

C. Al-Ekabi (✉)
ESPI, Schwarzenbergplatz 6, 1030 Vienna, Austria
e-mail: cenan.al-ekabi@espi.or.at

C. Al-Ekabi et al. (eds.), *Yearbook on Space Policy 2014*, Yearbook on Space Policy, DOI 10.1007/978-3-7091-1899-3_2

whether to continue with these three distinct launchers, or to rely solely on the two configurations of the Ariane 6 for future launches.[1]

On the issue of future international cooperation in space transportation and space exploration, the combined use of the ATV and other similar spacecraft from the U.S. (Dragon and Cygnus COTS missions), Japan (HTV) and Russia (Progress M) to serve ISS supply needs provides an example of the opportunities in the creation of common transportation policies among all participating space actors. While ESA began its final ATV-5 mission in mid-2014, to pay for its contribution to the operation of the ISS until 2017, ESA's dues for the utilization of the ISS on its extended operation will be paid by supplying NASA with an ATV-derived service module for NASA's Orion spacecraft.[2] The provision of ESA's ATV-technology for the Orion module will cover ESA's 8.3 % share of the ISS's annual operating costs for the period 2018–2020; estimated at a total cost of 455 million euros. At the International Astronautical Congress (IAC) held on 29 September to 3 October 2014, it was suggested that the Progress, ATV, and HTV cargo spacecraft would likely be welcome to fly to the upcoming Chinese space station; although some political and export control barriers need to be overcome before U.S. commercial supply services are able to launch to the Sino-space station.[3]

Seeking to increase its space exploration capability, ESA has been developing its entry, descent, and landing technology through its Intermediate Experimental Vehicle (IXV) since 2006. Having narrowly escaped cancellation following the previous financial crisis, the Italian led programme was scheduled to complete final integration in May 2014, and was planned to undergo a suborbital test launch on a Vega launcher by 18 November 2014.[4] However, the launch had to be pushed to 2015, to allow for additional analyses of the Vega flight trajectory.[5]

---

[1] Henry, Caleb. "ESA Hands Reins to Industry on New Launchers." 3 Dec. 2014. Satellite Today 22 June 2015. http://www.satellitetoday.com/launch/2014/12/03/esa-hands-reins-to-industry-on-new-launchers/.

[2] "ESA Workhorse to Power NASA's Orion Spacecraft." 16 Jan. 2013. ESA 8 Apr. 2014. http://www.esa.int/Our_Activities/Human_Spaceflight/Research/ESA_workhorse_to_power_NASA_s_Orion_spacecraft.

[3] Foust, Jeff. "SN Blog | The Role of International Cooperation in China's Space Station Plans." 14 Oct. 2014. SpaceNews 30 June 2015. http://spacenews.com/42183sn-blog-the-role-of-international-cooperation-in-chinas-space-station-plans/.

[4] Foust, Jeff. "65th International Astronautical Congress | European Re-entry Demonstrator Ready for November Test Flight." 3 Oct. 2014. SpaceNerws 14 July 2015. http://spacenews.com/4207765th-international-astronautical-congress-european-re-entry-demonstrator/.

[5] "Vega launch of ESA spaceplane postponed." 28 Oct. 2014. ESA 15 July 2015. http://www.esa.int/Our_Activities/Launchers/IXV/Vega_launch_of_ESA_spaceplane_postponed.

2.2.2 United States

Despite the success of NASA's Orion Multi-Purpose Crew Vehicle (MPCV) Exploration Flight Test 1 mission on the Space Launch System (SLS) on 5 December 2014, the ongoing White House and Congress debate on funding levels and prioritisation has resulted in missed opportunities, with the debut of the SLS probably delayed by a year. While NASA had aimed to have the SLS ready by December 2017, it seems more likely to have the launcher ready only in November 2018.[6] However a revised launch date awaits the completion of design reviews expected early in 2015. Other factors contributing to the expected delay come from the analyses of the 5 December launch, which tested many of the MPCV's key subsystems (e.g. its heat shield, electronics, and parachutes); and from ESA's inclusion into the programme by providing its ATV-derived service module.[7]

In broad policy matters, at the 30th Space Symposium in Colorado Springs, U.S., held on 19–22 May 2014, U.S. congressional staffers revealed that the U.S. government is preparing to update the Commercial Space Launch Act, last amended in 2004. Expected to come by 2015, the new legislation will address: whether the FAA will be allowed to begin writing human spaceflight safety regulations after October 2015, when the current regulatory grace period expires; in addition to changes to the formula FAA uses to determine how much insurance commercial launch providers must carry; and some changes to the National Oceanic and Atmospheric Administration's (NOAA) licensing regime for commercial remote sensing satellites. However, it not clear whether the new law will allow the FAA to regulate what commercially launched spacecraft can do while they are on orbit; that hot-button issue is currently being reviewed by congress.[8]

2.2.3 Russia

Russia is making progress in line with recent national space policies that have focused on improvement of the country's self-sufficiency, on technological and operational levels, to increase Russia's global market shares in the space sector. Development of the Vostochny Cosmodrome, located in the Far Eastern region of Russia, is continuing—to ensure the completion of the spaceport's Soyuz-2 and

---

[6] Foust, Jeff. "Looming SLS Delay To Rekindle Debate about NASA's Priorities." 29 Aug. 2014. SpaceNews 14 July 2015. http://spacenews.com/41710looming-sls-delay-to-rekindle-debate-about-nasas-priorities/.

[7] Foust, Jeff. "Next Orion Flight To Slip to 2018." 3 Dec. 2014. SpaceNews 14 July 2015. http://spacenews.com/42825next-orion-flight-to-slip-to-2018/.

[8] Leone, Dan. "Hill Staffers: Commercial Space Launch Bill Is Coming This Year." 26 May 2014. SpaceNews 14 July 2015. http://spacenews.com/40694hill-staffers-commercial-space-launch-bill-is-coming-this-year/.

Angara launch pads by 2016; which should also coincide with the completion of its Angara launcher.[9] The final construction stage of the Vostochny Cosmodrome, taking place between 2016 and 2018, will develop facilities for a super-heavy launch vehicle that will be capable of delivering 120–150 tons into space by 2020.[10]

On 9 July 2014, Russia succeeded in test launching its new indigenous Angara launcher, the first new rocket designed by Russia's space industry in over 20 years.[11] The inaugural sub-orbital flight of the Angara 1.2 was postponed from its intended 27 June 2014 launch date due to a drop in oxidizer pressure that stemmed from a poorly sealed valve on the launcher's liquid oxygen tank.[12] The Angara is a two stage launcher, with a main core that burns kerosene and liquid oxygen, which can be supplemented with additional boosters to increase power; while different upper stage designs can be used depending on the intended orbit altitude. The 9 July test launch was of the Angara 1—its simplest configuration—that should be capable of launching up to 4 tons to LEO orbit. The variants are planned to extend to a heavy-lift Angara 5 version intended to carry up to 7.5 tons to GEO orbit. The Angara will launch from both the Plesetsk spaceport and the soon to be built Vostochny Cosmodrome.[13] The Angara is meant to replace Russia's somewhat unreliable Proton M launcher to remain competitive with commercial launcher providers such as SpaceX.

2.2.4 *Japan*

Japan plans to develop a lower-cost, commercially viable successor to its H-2A rocket. On 17 May 2013, Japan's Space Transportation Systems Subcommittee of its Cabinet-level Office of National Space Policy (ONSP) presented a draft midterm report recommending an H-3 successor launcher.[14] On 24 December 2013, the

---

[9] Nowakowski, Tomasz. "Putin pledges $1.5 billion for completion of Vostochny Cosmodrome, OKs plans for creating super-heavy rockets." 9 Sept. 2014. Spaceflight Insider 16 July 2015. http://www.spaceflightinsider.com/organizations/roscosmos/putin-pledges-1-5-billion-completion-vostochny-cosmodrome-oks-plans-creating-super-heavy-rockets/.

[10] Bodner, Matthew. "Putin Pledges $1 Billion for Completion of New Cosmodrome." 2 Sept. 2014. The Moscow Times 16 July 2015. http://www.themoscowtimes.com/business/article/putin-pledges-1-billion-for-completion-of-new-cosmodrome/506321.html.

[11] De Selding, Peter B. "Russia's Angara 1.2 Rocket Succeeds in Inaugural Flight, Khrunichev Says." 9 July 2014. SpaceNews 16 July 2015. http://spacenews.com/41184russias-angara-12-rocket-succeeds-in-inaugural-flight-khrunichev-says/.

[12] Bodner, Matthew. "Angara Launch Troubles Reflect Russia's Struggling Space Industry." 1 July 2014. The Moscow Times 16 July 2015. http://www.themoscowtimes.com/business/article/angara-launch-troubles-reflect-russias-struggling-space-industry/502809.html.

[13] Amos, Jonathan. "Russia's Angara rocket 'makes debut'." 9 July 2014. BBC.com 16 July 2015. http://www.bbc.com/news/science-environment-28058633.

[14] Kallendar-Umezu, Paul. "Japanese Government Recommends Developing H-2A Successor." 27 May 2013. SpaceNews 30 May 2014. http://www.spacenews.com/article/launch-report/35499japanese-government-recommends-developing-h-2a-successor.

Japanese government approved initial funding of $70 million for 2014 for the development of the launcher, estimated to need $1.9 billion for full development. The two-stage H-3 is tentatively scheduled to have its first launch in 2020, and is projected to lift up to 6.5 metric ton payloads to GTO at a cost ranging between $50 million and $70 million per launch.[15] Mitsubishi Heavy Industries Corp. was selected by JAXA as the prime contractor, with the launcher expected to feature a liquid hydrogen/liquid oxygen core stage with up to six solid-fuel strap-on boosters to offer a wide range of payload-to-orbit capabilities.[16]

2.2.5 China

In line with its latest five-year plan for 2011–2016, released in December 2011,[17] China has continued to make incremental progress in the field of space transportation. It plans to develop three new launcher configurations by 2016, i.e. Long March 5 (with a 14 tons to GEO lift capacity), Long March 6 (1 ton to LEO), and Long March 7 (5.5 tons to LEO), using more efficient engines and an entirely new upper stage. China is also developing a new launch centre in the Northeastern corner of Hainan Island, far south of the mainland; its completion is expected to coincide with the completion of China's Long March 5, 6, and 7 launch systems. The use of the Hainan spaceport will allow China to free itself from the limitations of using its curving rail lines, and transporting vehicles through the narrow width of its train tunnels.[18]

The Long March 5 is expected to conduct its first launch sometime in 2015, from the Hainan spaceport. By the end of 2014, it was undergoing final testing and assembly.[19] The Long March 5 will be used to lift component modules of China's space station into orbit over the next decade. Comparable to other heavy launchers, such as the U.S. Delta IV Heavy and Europe's Ariane 5, the Long March 5 will be

[15] Onuki, Misuzi. "Japan Approves $1.9B for H-3 Rocket." 13 Jan. 2014. SpaceNews 30 May 2014. http://www.spacenews.com/article/civil-space/39069japan-approves-19b-for-h-3-rocket.

[16] Onuki, Misuzu. "MHI Formally Selected as H-X Prime Contractor, Operator." 25 Mar. 2014. SpaceNews 30 May 2014. http://www.spacenews.com/article/financial-report/39971mhi-formally-selected-as-h-x-prime-contractor-operator.

[17] White Papers of the Government of China. "China's Space Activities in 2011." Beijing 29 Dec. 2011. 6 Mar. 2012. http://www.china.org.cn/government/whitepaper/node_7145648.htm.

[18] David, Leonard. "China's New Spaceport to Launch Country's Largest Rocket Yet." 2 Apr. 2014. Space.com 16 July 2015. http://www.space.com/25323-china-new-spaceport-rocket-launches.html.

[19] Lin, Jeffrey and P.W. Singer. "China Is Building One Of The World's Largest Space Launch Vehicles." 11 Dec. 2014. Popular Science 16 July 2015. http://www.popsci.com/long-march-goes.

powerful enough to launch a lunar or Mars mission spacecraft with a total thrust at sea level of 1080 tons.

On 21 November 2014, China conducted the second launch of its Kuaizhou mobile satellite launch vehicle (SLV), meant to be a rapid satellite repopulation capability, following the first launch on 25 September 2013.[20] Also, China is developing a second emergency response launch vehicle, Long March 11, to allow it to rapidly enter space and meet the emergency launching demand in case of disasters and contingencies. It is expected to launch sometime before 2016.[21]

Moreover, China is also developing an even larger launcher, comparable to the U.S. SLS system, called the Long March 9. Up to December 2014, China had been conducting preliminary research, considering the technological feasibility and requirements needed for the launcher to be capable of exploration beyond the Moon. Similar to the SLS, the Long March 9 is expected to have a maximum payload of 130 tons, but its first launch will occur in the decade following the SLS's first launch.[22] At the International Astronautical Congress held in Beijing on 23–27 September 2013, the CALT published the main specifications of the two possible configurations. The first concept would have four YF-660 engines mounted in the core first stage and one in each of four side-mounted boosters. In the second concept, most of the initial thrust would come from four solid-propellant boosters, each generating 1000 tons of thrust, while four YF-220 concept engines would be mounted in the first stage. While work is underway on the engines, Chinese industry is still awaiting permission to begin developing the Long March 9.[23]

2.2.6 India

India's space launch programme primarily relies on the use of its Polar Satellite Launch Vehicle (PSLV), capable of carrying 3700 kg payloads to Low Earth Orbit and 800 kg to Geosynchronous Transfer Orbits (GTO).[24] However the successful launch of its Geosynchronous Satellite Launch Vehicle (GSLV)-Mark II on 5 January 2014 makes room for ISRO to develop its new GSLV Mark III rocket, capable of launching around 4000 kg into GTO. The GSLV-Mark II can launch up

[20] Fisher (Jr.), Richard D. "China launches second Kuaizhou mobile SLV." 26 Nov. 2014. HIS Jane's 360 16 July 2015. http://www.janes.com/article/46360/china-launches-second-kuaizhou-mobile-slv.

[21] Profiles of Government Space Programs. Paris: Euroconsult, 2014: 18.

[22] Lei, Zhao. "New Long March launcher on the drawing board." 9 Dec. 2014. SpaceDaily 16 July 2015. http://www.spacedaily.com/reports/New_Long_March_launcher_on_the_drawing_board_999.html.

[23] "Chinese Super-Heavy Launcher Designs Exceed Saturn V." 30 Sept. 2013. Aviation Week 2 May 2014. http://aviationweek.com/awin/chinese-super-heavy-launcher-designs-exceed-saturn-v.

[24] Federal Aviation Administration. Commercial Space transportation: 2011 Year in Review. Washington DC: FAA, Jan. 2012: 15.

to 2500 kg into GTO, using two liquid fuel stages, and a new indigenous cryogenic upper stage. Prior to the successful January 2014 launch of the GSAT 14 communications satellite, the Mark I and II had had mixed results over the past decade, exhibiting a trend of launch failures from 2006, and were not used following launch failures in 2010.[25] The GSLV-Mark III will share a number of components with its predecessor GSLV-Mark I and II launchers, but it will be the first of its series to field a second stage equipped with a restartable liquid fuel engine that should greatly improve the system's operational flexibility and commercial attractiveness.[26] The first experimental sub-orbital test of the GSLV-Mark III was conducted on 18 December 2014.[27] ISRO plans to market its GLSV to launch heavier commercial communications satellites through its Antrix commercial arm beginning in 2017.[28]

At the International Astronautical Congress held in Toronto from 29 September to 3 October 2014, India's intention to expand its presence in the global launch market became more apparent. Representatives from Antrix announced that India's PSLV had five slots available for commercial Earth observation and research customers, i.e. one in 2015, three in 2016, and one in 2017, to launch missions to sun-synchronous orbit. And one GSLV-Mark III launcher should be available per year for commercial sale starting in 2017.[29]

2.3 Space Science and Exploration

In this section, space science is understood to mean using mainly remote observation to make discoveries on the origin, evolution and future of the Universe, its galaxies, our Solar System, and other celestial bodies e.g. stars, exoplanets, comets, and asteroids. Space exploration, on the other hand, involves human and robotic spaceflight missions. While traditional governmental space agencies dominate in both these fields, progress in the latter category can be seen with the development of exploration involving commercial players, and with new space powers demonstrating the technology needed to carry out such missions.

---

[25] "GSLV." ISRO 19 Apr. 2013. http://www.isro.org/launchvehicles/GSLV/gslv.aspx.

[26] "GSLV MARK III." ISRO 19 Apr. 2013. http://www.isro.org/Launchvehicles/GSLVMARKIII/mark3.aspx.

[27] "First Experimental Flight of India's Next Generation Launch Vehicle GSLV Mk-III Successful." 18 Dec. 2014. ISRO 15 July 2015. http://www.isro.gov.in/update/18-dec-2014/first-experimental-flight-of-indias-next-generation-launch-vehicle-gslv-mk-iii.

[28] De Selding, Peter B. "65th International Astronautical Congress | India Poised To Expand Presence in Global Launch Market." 1 Oct. 2014. SpaceNews 14 July 2015. http://spacenews.com/4205565th-international-astronautical-congress-india-poised-to-expand-presence/.

[29] Ibid.

2.3.1 Human Spaceflight Activities

Human spaceflight was focused in Low Earth Orbit (LEO), with the International Space Station (ISS) at centre stage, following its formal extension to at least 2020. Following the retirement of NASA's Space Shuttle, Roscosmos is the sole launch provider relied upon to transport crew regularly to the ISS and, using Progress and Soyuz, it also provided ISS cargo resupply services with Europe's Automated Transfer Vehicle (ATV) and Japan's H-II Transfer Vehicle (HTV) providing auxiliary support.

On 22 November 2010, ESA's newest batch of astronauts, Luca Parmitano, Alexander Gerst, Samantha Cristoforetti, Timothy Peake, Andreas Mogensen, and Thomas Pesquet, graduated from the European Astronaut Centre (EAC). After Luca Parmitano completed his 5-month ISS Expeditions 36/37 mission on 11 November 2013, Alexander Gerst flew to the space station as a flight engineer for Expeditions 40/41 in May 2014. In November 2014, Samantha Cristoforetti began her mission to the ISS as the eighth ESA astronaut to participate in a long-duration visit.

Alexander Gerst began his 6-month 'Blue Dot' mission on the ISS on 28 May 2014. Serving as a flight engineer, Alexander assisted in the docking and unloading of five visiting cargo spacecraft, and was the prime operator for the final ESA Automated Transfer Vehicle (ATV), George Lemaître, where he monitored the arrival and docking of the vessel and took responsibility for the distribution of the contents of its cargo bay.[30] The German astronaut also tested the Electromagnetic Levitator, a furnace that can melt and solidify metal alloys away from the container's walls, helping to gain insight into the solidification and physical properties of molten alloys, and intended to improve industrial casting processes and permit finer and more delicate metal castings.[31] Gerst performed over 50 other experiments covering materials physics, human physiology, radiation biology, solar research, biotechnology, fluid physics and astrophysics, and a series of technology demonstrations.[32] Gerst also participated in NASA's Ocular Health research into how astronauts' eyes adapt to space conditions, and assisted with general maintenance on the station, such as repairing a xenon lamp in JAXA's SAIBO rack, for experiments aiming to grow more effective crops on Earth.[33] Further he participated in a 6.25 h spacewalk on 7 October with NASA astronaut Reid Wiseman, and

---

30 "ESA astronaut Alexander Gerst arrives at Space Station." 29 May 2014. ESA 26 May 2015. http://www.esa.int/Our_Activities/Human_Spaceflight/Blue_dot/ESA_astronaut_Alexander_Gerst_arrives_at_Space_Station.

31 "Ask astronaut Alexander anything." 11 Mar. 2014. ESA 26 May 2015. http://www.esa.int/Our_Activities/Human_Spaceflight/Blue_dot/Ask_astronaut_Alexander_anything.

32 "ESA astronaut Alexander Gerst arrives at Space Station." 29 May 2014. ESA 26 May 2015. http://www.esa.int/Our_Activities/Human_Spaceflight/Blue_dot/ESA_astronaut_Alexander_Gerst_arrives_at_Space_Station.

33 "Alexander's first week in space." 5 June 2014. ESA 26 May 2015. http://www.esa.int/Our_Activities/Human_Spaceflight/Blue_dot/Alexander_s_first_week_in_space.

choreographed a second 6.5 h spacewalk in the following week.[34] Gerst completed his Blue Dot mission on 10 November 2014, and returned to Earth along with Russian commander Maxim Suraev and NASA astronaut Reid Wiseman on the same Soyuz TMA-13M spacecraft that brought them to the station.[35]

Shortly afterward, on 24 November, ESA Astronaut Samantha Cristoforetti began a 5-month mission on the ISS under the mission banner 'Futura'.[36] In addition to helping to berth regular supply vessels such as the American Dragon, the Italian astronaut will be the prime operator for the undocking of ESA's final ATV from the ISS. Cristoforetti will run an extensive scientific programme including experiments in physical science, biology, human physiology, and radiation research and technology demonstrations, in addition to maintenance of the station.[37]

Although the life of the ISS has been extended by another 5 years, ESA is stopping the production of Automated Transfer Vehicles (ATVs) after 2015. The fifth and final ATV, George Lemaître, was launched to the ISS on 29 July 2014 on an Ariane 5 ES launcher from French Guiana. The ATV-5 carried on its exterior a Laser InfraRed Imaging Sensors (LIRIS) test demonstrator as a step toward future rendezvous with 'uncooperative' targets, such as orbiting debris or a Mars sample capsule. Using infrared cameras and lidar sensors (i.e. the light equivalent of radar), such future missions will scan the targets while onboard computers process the data using new guidance navigation and control software.[38] The ATV-5 was the heaviest of its progenitors, carrying over 6602 kg of experiments, spare parts, clothing, food, fuel, air, oxygen and water to the station—including the 400 kg Electromagnetic Levitator.[39] By the end of 2014, the ATV-5 had conducted more than three reboost/deboost manoeuvres,[40] including an unprecedented event in November where the ISS required urgent action to avoid a piece of Cosmos-2251 space debris.[41]

---

[34] "Wiseman and Gerst Complete First Spacewalk of Expedition 41." 7 Oct. 2014. NASA 27 May 2015. http://www.nasa.gov/content/wiseman-and-gerst-complete-first-spacewalk-of-expedition-41.

[35] "ESA astronaut Alexander Gerst returns to Earth." 10 Nov. 2014. ESA 26 May 2015. http://www.esa.int/Our_Activities/Human_Spaceflight/Blue_dot/ESA_astronaut_Alexander_Gerst_returns_to_Earth.

[36] "ESA astronaut Samantha Cristoforetti arrives at Space Station." 24 Nov. 2014. ESA 26 May 2015. http://www.esa.int/Our_Activities/Human_Spaceflight/Futura/ESA_astronaut_Samantha_Cristoforetti_arrives_at_Space_Station.

[37] "Futura mission in brief." 27 Oct. 2014. ESA 27 May 2015. http://www.esa.int/Our_Activities/Human_Spaceflight/Futura/Futura_mission_in_brief.

[38] "ATV-5 set to test new rendezvous sensors." 18 Mar. 2014. ESA 26 May 2015. http://www.esa.int/Our_Activities/Human_Spaceflight/ATV/ATV-5_set_to_test_new_rendezvous_sensors.

[39] "ATV-5: loaded and locked." 23 July 2014. ESA 26 May 2015. http://www.esa.int/Our_Activities/Human_Spaceflight/ATV/ATV-5_loaded_and_locked.

[40] "ATV-5 Mission Updates." 1Oct. 2014. Spaceflight101.com 27 May 2015. http://www.spaceflight101.com/atv-5-mission-updates.html.

[41] "ESA space ferry moves Space Station to avoid debris." 4 Nov. 2014. ESA 26 May 2015. http://www.esa.int/Our_Activities/Human_Spaceflight/ATV/ESA_space_ferry_moves_Space_Station_to_avoid_debris.

DLR continued to develop and implement the German Space Programme, based on its National Space Strategy published in 2010, with strong involvement in the ISS, ExoMars and lunar preparatory activities at ESA level, and scientific instrumentation, robotics contributions and data analysis in the National Programme and cooperatively for international exploration missions. The flight of astronaut Alexander Gerst to ISS in May 2014 was strongly supported by Germany's national research and outreach programme.[42]

In 2014, Russia launched four expeditions to the ISS on its Soyuz launcher: TMA-12M on 26 March 2014 with the Expedition 39/40 crew of Steve Swanson, Alexander Skvortsov, and Oleg Artemyev[43]; TMA-13M on 28 May 2014 with the Expedition 40/41 crew of Reid Wiseman, Maxim Suraev, and Alexander Gerst[44]; TMA-14M on 26 September 2014 with the Expedition 41/42 crew of Aleksandr Samoukutyaev, Elena Serova, and Barry Wilmore[45]; and TMA-15M on 24 November 2014 with the Expedition 42/43 crew of Terry Virts, Anton Shkaplerov, and Samantha Cristoforetti.[46] In addition to conducting a 6.5 h space-walk outside the ISS, and launching a Russo-Peruvian nanosatellite, Chasqui I (Messenger), Russian cosmonauts Aleksandr Skvortsov and Oleg Artemyev carried out a number of technical operations on the ISS.[47] Elena Serova is the first Russian woman to work on the ISS, and the fourth Russian woman to fly in space; Serova will work on or help to plan a variety of ongoing experiments on the station.[48] Russia also conducted four successful resupply launches to the ISS with its Progress cargo transfer vehicles: M-22M on 5 February 2014, M-23M on 9 April 2014, M-24M on 24 July 2012, and M-25M on 29 October 2014.[49]

[42] Annual Report 2013 of the International Space Exploration Coordination Group." 20 Dec. 2013. International Space Exploration Coordination Group (ISECG) 14 Apr. 2014. http://www.globalspaceexploration.org/wordpress/wp-content/uploads/2013/12/Annual-Report_2013_FINAL.pdf.

[43] "Russia launches Expedition 39 to ISS." Russian Space Web 26 May 2015. http://www.russianspaceweb.com/iss_soyuz_tma12m.html.

[44] "Russia launches fresh crew to ISS." Russian Space Web 26 May 2015. http://www.russianspaceweb.com/iss_soyuz_tma13m.html.

[45] "Mission of Soyuz TMA-14M." Russian Space Web 26 May 2015. http://www.russianspaceweb.com/iss_soyuz_tma14m.html.

[46] "Mission of Soyuz TMA-15M." Russian Space Web 26 May 2015. http://www.russianspaceweb.com/iss_soyuz_tma15m.html.

[47] Burke, Myles. "Russian cosmonauts make six-and-a-half-hour spacewalk outside ISS." 18 Aug. 2014. The Telegraph 21 Aug. 2015. http://www.telegraph.co.uk/news/science/space/11041065/Russian-cosmonauts-make-six-and-a-half-hour-spacewalk-outside-ISS.html.

[48] Allen, Nick. "Zebrafish, athlete's foot and 3D printing: 10 far out experiments on the International Space Station." 26 Sept. 2014. The Telegraph 21 Aug. 2015. http://www.telegraph.co.uk/news/science/space/11122637/Zebrafish-athletes-foot-and-3D-printing-10-far-out-experiments-on-the-International-Space-Station.html.

[49] "A complete chronology of ISS missions." Russian Space Web 26 May 2015. http://www.russianspaceweb.com/iss_chronology_flights.html.

Following the success of the Tiangong 1 space laboratory and associated milestones in docking its Shenzhou spacecraft to the station, and repeatedly sending Chinese astronauts to the station in 2012 and 2013, China did not conduct human spaceflight missions in 2014. China plans to complete the construction and launch of its Tiangong 2 space lab in 2015, followed by another experimental core module of the future space station around 2018.[50] However, the launch of the Tiangong 2 space lab has been pushed to 2016. The Shenzhou-11 spacecraft and Tianzhou-1 cargo spacecraft will be launched shortly afterward to dock with the space lab. China's full space station is to be completed by 2022.[51]

2.3.2 Lunar Science

The Moon stimulated a great deal of interest in this reporting period in regard to the science and exploration activities that were planned or already underway. Europe, China, Japan and India made progress toward a robotic and human lunar presence, however budget constraints had the potential to delay well-intentioned initiatives. The U.S. and Russia also maintained their interest in exploring the Moon, and its potential benefit for future Mars exploration.

At ESA's Ministerial Council meeting held on 2 December 2014, the agency gained approval and funding to investigate participation with Russia in robotic missions for the exploration of the Moon; Russia intends to place a lander and a rover on the Moon's South Pole. Discussions of a partnership with Russia's Roscosmos on the project have been ongoing following ESA's inability to gather enough support for its own proposed ESA Lunar Lander mission at the Ministerial Council held in 2012. The project had been passed over in favour of launcher development, EO, ISS operations, and the joint ExoMars mission with Russia.[52] In the new proposed partnership, ESA would contribute technologies to Roscosmos' Luna-Resource Lander, also known as Luna 27, which is scheduled for launch in 2019, as well as the Lunar Sample Return, planned for the early 2020s. Pursuing lunar missions is strategically important for ESA, not only to secure access to the Moon's surface for European scientists, but also to ensure that European expertise and technology is involved in future lunar exploration—including, ultimately, international crewed missions and even a permanent lunar base.[53]

---

[50] "China plans to launch Tiangong-2 space lab around 2015." 27 June 2013. Space Daily 8 Apr. 2014. http://www.spacedaily.com/reports/China_plans_to_launch_Tiangong_2_space_lab_around_2015_999.html.

[51] "China's space station to be established around 2022." 10 Sept. 2014. English.news.cn 11 June 2015. http://news.xinhuanet.com/english/china/2014-09/10/c_133633441.htm.

[52] Clark, Stephen. "ESA lunar lander shelved ahead of budget conference." 20 Nov. 2012. Spaceflight Now 21 Apr. 2013. http://spaceflightnow.com/news/n1211/20moonlander/.

[53] Gibey, Elizabeth. "Europe proposes joint Moon trips with Russia." 9 Dec. 2014. Nature 28 Feb. 2015. http://www.nature.com/news/europe-proposes-joint-moon-trips-with-russia-1.16517.

NASA's Lunar Reconnaissance Orbiter (LRO), launched in June 2009, is scouting the Moon in preparation for future lunar exploration, including finding landing sites, locating resources such as water, ice and hydrogen, and investigating the long-term effects of the lunar environment. The mission has created the most precise and complete topographic maps of the Moon, including the largest high resolution mosaic of the Moon's North polar region to date,[54] and has determined areas of the Moon that are in perpetual darkness or in near-continuous sunlight. The LRO helps scientists study the deformation of the Moon due to Earth's pull, which provides clues about the Moon's interior. Earth's distorting effect on the Moon, i.e. the lunar body tide, provides enough force to raise a 51 cm bulge on the near side of the Moon and a corresponding dip on the far side, with the bulge shifting according to the Earth's movements.[55]

In a study published in August 2014, it was proposed that high-energy particles from uncommon, large solar storms penetrate the Moon's polar regions and electrically charge the soil potentially significantly altering soil properties, which could have important implications for current understanding of the evolution of planetary surfaces in extremely cold regions that are exposed to harsh radiation from space.[56] Moreover, the LRO has provided researchers strong evidence that the moon's volcanic activity slowed gradually instead of stopping abruptly a billion years ago, having observed distinctive rock deposits that are estimated to be less than 100 million years old, with some areas appearing less than 50 million years old.[57] Also, through measuring the Moon's radiation environment, the spacecraft's CRaTER instrument was able to show that in shielding against the radiation faced by astronauts in deep-space missions, tissue-equivalent plastics and other lightweight materials can provide even more effective protection than standard shielding, such as aluminium.[58] The LRO completed its first extended science mission on 15 September 2014, beginning its second 2 year extended science mission slated

[54] Press Release. "NASA Releases First Interactive Mosaic of Lunar North Pole." 18 Mar. 2014. NASA 28 Feb. 2015. http://www.nasa.gov/press/2014/march/nasa-releases-first-interactive-mosaic-of-lunar-north-pole/.

[55] Zubritsky, Elizabeth. "NASA Missions Let Scientists See Moon's Dancing Tide From Orbit." 29 May 2014. NASA 28 Feb. 2015. http://www.nasa.gov/content/goddard/nasa-missions-let-scientists-see-moon-s-dancing-tide-from-orbit/.

[56] "Electric Sparks May Alter Evolution of Lunar Soil." 21 Aug. 2014. NASA 28 Feb. 2015. http://www.nasa.gov/content/goddard/electric-sparks-may-alter-evolution-of-lunar-soil/.

[57] Release. "NASA Mission Finds Widespread Evidence of Young Lunar Volcanism." 12 Oct. 2014. NASA 28 Feb. 2015. http://www.nasa.gov/press/2014/october/nasa-mission-finds-widespread-evidence-of-young-lunar-volcanism/.

[58] Zubritsky, Elizabeth. "Internet Radio Provides Musical Space-Weather Reports from NASA's LRO Mission." 9 Jan. 2014. NASA 28 Feb. 2015. http://www.nasa.gov/content/goddard/musical-space-weather-reports-from-nasas-lro/.

for completion in September 2016[59]; even so, the LRO retains sufficient fuel to maintain its current orbit for around more 8 years.[60]

NASA's newly launched Lunar Atmosphere and Dust Environment Explorer (LADEE) lifted into space on 6 September 2013. After completing its development phase in reaching the Moon at the end of 2013, the satellite began its 100-day mission to study the Moon's exospheric dust environment and determine the composition of the lunar atmosphere, including the processes that control its distribution and variability.[61] LADEE uses a set of three instruments to measure the chemical composition of the atmosphere, and collect and analyse samples of lunar dust particles in the atmosphere. Its Neutral Mass Spectrometer, measuring atoms and molecules in the lunar atmosphere, has already recorded noble gases helium, neon, and argon-40. The Ultraviolet-Visible Spectrometer views the lunar horizon seeking the glow of atoms, molecules and dust in the lunar atmosphere and has made measurements of atmospheric sodium and potassium at lunar sunset, sunrise and noon. LADEE's Lunar Dust Experiment (LDEX) has measured the dust tossed up by a fairly steady "rain" of meteoroids on the lunar surface, due to meteoroid showers and "dust bursts" that may be from LADEE flying through plumes kicked up from nearby meteoroid impacts. [62] While the 100-day mission was completed on 1 March 2014, it had been extended for an additional lunar day (29.5 Earth days) to gain a further understanding of the sources and drivers of the tenuous lunar atmosphere, as they are subject to change from month to month.[63] LADEE's final manoeuvre was performed on 11 April, adjusting its trajectory to avoid any previous lunar mission landings; and following a total lunar eclipse on April 14 to 15, LADEE impacted the surface of the moon on 17 April 2014.[64]

China's Chang'e 3 Lunar lander and rover mission successfully soft-landed on the Moon on 14 December 2013, making it the third country in the world to soft land a spacecraft on the Moon. Shortly afterward, the Yutu rover, 'Jade Rabbit', began its planned 3-month mission, using ground-penetrating radar and spectrometers to inspect the composition of the Moon's soil and the structure of the much

[59] Keller, J. "The Lunar Reconnaissance Orbiter and the New Moon: Mission Highlights and Two More Years of Science From Lunar Orbit!" 22 Nov. 2014. LRO—Lunar Planetary Institute 16 Mar. 2015. http://www.lpi.usra.edu/meetings/leag2014/presentations/keller.pdf.

[60] Petro, Noah E., and John W. Keller. "The second extended science mission for the lunar reconnaissance orbiter: Status, science goals, and data deliveries." 15 Sept. 2014. LPSC 16 Mar. 2015. http://www.hou.usra.edu/meetings/lpsc2015/pdf/2278.pdf.

[61] "Missions—LADEE—NASA Science." NASA 16 Apr. 2014. http://science.nasa.gov/missions/ladee/.

[62] Hoover, Rachel. "NASA Extends Moon Exploring Satellite Mission." 31 Jan. 2014. NASA 28 Feb. 2015. http://www.nasa.gov/ames/nasa-extends-moon-exploring-satellite-mission/.

[63] Elphic, Rick. "LADEE Project Scientist Update: Citius, propius, occupatus." 10 Apr. 2014. NASA 28 Feb. 2015. http://www.nasa.gov/ames/ladee-project-scientist-update/.

[64] "NASA Completes LADEE Mission with Planned Impact on Moon's Surface." 18 Apr. 2014. NASA 28 Feb. 2015. http://www.nasa.gov/ames/nasa-completes-ladee-mission-with-planned-impact-on-moons-surface/.

deeper lunar crust.[65] While the rover was able to activate all of the scientific tools within days of deployment, it experienced a myriad of challenges from the outset of its mission. Those challenges stemmed from the extreme lunar environment and mechanical control problems that eventually kept the rover from entering normal dormancy and folding its mast and solar panels by its second 14-day lunar night. The rover completed its first examination of the lunar soil on 16 January 2014.[66] In the following weeks, communication with the rover became more difficult as the rover's solar panels began to degrade due to the colder than anticipated temperatures on the Moon; the rover became immobile from 25 January 2014,[67] however it was still able to send signals to Earth for the remainder of the year.

Following the continued success of Chang'e 3 lander and the Jade Rabbit rover, the country pressed on with the 23 October 2014 launch of its test capsule Chang'e 5-TI, which looped around the Moon as a test-run for an eventual lunar sample return mission.[68] The mission was to verify technologies for the sample return initiative in the third phase of China's lunar exploration programme; it also accommodated a mini-satellite called the Manfred Memorial Moon Mission (4M) developed by LuxSpace.[69] China plans to return a sample of lunar soil to Earth with its Chang'e 5 to be launched in 2017,[70] in addition to sketching plans for a manned lunar landing sometime between 2025 and 2030.[71]

The Google Lunar X PRIZE is a competition for a total of $30 million in prizes for the first privately funded team to safely land a rover on the Moon. To win, the rover must travel at least 500 m on the Moon's surface and send high-definition video, images, and data back to the Earth. To provide additional incentive for accelerated development, the prize will reduce in value after a government-funded mission explores the lunar surface. Early in November 2013, in an effort to assist the competing teams by allowing them to access financing at a critical point in their mission timeline and raise public excitement and support for the teams, X PRIZE and Google announced a series of Milestone Prizes available to the competing

[65] "Chang E III launch next year will achieve lander and rover joint probe." 14 June 2012. News. Xinhuanet.Com 21 Aug. 2015. http://news.xinhuanet.com/politics/2012-06/14/c_112217044.htm.

[66] "China's Jade Rabbit rover explores Moon soil." 16 Jan. 2014. BBC News 21 Aug. 2015. http://www.bbc.com/news/science-environment-25763168.

[67] Chen, Stephen. "Last-ditch efforts to salvage mission of China's stricken Jade Rabbit lunar rover." 18 Apr. 2014. South China Morning Post 21 Aug. 2015. http://www.scmp.com/news/china/article/1486425/last-ditch-efforts-salvage-mission-chinas-stricken-jade-rabbit-lunar.

[68] Spudis, Paul D. "China is Now Positioned to Dominate the Moon." 6 Nov. 2014. Air & Space Smithsonian 28 Feb. 2015. http://www.airspacemag.com/daily-planet/china-now-positioned-dominate-moon-180953267/?no-ist.

[69] Lakdawalla, Emily. "UPDATED: China successfully launched test mission for Chang'e 5 program today." 23 Oct. 2014. Plantary.org 28 Feb. 2015. http://www.planetary.org/blogs/emily-lakdawalla/2014/10230750-china-to-launch-test-mission.html.

[70] "China Targets Moon Sample-Return Mission in 2017." 26 Dec. 2013. Space.com 7 Apr. 2014. http://www.space.com/24055-china-moon-sample-return-mission.html.

[71] "China considering manned lunar landing in 2025–2030." 24 May 2009. China View 21 Apr. 2013. http://news.xinhuanet.com/english/2009-05/24/content_11425131.htm.

teams. With amounts ranging from between $250,000 to $1 million available to teams that demonstrate (via actual testing and analysis) robust hardware and software to combat key technical risks in the areas of imaging, mobility and lander systems, the awards can be won through the end of September 2014.[72] With 25 teams registered for the competition on 31 December 2010,[73] by the end of 2013, that number had reduced to 18 active teams, with several competitors leaving to pursue other business interests.[74] On 16 December 2014, the Lunar X Prize was competition deadline was extended to 2016.[75]

2.3.3 *Mars Science*

The focus for Mars science has for decades remained the investigation of the planet's habitability, in a search for the presence of water. The collected data continues to suggest that Mars was once partially covered by large oceans, and that life could have been possible in many locations on the planet's surface.

ESA's Mars Express orbiter, launched in June 2003, continued its mission imaging the entire surface of the planet at high resolution, including maps of the mineral composition and atmosphere, and determining the structure of the sub-surface to a depth of a few kilometres, the effect of the atmosphere on the surface, and the interaction of the atmosphere with the solar winds. In October 2014, the orbiter was repositioned to view the comet Siding Spring as it streaked past Mars at a distance reaching 139,500 km; i.e. just over a third of the distance from the Moon to the Earth. The flyby presented the opportunity for close-up observations of the comet, the influence of the comet's gas and dust on Mars atmosphere, and the complex three-way interaction between Mars, the comet and the solar wind.[76]

The ExoMars mission, now a joint endeavour between ESA and Russia, continued its development into 2014. ESA will provide the Trace Gas Orbiter (TGO) and the Entry, Descent and Landing Demonstrator Module (EDM) 'Schiaparelli' in

---

72 "Recognizing Giant Leaps: Google Lunar XPRIZE Establishes Milestone Prizes (Op-Ed)." 7 Nov. 2013. Space.com 7 Apr. 2014. http://www.space.com/23503-google-lunar-xprize-milestone-prizes.html.

73 "Google Lunar X PRIZE." Google Lunar XPRIZE 26 Mar. 2013. http://www.googlelunarxprize.org/.

74 "As 2013 Comes To An End, Competition Intensifies In Private Race To The Moon." 19 Dec. 2013. Google Lunar X Prize 7 Apr. 2014. http://www.googlelunarxprize.org/blog/2013-comes-end-competition-intensifies-private-race-moon.

75 Foust, Jeff. "Google Lunar X Prize Extends Competition Deadline." 17 Dec. 2014. SpaceNews 23 July 2015. http://spacenews.com/google-lunar-x-prize-extends-competition-deadline/.

76 "Mars Express ready for comet encounter." 17 Oct. 2014. ESA 1 Mar. 2015. http://www.esa.int/Our_Activities/Operations/Mars_Express_ready_for_comet_encounter.

201,[77] and the carrier and rover in 2018; while Roscosmos is responsible for the 2018 descent module and surface platform, and will provide launchers for both missions. The ExoMars mission will address whether life has ever existed on Mars by drilling the surface of the planet and analyzing the samples in situ. By February 2014, the orbiter's core TGO module was completed, and will be launched in the 2016 mission to search for evidence of methane and other atmospheric gases that could be signatures of active biological or geological processes.[78] The ExoMars rover, to be launched in 2018, will search the planet's surface for signs of life, past and present, and will be able to drill to a depth of 2 m.[79] In October 2014, an ESA-appointed panel formally recommended further analysis of four possible landing sites for the rover, i.e. Mawrth Vallis, Oxia Planum, Hypanis Vallis, and Aram Dorsum—each located relatively close to Mars' equator. The candidate sites will need to satisfy the technical constraints of the mission, meeting the operational and engineering requirements for safe landing and roving on the surface. The first certification is expected by the end of 2016, while a final decision on the landing site will be taken sometime in 2017.[80]

NASA's Mars Odyssey mission, launched on 7 April 2001, is the longest-operating spacecraft to be sent to Mars. The satellite orbited the planet's poles at about 5 am/pm local Mars time for the first 6 years of its mission, providing optimal temperature for its Gamma Ray Spectrometer search for evidence of water near the Martian surface (e.g. how widely water ice and other elements are distributed on Mars). In the next 3 years, a 4 am/pm orbit allowed the orbiter's Thermal Emission Imaging System (THEMIS) to more easily identify the infrared signatures of minerals. By 2012, after providing radio-relay support for the landing of the Curiosity Mars rover, the orbit was manoeuvred on a slow drift to later times of the day to aid in preserving battery life. In February 2014, it was proposed to have the orbiter moved to a 6:45 am/pm orbit that will have the Odyssey make daylight observations from South-to-North, which could present seasonal differences; the orbit-adjustment locking manoeuvre will be conducted on 15 November 2015.[81] On 5 August 2014, the orbiter was manoeuvred to place it behind Mars during the 30 min when the comet Siding Spring's dust particles were most likely to reach Mars, avoiding a potential impact that might have ended the Odyssey mission a

---

[77] "ExoMars Lander Module Named Schiaparelli." 8 Nov. 2013. ESA 28 Mar. 2014. http://exploration.esa.int/mars/53145-exomars-lander-module-named-schiaparelli/.

[78] "ExoMars orbiter core module completed." 3 Feb. 2014. ESA 1 Mar. 2015. http://exploration.esa.int/mars/53642-exomars-orbiter-core-module-completed/.

[79] "ExoMars: ESA and Roscosmos for Mars Missions." 14 Mar. 2013. ESA 28 Mar. 2014. http://exploration.esa.int/mars/51495-exomars-esa-and-roscosmos-set-for-mars-missions/.

[80] "Four candidate landing sites for ExoMars 2018." 1 Oct. 2014. ESA 1 Mar. 2015. http://exploration.esa.int/mars/54708-four-candidate-landing-sites-for-exomars-2018/.

[81] Press Release. "NASA Moves Longest-Serving Mars Spacecraft for New Observations." 12 Feb. 2014. NASA 1 Mar. 2015. http://mars.jpl.nasa.gov/odyssey/news/whatsnew/index.cfm?FuseAction=ShowNews&NewsID=1602.

decade too early.[82] On 19 October 2014, the orbiter's THEMIS was used to observe the comet's arrival, while its Neutron Spectrometer and High Energy Neutron detector were used to assess possible effects on Mars' atmosphere of dust and gas from the comet.[83]

NASA's Mars Reconnaissance Orbiter (MRO) continued to provide valuable data for the purpose of determining whether or not life has existed on Mars, characterising the climate and geology, and preparing for future human exploration. In previous years, it returned data that suggested that water still flows in some places on Mars, depicted as dark, finger-like features that appear and extend down some Martian slopes that change during the seasons,[84] in addition to finding evidence of carbon-dioxide snowfalls (i.e. 'dry ice') occurring around Mars' South Pole in its winter season.[85] In February 2014, a possible explanation for the dark, finger-like slopes came from clues corresponding to the seasonal changes in iron ore minerals and a survey of ground temperatures and other traits that supported a suggestion that the slopes may be brines with an iron-mineral anti-freeze, such as ferric sulfate, that flow seasonally.[86] Images from the spacecraft's High Resolution Imaging Science Experiment (HiRISE) camera also helped to reveal more detail about the changing Martian seasons, such as showing the formation of a new gully channel that broke out from an older route and eroded its new path; an activity that generally tends to occur in the planet's winter period where temperatures are cold enough to keep carbon dioxide in its liquid state.[87] The MRO was also manoeuvred on 2 July and 27 August to place the satellite in the right position to view the gases in comet Siding Spring's coma, in addition to detailed views of the comet's small nucleus and potentially its rotation rate and surface features.[88]

---

[82] "Orbiter Completes Maneuver to Prepare for Comet Flyby." 6 Aug. 2014. NASA 1 Mar. 2015. http://mars.jpl.nasa.gov/odyssey/news/whatsnew/index.cfm?FuseAction=ShowNews&NewsID=1689.

[83] "NASA's Mars Odyssey Orbiter Watches Comet Fly Near." 19 Oct. 2014. NASA 1 Mar. 2015. http://mars.jpl.nasa.gov/odyssey/news/whatsnew/index.cfm?FuseAction=ShowNews&NewsID=1736.

[84] "Mars Reconnaissance Orbiter." NASA Jet Propulsion Laboratory, California Institute of Technology 26 Mar. 2013. http://marsprogram.jpl.nasa.gov/mro/.

[85] "NASA Orbiter Observations Point to 'Dry Ice' Snowfall on Mars." 11 Sept. 2012. Jet Propulsion Laboratory 28 Mar. 2014. http://mars.jpl.nasa.gov/mro/news/whatsnew/index.cfm?FuseAction=ShowNews&NewsID=1341.

[86] "NASA Mars Orbiters See Clues to Possible Water Flows." 10 Feb. 2014. NASA 1 Mar. 2015. http://mars.jpl.nasa.gov/mro/news/whatsnew/index.cfm?FuseAction=ShowNews&NewsID=1599.

[87] "NASA Orbiter Finds New Gully Channel on Mars." 19 Mar. 2014. NASA 1 Mar. 2015. http://mars.jpl.nasa.gov/mro/news/whatsnew/index.cfm?FuseAction=ShowNews&NewsID=1613.

[88] "NASA's Mars Spacecraft Manoeuvres to Prepare for Close Comet Flyby." 25 July 2014. NASA 1 Mar. 2014. http://mars.jpl.nasa.gov/mro/news/whatsnew/index.cfm?FuseAction=ShowNews&NewsID=1675.

The NASA Mars Science Laboratory (MSL) rover, nicknamed Curiosity, reached Mars on 5 August 2012, and is currently studying Mars's habitability.[89] The rover has eight scientific objectives, i.e. determining the nature and inventory of organic carbon compounds; conducting an inventory of the chemical building blocks of life; identifying features that may represent the effects of biological processes; investigating the chemical, isotopic, and mineralogical composition of Martian geological materials; interpreting the processes that have formed and modified rocks and soils; assessing 4-billion-year timescale atmospheric evolution processes; determining the present state, distribution, and cycling of water and carbon dioxide; and characterizing the broad spectrum of surface radiation, including galactic cosmic radiation, solar proton events, and secondary neutrons.[90] By 6 May 2014, Curiosity had drilled into its third rock sample, a sandstone slab designated as Windjana, to be studied by the rover's internal instruments. Whereas the two previous rock-drilling sites, conducted on mudstone targets, revealed evidence of an ancient lakebed environment that held key chemical elements and a chemical energy source that long ago provided conditions favourable for microbial life, the Windjana site is expected to reveal the cementing material that holds together sand-size grains in sandstone.[91] On 5 August 2014, Curiosity celebrated its second year on Mars; while in its first year, Curiosity's major science goal was in determining whether Mars ever offered environmental conditions favourable for microbial life, its second year was meant to determine how the environmental conditions on Mars evolved.[92] After bypassing a fourth drilling target for safety concerns in August, the rover reached Mount Sharp on 24 September, where it began to sample the base layer of the mountain.[93] There, Curiosity confirmed the presence of the mineral hematite, an iron-oxide mineral that gives clues about ancient environmental conditions from when it formed, first mapped from orbit by the NASA MRO in 2010.[94] On 16 December 2014, after a year of analysing a drilled sample of the Sheepbed mudstone in the Gale crater, the team responsible for the Sample Analysis at Mars (SAM) instrument made the first definitive detection of organic material on Mars—i.e. short of finding signs of life at the Gale crater, the discovery shows that the ancient environment offered a supply of

[89] "NASA Lands Car-Size Rover Beside Martian Mountain." 5 Aug. 2012. NASA 31 Mar. 2014. http://www.nasa.gov/mission_pages/msl/news/msl20120805c.html.

[90] Mars Science Laboratory (MSL). 14 May. 2012. NASA NSSDC 14 Jan. 2013. http://nssdc.gsfc.nasa.gov/nmc/spacecraftDisplay.do?id=2011-070A.

[91] "NASA's Curiosity Rover Drills Sandstone Slab on Mars." 6 May 2014. NASA 1 Mar. 2015. http://www.nasa.gov/jpl/msl/drill-hole-20140506/.

[92] "NASA Mars Curiosity Rover: Two Years and Counting on Red Planet." 5 Aug. 2014. NASA 1 Mar. 2015. http://www.nasa.gov/jpl/msl/curiosity-20140805/.

[93] "NASA Rover Drill Pulls First Taste From Mars Mountain." 25 Sept. 2014. NASA 1 Mar. 2015. http://www.nasa.gov/jpl/msl/nasa-rover-drill-pulls-first-taste-from-mars-mountain/.

[94] "NASA's Curiosity Mars Rover Finds Mineral Match." 4 Nov. 2014. NASA 1 Mar. 2015. http://www.nasa.gov/jpl/msl/nasas-curiosity-mars-rover-finds-mineral-match/.

reduced organic molecules for use as building blocks for life and an energy source for life.[95]

NASA's Mars Atmosphere and Volatile EvolutioN (MAVEN) mission, launched on 18 November 2013, successfully entered Mars's orbit on 21 September 2014.[96] MAVEN aims to explore the planet's upper atmosphere, ionosphere and interactions with the Sun and solar wind, which will be used to determine the role that the loss to space of volatile compounds from the Mars atmosphere has played in the history of Mars' habitability.[97] In other words, by studying the planet's upper atmosphere and measuring current rates of atmospheric loss, MAVEN scientists hope to understand how Mars transitioned from a warm, wet planet to its current dry desert state.[98] The satellite was also manoeuvred to avoid the high-velocity dust released by comet Siding Spring, while its Imaging Ultraviolet Spectrograph observed intense ultraviolet emissions from magnesium and iron ions high in the atmosphere in the aftermath of the meteor shower, and its Neutral Gas and Ion Mass Spectrometer detected eight different types of metal ions, including sodium, magnesium and iron.[99] By 16 November 2014, MAVEN had completed its commissioning activities, and formally began its 1-year primary science mission[100]; and by the end of the year new insight was gained about the depth of penetration solar wind makes in a planet's atmosphere.[101]

ISRO's Mars Orbiter Mission (MOM) lifted into space on 5 November 2013, entering Mars' orbit on 24 September 2014.[102] Shortly thereafter, the spacecraft began its science activities fully, activating its five indigenous scientific payloads

[95] "NASA Goddard Instrument Makes First Detection of Organic Matter on Mars." 16 Dec. 2014. NASA 1 Mar. 2015. http://www.nasa.gov/content/goddard/mars-organic-matter/.

[96] "MAVEN spacecraft enters orbit around Mars." 22 Sept. 2014. University of Colorado Boulder 1 Mar. 2015. http://lasp.colorado.edu/home/maven/2014/09/22/maven-spacecraft-enters-orbit-around-mars/.

[97] "MAVEN." University of Colorado Boulder 14 Jan. 2013. http://lasp.colorado.edu/home/maven/.

[98] "NASA Launches Mission to Study Upper Atmosphere of Mars." 18 Nov. 2013. NASA 31 Mar. 2014. http://www.nasa.gov/press/2013/november/nasa-launches-mission-to-study-upper-atmosphere-of-mars/#.Uzl3uvmSwj5.

[99] "Mars Spacecraft Reveal Comet Flyby Effects on Martian Atmosphere." 7 Nov. 2014. University of Colorado Boulder 1 Mar. 2015. http://lasp.colorado.edu/home/maven/2014/11/07/mars-spacecraft-reveal-comet-flyby-effects-on-martian-atmosphere/.

[100] "MAVEN Completes Commissioning And Begins Its Primary Science Mission." 17 Nov. 2014. University of Colorado Boulder 1 Mar. 2015. http://lasp.colorado.edu/home/maven/2014/11/17/maven-completes-commissioning-and-begins-its-primary-science-mission/.

[101] "MAVEN Identifies Links in Chain Leading to Atmospheric Loss." 15 Dec. 2014. University of Colorado Boulder 1 Mar. 2015. http://lasp.colorado.edu/home/maven/2014/12/15/maven-identifies-links-in-chain-leading-to-atmospheric-loss/.

[102] "Mars Mission: India creates history as Mangalyaan successfully enters Mars orbit in first attempt." 24 Sept. 2014. The Economic Times 22 Mar. 2015. http://economictimes.indiatimes.com/news/science/mars-mission-india-creates-history-as-mangalyaan-successfully-enters-mars-orbit-in-first-attempt/articleshow/43299562.cms.

consisting of the Mars Colour Camera, the Methane Sensor For Mars, the Thermal Infrared Imaging Spectrometer, the Mars Exospheric Neutral Composition Analyser, and the Lyman Alpha Photometer. Also, at the end of September, NASA and ISRO officials signed an agreement to establish a working group to explore potential coordinated observations and science analysis between the MAVEN orbiter and MOM.[103] MOM will observe the surface of Mars and its atmosphere and exosphere extending up to 80,000 km for a detailed understanding of the planet's evolution, especially its related geological and possible biogenic processes.[104]

2.3.4 Saturn Science

The Cassini-Huygens mission, a joint NASA, ESA and ASI mission, was launched in 1997. Reaching Saturn in 2004, Cassini went on to drop the Huygens probe onto Saturn's moon, Titan. The renamed Cassini Solstice Mission was supposed to end in June 2008, however, funding was provided to allow continued operation to provide new insights on Saturn and its moons; it is now slated to explore Saturn until 2017.[105] In May 2014, data from the now 17-year old mission provided a new technique to understand the atmospheres of exoplanets by comparing sunsets seen through Titan's hazy atmosphere; due to a similarity between exoplanet transits and Titan's sunsets as seen by the Cassini spacecraft, the solar occultations effectively allowed the scientists to observe Titan as a transiting exoplanet without having to leave the solar system. With Titan as an example, scientists found that hazes high above some transiting exoplanets might strictly limit what their spectra can reveal to planet transit observers. Moreover, as with Titan's hazes, which more strongly affect shorter wavelength colours of light, exoplanet spectra hazes may not affect colours of light uniformly.[106] Titan's ocean beneath its outer ice shell showed that it was relatively high in density, and likely to be an extremely salty brine of water mixed with dissolved salts likely composed of sulphur, sodium and potassium.[107] On Titan's south pole, the emergence of a giant polar vortex containing frozen particles of the toxic compound hydrogen cyanide (HCN) suggests that the

[103] Laxman, Srinivas. "Mars Orbiter Mission activates all science instruments as NASA, ISRO form joint Mars working group." 1 Oct. 2014. Planetary.org 22 Mar. 2015. http://www.planetary.org/blogs/guest-blogs/2014/10010914-mars-orbiter-mission.html.

[104] "Indian Space Research Organisation | Mars Orbiter Mission." Indian Space Research Organisation 31 Mar. 2014. http://www.isro.org/mars/home.aspx.

[105] Mason, Betsy. "Cassini Gets Life Extension to Explore Saturn Until 2017." 3 Feb. 2010. WIRED 18 Dec. 2012. http://www.wired.com/wiredscience/2010/02/cassini-life-extension-2017/.

[106] "Sunsets on Titan Reveal the Complexity of Hazy Exoplanets." 27 May 2014. NASA 1 Mar. 2015. http://saturn.jpl.nasa.gov/news/newsreleases/newsrelease20140527/.

[107] "Ocean on Saturn Moon Could be as Salty as the Dead Sea." 2 July 2014. NASA 1 Mar. 2015. http://saturn.jpl.nasa.gov/news/newsreleases/newsrelease20140702/.

nitrogen-rich atmosphere of Titan's southern hemisphere is cooling much faster than expected; HCN forms frozen particles only if the atmospheric temperature is as cold as minus 148 °C—while predictions from current theoretical models of Titan's upper atmosphere were anticipated to be about minus 48 °C.[108] The nitrogen in Titan's atmosphere appears to have originated in conditions similar to the comets from the Oort cloud rather than from the relatively warmer disk of material surrounding Saturn during its formation or from comets born in the Kuiper belt.[109]

By July 2014, scientists using Cassini mission data had identified 101 distinct geysers erupting from the four tiger stripe fractures on the icy Enceladus. Moreover, the location of the greatest geyser activity coincided with the location of greatest thermal emissions. The size of hotspots was too small to be caused by the rubbing of opposing walls of the fractures; however they were the right size to come from the condensation of vapour on the near surface walls of the fractures, and thus the geyser process was actually causing the heat. As a sea is known to exist beneath the ice shell, and narrow pathways through the ice shell can remain open from the sea all the way to the surface, if filled with liquid water, it is the only plausible source of the material forming the geysers.[110]

2.3.5 Venus Science

ESA's Venus Express mission was launched in 1995 and reached Venus in April 2006. It studies Venus's atmosphere, including its dynamics and chemistry, atmosphere-surface interactions, and interactions with solar wind, to address open questions such as the workings of the complex global dynamics of the planet, its cloud system, processes that govern the chemical state of the atmosphere, and the 'green-house effect' in its global climate. On 16 May 2014, after 8 years in orbit around Venus, and after concluding its routine science operations, the satellite began preparations for a final mission involving aerobraking in Venus' atmosphere, conducted from 18 June through 11 July 2014. Through short drag campaigns in the upper atmosphere layers, progressively dipping to an altitude of around 130 km, researchers gained new insights into the inaccessible regions of Venus's atmosphere, and how the spacecraft and its components respond to such a hostile environment. Moreover, the spacecraft conducted some limited science measurements with its magnetic field, solar wind and atom analysing instruments, while temperature and pressure sensors recorded the conditions that the spacecraft

[108] "Swirling Cloud at Titan's Pole is Cold and Toxic." 1 Oct. 2014. NASA 1 Mar. 2015. http://saturn.jpl.nasa.gov/news/newsreleases/newsrelease20141001/.

[109] "Titan's Building Blocks Might Pre-date Saturn." 23 June 2014. NASA 1 Mar. 2015. http://saturn.jpl.nasa.gov/news/newsreleases/newsrelease20140623/.

[110] "Cassini Spacecraft Reveals 101 Geysers and More on Icy Saturn Moon." 28 July 2014. NASA 1 Mar. 2015. http://saturn.jpl.nasa.gov/news/newsreleases/newsrelease20140728/.

experienced.[111] Full contact with the spacecraft was lost on 28 November 2014, as it had exhausted its propellant and only limited information could be retrieved from the telemetry and telecommand links; the mission came to an end by December 2014.[112] Through its eight-year mission, its major discoveries included the further understanding of Venus' shape-shifting polar vortices; thousands of volcanoes and widespread volcanic plains suggesting a planet-wide resurfacing during the last 500 million years; that the planet appears to be slowing in its rotation, while the atmosphere rotation is accelerating; that a region in Venus' atmosphere is cold enough for dry-ice snow to form; the existence of an ozone layer on the planet; the loss of water from the planet into space over billions of years; and instead of an internal magnetic field, the existence of an elongated magnetotail on the side of Venus that points away from the Sun.[113]

2.3.6 Mercury Science

Running as a partnership between ESA and JAXA, BepiColombo will be Europe's first mission to Mercury. The mission's targeted launch date has been moved to July 2016 with arrival at Mercury in January 2024 for a one-year mission, with a possible 1-year extension. To be executed under ESA leadership, BepiColumbo is currently in the implementation stage, to be eventually launched on the Ariane 5 launch vehicle. The mission is made up of two spacecraft: the Mercury Planetary Orbiter (MPO), a three-axis stabilized spacecraft provided by ESA that will study the planet's geology, composition, inner structure, and exosphere, and the Mercury Magnetospheric Orbiter (MMO), a spin-stabilised spacecraft provided by JAXA that will study the planet's magnetic field, atmosphere, magnetosphere and inner interplanetary space. Enduring temperatures in excess of 350 °C, BepiColombo has been designed to provide the measurements necessary to study and understand the composition, geophysics, atmosphere, magnetosphere and history of Mercury.[114] On 21 July 2014, the protoflight models of the Mercury Transfer Module (MTM), the MPO, and the MMO were completed at the Thales Alenia Space facility in Turin, Italy, and were sent to ESTEC where final integration tasks, and environmental testing will be performed.[115]

[111] "Venus Express gets ready to take the plunge." 16 May 2014. ESA 1 Mar. 2015. http://www.esa.int/Our_Activities/Space_Science/Venus _Express/Venus_Express_gets_ready_to_take_the_plunge.

[112] "Venus Express goes gently into the night." 16 Dec. 2014. ESA 1 Mar. 2015. http://www.esa.int/Our_Activities/Space_Science/Venus_Express/Venus_Express_goes_gently_into_the_night.

[113] "Major Discoveries by Venus Express: 2006–2014." 4 July 2014. ESA 1 Mar. 2015. http://sci.esa.int/venus-express/54062-1-shape-shifting-polar-vortices/.

[114] "BepiColombo Fact Sheet." 3 Dec. 2013. ESA 27 Mar. 2014. http://sci.esa.int/bepicolombo/47346-fact-sheet/.

[115] "#13: BepiColombo integration and functional testing completed at Thales Alenia Space in Turin." 21 July 2014. ESA 1 Mar. 2015. http://sci.esa.int/bepicolombo/54364-13-bepicolombo-integration-and-functional-testing-completed-at-thales-alenia-space-in-turin/.

MErcury Surface, Space ENvironment, GEochemistry and Ranging (MESSENGER), a NASA discovery-class mission, was launched in August 2004. On 18 March 2011, it became the first spacecraft to orbit Mercury, following three flybys. After completing its year-long task to perform the first complete reconnaissance of the geochemistry, geophysics, geological history, atmosphere, magnetosphere, and plasma environment of Mercury by 17 March 2012, MESSENGER began its extended mission to build on its discoveries.[116] On 16 March 2014, based on global imaging and topographic data from MESSENGER, it was determined that Mercury had contracted far more than had been previously estimated.[117] In the course of 3 years, MESSENGER's +Mercury Dual Imaging System (MDIS) had imaged 100 % of the planet[118]; and in June 2014, its orbit was modified to prepare for a low-altitude campaign that would extend its mission. Two additional manoeuvres were conducted in September and October, along with an additional manoeuvre early in 2015, to raise the minimum altitude above the planet.[119] In previous years, a long-held hypothesis that Mercury harbours abundant water ice and other frozen volatile materials in its permanently shadowed polar craters gained compelling support from three independent lines of evidence: i.e. excess hydrogen at Mercury's north pole was measured with MESSENGER's Neutron Spectrometer; reflectance of Mercury's polar deposits at near-infrared wavelengths was measured with the Mercury Laser Altimeter (MLA); and the first detailed models of the surface and near-surface temperatures of Mercury's north polar regions that utilize the actual topography of Mercury's surface were measured by the MLA.[120] On 15 October 2014, the first optical images of ice and other frozen volatile materials within permanently shadowed craters in Mercury's north pole were revealed, which also hold clues as to when the ices were trapped and how they have evolved.[121]

[116] "MESSENGER Completes Primary Mission at Mercury, Settles in for Another Year." 19 Mar. 2012. MESSENGER 13 Jan. 2013. http://messenger.jhuapl.edu/news_room/details.php?id=197.

[117] "Mercury's Contraction Much Greater Than Thought." 16 Mar. 2014. MESSENGER 1 Mar. 2015. http://messenger.jhuapl.edu/news_room/details.php?id=254.

[118] "MESSENGER Has Imaged 100 Percent of Mercury." 6 Mar. 2013. NASA 27 Mar. 2014. http://www.nasa.gov/mission_pages/messenger/media/Imaged100Percent.html#.UzQGLPldUj4.

[119] "Third of Four Planned Maneuvers Extends MESSENGER Orbital Operations." 24 Oct. 2014. MESSENGER 1 Mar. 2015. http://messenger.jhuapl.edu/news_room/details.php?id=267.

[120] "MESSENGER Finds New Evidence for Water Ice at Mercury's Poles." 29 Nov. 2012. NASA 27 Mar. 2014. http://www.nasa.gov/mission_pages/messenger/media/PressConf20121129.html#.UzQGLPldUj4.

[121] "MESSENGER Provides First Optical Images of Ice Near Mercury's North Pole." 15 Oct. 2014. MESSENGER 1 Mar. 2015. http://messenger.jhuapl.edu/news_room/details.php?id=266.

2.3.7 Jupiter Science

In May 2012, the proposed ESA Jupiter Icy moon Explorer (JUICE)[122] mission was selected by ESA's Space Programme Committee (SPC) as the first large (L-class) mission opportunity in ESA's Cosmic Vision 2015–2025 plan, with a foreseen launch date of 2022 and arrival in 2030. The proposed nearly 5 tonne spacecraft will make a careful investigation of Jupiter's three biggest moons, i.e. it will use the gravity of Jupiter to initiate a series of close fly-bys around Callisto and Europa, and then finally to put itself in a settled orbit around Ganymede. As all three moons are suspected of having oceans of water beneath their icy crusts, scientists are trying to understand whether there is any possibility that these moons could host microbial life.[123] On 20 November 2014, the mission gained approval from ESA's Science Programme Committee to begin its implementation phase. Moreover, the multilateral agreement for JUICE was also approved, which provides the legal framework for provision of payload equipment and ongoing mission support between funding agencies.[124]

2.3.8 Solar Observation

Continued observation of the Sun's external activity has the benefit of improving our understanding of its interior, its corona, the monitoring of solar wind and its consequences on Earth and its neighbouring planets. Coronal mass ejections (CMEs) from the Sun emit surges of charged particles in directions that may cross Earth's path and can damage satellites, impede space-based services and affect the terrestrial electrical infrastructure.

ESA's PRoject for OnBoard Autonomy (PROBA)-2 microsatellite continued its solar observation activity, with its programme extended again on 20 November 2014 by ESAs SPC, with the mission now operating until at least the end of 2016.[125] PROBA-2 tracks spikes in CMEs ejecting from the Sun that have previously been seen to just skim Earth, typically bringing with them a burst of radio energy.[126]

[122] JUICE was renamed during its reformulation exercise from the designation Europa Jupiter System Mission (EJSM)–Laplace in 2011.

[123] "ESA Selects 1bn-Euro JUICE Probe to Jupiter." 2 May 2012. BBC News 27 Mar. 2014. http://www.bbc.com/news/science-environment-17917102.

[124] "JUICE mission gets green light for next stage of development." 27 Nov. 2014. ESA 1 Mar. 2015. http://sci.esa.int/juice/55055-juice-mission-gets-green-light-for-next-stage-of-development/.

[125] "Working life extensions for ESA's science missions." 20 Nov. 2014. ESA 1 Mar. 2015. http://sci.esa.int/director-desk/54999-working-life-extensions-for-esas-science-missions/.

[126] "Small Sun-Watcher Proba-2 Offers Detailed View of Massive Solar Eruption." 9 June 2011. ESA 17 Apr. 2013. http://www.esa.int/Our_Activities/Technology/Small_Sun-watcher_Proba-2_offers_detailed_view_of_massive_solar_eruption.

The mission has gathered over 400,000 images of the Sun, and made nearly 20 million in-situ ionospheric observations.[127] ESA is also developing the Proba-3, as a pair of satellites maintaining a fixed configuration to form a 150 m long solar chronograph to study the Sun's faint corona closer to the solar rim than previously achieved.[128]

CNES' solar metrology mission 'PICARD', launched on 15 June 2010, aims to improve knowledge of how the Sun functions and the influence of solar activity on the Earth's climate. PICARD will accomplish this by measuring absolute total and spectral solar irradiance, solar diameter and shape, and by probing the interior of the Sun using the helioseismology method. PICARD received is final telecommand on 4 April 2014.[129]

The Solar Dynamics Observatory (SDO) is the first NASA mission to operate under its Living With a Star (LWS) programme. Launched on 11 February 2010, its objectives are to determine how the Sun's magnetic field is generated and structured, and how this stored magnetic energy is released in the form of the solar wind, energetic particles and variations in the solar irradiance. The spacecraft is comprised of three scientific experiments: the Atmospheric Imaging Assembly (AIA), the EUV Variability Experiment (EVE) and the Helioseismic and Magnetic Imager (HMI).[130] The SDO's global view of the Sun facilitates research that focuses on the previously unrecorded real fine structure of the star.[131] A new technique to probe the Sun's interior came from using the HMI instead of the more commonly applied technique of tracking the time it takes for waves to travel from one side of the Sun to the other. The research published in April 2014 showed that it is possible to track the size of different magnetically-balanced areas on the Sun, which corresponded to the size of the Sun's granules and supergranules, and reveal spotted areas much larger than previously noted.[132]

The SOlar and Heliospheric Observatory (SOHO) continued to operate during this reporting period. As an international cooperation project between ESA and NASA, this EADS Astrium-*et al*-built spacecraft was launched on 2 December 1995. The spacecraft orbits around the Sun in step with the Earth, at a distance of

---

[127] "ESA's Space Weather Station Proba-2 Tracks Stormy Sun." 2 Dec. 2011. ESA 17 Apr. 2013. http://www.esa.int/Our_Activities/Technology/ESA_s_space_weather_station_Proba-2_tracks_stormy_Sun.

[128] "About PROBA3." 19 Nov. 2012. ESA 1 Apr. 2014. http://www.esa.int/Our_Activities/Technology/Proba_Missions/About_Proba-3.

[129] "PICARD News." Cnes 1 Mar. 2015. http://smsc.cnes.fr/PICARD/GP_actualites.htm.

[130] "SDO | Solar Dynamics Observatory." NASA Goddard Space Flight Center 4 Mar. 2013. http://sdo.gsfc.nasa.gov/.

[131] SpaceNews Staff. "NASA Boasts Big Results from 5-minute Spaceflight." 28 Jan. 2013 SpaceNews 4 Mar. 2013. http://www.spacenews.com/article/nasa-boasts-big-results-from-5-minute-spaceflight.

[132] "Bright Points in Sun's Atmosphere Mark Patterns Deep In Its Interior." 17 Apr. 2014. NASA 1 Mar. 2015. http://www.nasa.gov/content/goddard/bright-points-mark-patterns-inside-sun/.

1.5 million kilometres from Earth, enabling an uninterrupted view of the star.[133] Its scientific objectives are to investigate the solar interior and explain the extreme heating of the solar corona and the mechanism by which the solar wind is produced and accelerated. The SOHO mission has very significantly exceeded its expected lifetime of two years and on 19 June 2013, it was extended until 31 December 2016.[134] In 2014, new analysis of data from SOHO's Solar Ultraviolet Measurements of Emitted Radiation (SUMER) instrument revealed that the comet ISON had stopped producing dust and gas, and disintegrated shortly prior to its closest approach to the Sun, which provided an exceptional chance to see the entirety of the comet.[135]

NASA's Solar TErrestrial RElations Observatory (STEREO) is made up of two space-based observatories, i.e. STEREO-A travelling in a smaller and faster orbit (ahead of Earth's orbit), and STEREO-B trailing behind with a larger and slower orbit.[136] They provide new insights into CMEs, including detecting and processing data that enables the tracking of CMEs headed toward Earth. In June 2014, STEREO showed that the corona (or atmosphere) of the Sun extended further than previously thought, i.e. around 8 million km from the Sun's surface, which will have implications for NASA's close-orbiting Solar Probe Plus mission that is due to launch in 2018.[137] STEREO continued to operate for the remainder of 2014. On 20 August and 1 December 2014, the mission underwent side lobe repointing operations that will provide lower resolution data; the mission will remain in this state until January 2016.[138]

NASA's Interface Region Imaging Spectrograph (IRIS) satellite was launched on 26 June 2013, with the purpose of observing how solar material moves, gathers energy and heats up as it travels through the Sun's lower atmosphere. In addition to being where most of the Sun's ultraviolet emission is generated, this region between the Sun's photosphere and corona powers the Sun's million-degree atmosphere, and drives the solar wind. For its 2-year mission, IRIS will enter a sun-synchronous polar orbit, making continuous solar observations throughout this time.[139] IRIS

---

[133] About the SOHO Mission. "SOHO Fact Sheet." SOHO—Solar and Heliospheric Observatory 4 Mar. 2013. http://sohowww.nascom.nasa.gov/about/docs/SOHO_Fact_Sheet.pdf.

[134] "ESA Science Missions Continue in Overtime." 20 June 2013. ESA 27 Mar. 2014. http://sci.esa.int/director-desk/51944-esa-science-missions-continue-in-overtime/.

[135] "Comet ISON's dramatic final hours." 16 July 2014. ESA 1 Mar. 2015. http://sci.esa.int/soho/54344-comet-ison-dramatic-final-hours/.

[136] "First Ever STEREO Images of the Entire Sun." 6 Feb. 2011. NASA 5 Mar. 2013. http://www.nasa.gov/mission_pages/stereo/news/entire-sun.html

[137] "NASA's STEREO Maps Much Larger Solar Atmosphere Than Previously Observed." 25 June 2014. NASA 1 Mar. 2015. http://www.nasa.gov/content/goddard/nasas-stereo-maps-much-larger-solar-atmosphere-than-previously-observed/.

[138] "Q & A for Operations of STEREO During Superior Conjunction." 3 July 2014. NASA 1 Mar. 2015. http://www.nasa.gov/content/goddard/q-a-for-stereo-during-superior-conjunction/.

[139] "IRIS Solar Observatory Launches, Begins Mission." 28 June 2013. NASA 1 Apr. 2014. http://www.nasa.gov/content/iris-solar-observatory-launches-begins-mission/#.UzqYe_mSwj4.

caught its first CME on 9 May 2014, which hurled a giant cloud of particles, several times the size of Earth into space.[140] It also witnessed dozens of solar flares in 2014, including two X-class flares and numerous M-class flares that are a tenth as strong. One X-class flare was fortuitously caught by both IRIS and SDO, allowing IRIS to view the relatively lower temperature lower atmosphere of the Sun, called the transition region, while SDO viewed the hotter material higher up in the Sun's corona.[141]

The Reuven Ramaty High Energy Solar Spectroscopic Imager (RHESSI), a NASA SMall EXplorer (SMEX) mission with the objective of exploring the basic physics of particle acceleration and energy release in solar flares also continued to operate. By 1 September 2014, the spacecraft had completed its fourth annual procedure to rejuvenate its detectors from the effects of radiation damage, and had resumed collecting solar X-ray and gamma-ray data.[142]

2.3.9 *Solar Science*

Near-Earth Objects continue to be widely discussed as a target for human exploration, with renewed interest in the study of comets and asteroids.

ESA's Rosetta mission turned out to be a stunning success, following its mid-2014 rendezvous with the comet 67P/Churyumov-Gerasimenko, and the release of its lander Philae which accomplished the first soft-landing on a comet's nucleus. While en route to the comet, Rosetta's Optical, Spectroscopic and Infrared Remote Imaging System (OSIRIS) camera observed variable activity around the comet, as its coma rapidly increased in brightness and then died down during that period.[143] Rosetta made its first observations of water vapour streaming out on 6 June; using its Microwave Instrument for Rosetta Orbiter (MIRO) it was estimated that the water vapour was being ejected into space at about 300 ml/s.[144] And between 13 and 21 July, at a distance of over 5000 km from the comet, Rosetta's visible, infrared and thermal imaging spectrometer (VIRTIS) collected the infrared light emitted by the comet, which enabled scientists to determine that its average

[140] "A First for NASA's IRIS: Observing a Gigantic Eruption of Solar Material." 30 May 2014. NASA 1 Mar. 2015. http://www.nasa.gov/content/goddard/a-first-for-nasas-iris-observing-a-gigantic-eruption-of-solar-material/.

[141] "NASA Releases IRIS Footage of X-class Flare." 17 Sept. 2014. NASA. http://www.nasa.gov/content/goddard/nasa-releases-iris-footage-of-x-class-flare/.

[142] "RHESSI Resumes Observations." 1 Sept. 2014. RHESSI Science Nuggets 24 Mar. 2015. http://sprg.ssl.berkeley.edu/~tohban/wiki/index.php/RHESSI_Resumes_Observations.

[143] "Rosetta arrives at comet destination." 6 Aug. 2014. ESA 9 Mar. 2015. http://www.esa.int/Our_Activities/Space_Science/Rosetta/Rosetta_arrives_at_comet_destination.

[144] "Rosetta's comet 'sweats' two glasses of water a second." 30 June 2014. ESA 9 Mar. 2015. http://www.esa.int/Our_Activities/Space_Science/Rosetta/Rosetta_s_comet_sweats_two_glasses_of_water_a_second.

surface temperature is about –70 °C, providing clues about the composition and physical properties of the comet's dusty surface. While comet 67P/Churyumov-Gerasimenko was already known to have a low reflectance from ground-based observations, which indicated that it lacked a 'clean' icy surface, the surface temperature was found to be around 20–30 °C warmer than for a comet at that distance covered exclusively in ice, as darker material heats up and emits heat more readily than ice when it is exposed to sunlight.[145] By November, following Philae's successful landing on the comet, and prior to its hibernation, the lander had utilized its instruments including a sampling drill to penetrate the dust covered surface ice and collect a soil sample. However, further analysis of the data is needed before it can be confirmed whether a sample was collected and put into Philae's micro-oven for analysis.[146] Nevertheless, some findings from the Rosetta mission are providing insight into the origin of Earth's oceans, as Rosetta's Orbiter Spectrometer for Ion and Neutral Analysis (ROSINA) measured the comet's deuterium/hydrogen (D/H) ratio to be more than three times greater than for Earth's oceans and for other recorded Kuiper Belt comets; it also measured higher than previous Oort cloud comets as well, potentially indicating that comets in the Kuiper belt may have formed over a wider range of distances than previously thought.[147]

NASA's reboot of the Wide-field Infrared Survey Explorer (WISE) mission, this time to discover and characterize near-Earth objects (NEOs) with infrared light, was reactivated in September 2013 for three additional years of service.[148] The original WISE mission was completed early in 2011, following the completion of surveying the whole sky twice in infrared light the space telescope had been placed in 31 months of hibernation.[149] Now dubbed NEOWISE, the new mission commenced at the start of 2014, detecting 857 minor bodies in our solar system, including 22 NEOs and four comets in the first 25 days of its operation.[150] NEOWISE was also used to observe the active and dusty comet Siding Spring prior to its dramatically-close flyby of Mars, both for characterization and to study dust particle sizes and amounts produced by the comet to understand the potential

[145] "Rosetta takes comet's temperature." 1 Aug. 2014. ESA 9 Mar. 2015. http://www.esa.int/Our_Activities/Space_Science/Rosetta/Rosetta_takes_comet_s_temperature.

[146] De Selding, Peter B. "Philae Managers Say Recovery Possible as Comet Approaches Sun." 18 Nov. 2014. SpaceNews 14 May 2015. http://spacenews.com/42586philae-managers-say-recovery-possible-as-comet-approaches-sun/.

[147] "Rosetta fuels debate on origin of Earth's oceans." 10 Dec. 2014. ESA 9 Mar. 2015. http://www.esa.int/Our_Activities/Space_Science/Rosetta/Rosetta_fuels_debate_on_origin_of_Earth_s_oceans.

[148] "NASA Spacecraft Reactivated to Hunt for Asteroids." 21 Aug. 2013. NASA 2 Apr. 2014. http://www.jpl.nasa.gov/news/news.php?release=2013-257.

[149] "NASA's Asteroid Hunter Spacecraft Returns First Images after Reactivation." 19 Dec. 2013. NASA 9 Mar. 2015. http://www.jpl.nasa.gov/wise/newsfeatures.cfm?release=2013-373.

[150] "NEOWISE Celebrates First Month of Operations After Reactivation." 23 Jan. 2014. NASA 9 Mar. 2015. http://www.jpl.nasa.gov/wise/newsfeatures.cfm?release=2014-023.

risks to the Mars orbiters.[151] NEOWISE also discovered its first new comet since coming out of hibernation, officially named "C/2014 C3 (NEOWISE)" at a distance of 230 million kilometres from Earth; however its orbit is still a bit uncertain.[152] Another comet, "C/2013 UQ4 (Catalina)" was observed by NEOWISE a day after passing through its closest approach to the sun; due to its inactivity when observed in 2013, it was originally thought to be an asteroid, but it became highly active as it closed in on the Sun in 2014. Its next approach to the Sun will be in 450 years.[153]

2.3.10 Outer Solar Science

ESA launched its Gaia mission on 19 December 2013 to measure the positions and motions of a billion stars of the roughly 100 billion stars in our galaxy to create the most accurate map yet of the Milky Way.[154] Gaia followed a trajectory to the L2 Lagrangian point 1.5 million km beyond Earth's orbit, where it is spared from sweeping in and out Earth's shadow, heating up and cooling down the telescope, and distorting its view. It also has a sunshield to protect its instruments from the light and heat that comes from the Sun, Earth and Moon simultaneously.[155] It shares this position with the Herschel infrared telescope, and the Planck microwave observatory. After a series of tests and calibrations, Gaia began its 5-year scientific mission on 25 July 2014, spinning slowly as it repeatedly scans the sky with its two telescopes, and focuses their light simultaneously onto a single digital camera to create an extraordinarily precise 3D map of these stars, also mapping their motions, luminosity, temperature and composition.[156] Already on 30 August, Gaia observed its first super nova, a Type Ia supernova that was detected in the form of a sudden spike in the brightness of light coming from another distant galaxy that had appeared much dimmer when Gaia first looked at it in July.[157]

While ESA's Herchel science mission ended on 29 April 2013, following the exhaustion of its helium coolant reserves, data from Herchel's mission is continuing

[151] "NASA Preparing for 2014 Comet Watch at Mars." 28 Jan. 2014. NASA 9 Mar. 2015. http://www.jpl.nasa.gov/wise/newsfeatures.cfm?release=2014-026.

[152] "NEOWISE Spies Its First Comet." 28 Feb. 2014. NASA 9 Mar. 2015. http://www.jpl.nasa.gov/wise/newsfeatures.cfm?release=2014-067.

[153] "NEOWISE Spots a Comet That Looked Like an Asteroid." 23 July 2014. NASA 9 Mar. 2015. http://www.jpl.nasa.gov/wise/newsfeatures.cfm?release=2014-241.

[154] "ESA PR 44-2013: Liftoff for ESA's billion-star surveyor." 19 Dec. 2013. ESA 9 Mar. 2015. http://sci.esa.int/gaia/53536-esa-pr-44-2013-liftoff-for-esas-billion-star-surveyor/.

[155] "The experts behind Gaia's arrival at nothingness." 15 Jan. 2014. ESA 9 Mar. 2015. http://www.esa.int/Our_Activities/Operations/The_experts_behind_Gaia_s_arrival_at_nothingness.

[156] "Gaia overview." 23 Apr. 2014. ESA 9 Mar. 2015. http://www.esa.int/Our_Activities/Space_Science/Gaia/Gaia_overview.

[157] "Gaia discovers its first supernova." 12 Sept. 2014. ESA 9 Mar. 2015. http://sci.esa.int/gaia/54630-gaia-discovers-its-first-supernova/.

to make significant findings. In January 2014, the first unambiguous detection of water vapour was discovered around Ceres, the dwarf planet orbiting between Mars and Jupiter, providing proof that Ceres has an icy surface and an atmosphere; estimates approximate 6 kg of water vapour being produced per second.[158] By March 2014, researchers had produced the largest census of dust in local galaxies, wherein observing the light emitted by the dust as a function of wavelength provides a means to study the physical properties of the dust, and provides a local benchmark for quantifying the role dust has played in galaxy evolution throughout the history of the Universe.[159] Moreover, Herschel helped researchers to discover that the molecule, OH+ (a positively charged combination of single oxygen and hydrogen atoms), vital for creating water, can be formed by the ultraviolet radiation poured out from white dwarfs even while the same radiation may destroy molecules that had previously been ejected by the star.[160]

While ESA's Planck microwave observatory mission to study the Cosmic Microwave Background (CMB) ended on 23 October 2013, research continued on into the following year. Scientists are now assessing CMB data to measure the polarisation of the light, which could provide evidence of gravitational waves generated in the Universe immediately after its birth.[161] After 4.5 years of operation where Planck completed five full-sky surveys with both its Low Frequency Instrument (LFI) and High Frequency Instrument (HFI), the spacecraft was put into permanent hibernation. The HFI exhausted its liquid helium coolant in January 2012, whereas the LFI was able to operate until 3 October 2013 at higher temperatures, before being switched off shortly thereafter.[162]

CNES' COnvection, ROtation and planetary Transits (COROT) space telescope ended its mission on 17 June 2014. The mission, operating for more than 7 years, probed the inner structure of stars using stellar seismology and was used to detect extrasolar planets. This astronomy mission, launched on 27 December 2006, announced the discovery of ten new planets at the beginning of this reporting period.[163, 164] While in operation, the COROT mission was the first to discover a

---

[158] "Herschel discovers water vapour around dwarf planet Ceres." 22 Jan. 2014. ESA 9 Mar. 2015. http://www.esa.int/Our_Activities/Space_Science/Herschel/Herschel_discovers_water_vapour_around_dwarf_planet_Ceres.

[159] "Herschel completes largest survey of cosmic dust in local Universe." 18 Mar. 2014. ESA 9 Mar. 2015. http://www.esa.int/Our_Activities/Space_Science/Herschel/Herschel_completes_largest_survey_of_cosmic_dust_in_local_Universe.

[160] "New molecules around old stars." 17 June 2014. ESA 9 Mar. 2015. http://www.esa.int/Our_Activities/Space_Science/Herschel/New_molecules_around_old_stars.

[161] "Planck takes magnetic fingerprint of our Galaxy." 6 May 2014. ESA 9 Mar. 2015. http://www.esa.int/Our_Activities/Space_Science/Planck/Planck_takes_magnetic_fingerprint_of_our_Galaxy.

[162] "Last Command Sent to ESA's Planck Space Telescope." 23 Oct. 2013. ESA 2 Apr. 2014. http://www.esa.int/Our_Activities/Space_Science/Planck/Last_command_sent_to_ESA_s_Planck_space_telescope.

[163] "CaRoT discovers 10 new extra-solar planets." 14 June 2011. CNES 17 Apr. 2013. http://smsc.cnes.fr/COROT/GP_actualite.htm.

[164] "CoRoT Events." Cnes 2 Mar. 2015. http://smsc.cnes.fr/COROT/GP_actualite.htm.

confirmed Earth-like exoplanet orbiting a star similar to the Sun, and has since revealed 32 planets, while 100 more await confirmation.[165]

NASA's Kepler space telescope mission was recently extended to 2016 with the continued aim of finding Earth-sized planets in the habitable zone of other solar-like oscillating stars, where liquid water could exist on their surfaces.[166] In February 2014, Kepler-413b was found to wobble wildly on its spin axis, varying as much as 30° over 11 years; by comparison, Earth's spin axis varies by 23.5° over 26,000 years, resulting in much less rapid and erratic changes in weather patterns.[167] In April 2014, the first Earth-sized planet, Kepler-186f, was validated to orbit a distant star in the habitable zone. Whereas Kepler-186f is less than 10 % larger than Earth, the next closest Earth-like planet is Kepler-62f, which is 40 % larger than Earth and also orbits its habitable zone.[168] By July 2014, the number of confirmed exoplanets exceeded 1800,[169] with 715 exoplanets verified to orbit 305 stars.[170] Through the use of Kepler and Spitzer space telescope data, scientists have made the most precise measurement to date of an exoplanet, Kepler-93b, with an uncertainty of 119 km on either side of the "super-Earth" planetary body.[171] And data from the duo, in addition to the Hubble telescope's Wide Field Camera 3, helped to identify that exoplanet HAT-P-11b, a Neptune-sized planet that orbits the star HAT-P-11, had clear skies and water vapour, through a technique called transmission spectroscopy which observed the distinct signatures made by starlight as it is filtered through the rim of a planet's atmosphere containing the vapour.[172] And after losing two of its four reaction wheels in May 2013, resulting in the spacecraft losing its ability to precisely point at the original field of view, plans

[165] "Mission Accomplished for CoRoT." 24 June 2013. CNES 2 Apr. 2014. http://smsc.cnes.fr/COROT/PDF/CP039-2013_mission_CoRoT_va.pdf.

[166] "NASA Approves Kepler Mission Extension." 4 Apr. 2012. NASA 16 Apr. 2013. http://kepler.nasa.gov/news/nasakeplernews/index.cfm?FuseAction=ShowNews&NewsID=199.

[167] "Kepler Finds a Very Wobbly Planet." 4 Feb. 2014. NASA 9 Mar. 2015. http://kepler.nasa.gov/news/nasakeplernews/index.cfm?FuseAction=ShowNews&NewsID=319.

[168] "First Earth-Size Planet in 'Habitable Zone'." 17 Apr. 2014. NASA 9 Mar. 2015. http://kepler.nasa.gov/news/nasakeplernews/index.cfm?FuseAction=ShowNews&NewsID=330.

[169] "Kepler-421b: An exoplanet orbiting near its star's "snow line"." 21 July 2014. NASA 9 Mar. 2015. http://kepler.nasa.gov/news/nasakeplernews/index.cfm?FuseAction=ShowNews&NewsID=347.

[170] "715 Newly Verified Planets More Than Triples the Number of Confirmed Kepler Planets." 26 Feb. 2014 NASA 9 Mar. 2015. http://kepler.nasa.gov/news/nasakeplernews/index.cfm?FuseAction=ShowNews&NewsID=324.

[171] "Kepler-93b: The Most Precise Measurement of an Alien World's Size." 23 July 2014. NASA 9 Mar. 2015. http://kepler.nasa.gov/news/nasakeplernews/index.cfm?FuseAction=ShowNews&NewsID=348.

[172] "NASA Telescopes Find Clear Skies and Water Vapor on Exoplanet." 24 Sept. 2014. NASA 9 Mar. 2015. http://kepler.nasa.gov/news/nasakeplernews/index.cfm?FuseAction=ShowNews&NewsID=355.

were underway to recover stability in the spacecraft using the Sun to maintain stability, under the designation Kepler 2 (K2)[173] In May 2014, the K2 mission was approved by NASA, which repurposed Kepler to accurately point at target sky fields along the ecliptic plane of Earth's orbit.[174] And by year end, the K2 mission succeeded in discovering its first exoplanet HIP 116454b, in addition to having observed more than 35,000 stars and collected data on star clusters, dense star-forming regions, and several planetary objects within our own solar system.[175]

After running out of the coolant needed to chill its longer-wavelength instruments in 2009, NASA's Spitzer Space Telescope was repurposed to track exoplanets around other stars with the use of infrared light. As exoplanets cross in front of their stars, they block out a fraction of the light, allowing the size of the planet to be revealed, in addition to giving clues about the planet's atmosphere by the infrared light that they also emit.[176] Spitzer has already observed infrared light emanating from a "super-Earth" planet in another solar system[177] and observed an unprecedented simultaneous elliptical galaxy with another thin disk existing in its interior.[178] By the beginning of 2014, with the use of NASA's Spitzer and ESA's Herschel Space Observatory, the evolutionary sequence of compact elliptical galaxies that erupted and burned out early in the history of the universe was pieced together; i.e. from their initial burst of star formation through their development of dense stellar cores to their ultimate reality as giant ellipticals. It was determined that compact ellipticals voraciously consumed the gas available for star formation, to the point where they could not create new stars, and then merged with smaller galaxies to form giant ellipticals.[179] In August 2014, Spitzer observed the eruption of dust around the star NGC 2547-ID8, when it surged in the amount of observable

[173] "NASA STATEMENT: Two-Wheel Kepler Mission Invited to 2014 Senior Review." 4 Dec. 2013. National Aeronautics and Aerospace Administration 2 Apr. 2014. http://www.nasa.gov/kepler/nasa-statement-two-wheel-kepler-mission-invited-to-2014-senior-review/.

[174] "Kepler Begins K2 Mission Field 1 Observing." 30 May 2014. NASA 9 Mar. 2015. http://kepler.nasa.gov/news/nasakeplernews/index.cfm?FuseAction=ShowNews&NewsID=341.

[175] "NASA's Kepler Reborn, Makes First Exoplanet Find of New Mission." 18 Dec. 2014. NASA 9 Mar. 2015. http://www.nasa.gov/press/2014/december/nasa-s-kepler-reborn-makes-first-exoplanet-find-of-new-mission/.

[176] "How Engineers Revamped Spitzer to Probe Exoplanets." 24 Sept. 2013. NASA 3 Apr. 2014. http://www.spitzer.caltech.edu/news/1560-feature13-07-How-Engineers-Revamped-Spitzer-to-Probe-Exoplanets.

[177] "NASA's Spitzer Sees The Light of Alien 'Super Earth'." 8 May 2012 NASA 16 Apr. 2013. http://www.spitzer.caltech.edu/news/1419-ssc2012-07-NASA-s-Spitzer-Sees-The-Light-of-Alien-Super-Earth-.

[178] "NASA's Spitzer Finds Galaxy with Split Personality." 24 Apr. 2012. NASA 16 Apr. 2013. http://www.spitzer.caltech.edu/news/1412-ssc2012-06-NASA-S-Spitzer-Finds-Galaxy-with-Split-Personality.

[179] "NASA and ESA Telescopes Help Solve Mystery of Ultra-Compact, Burned-Out Galaxies." 29 Jan. 2014. NASA 9 Mar. 2015. http://www.spitzer.caltech.edu/news/1607-ssc2014-01-NASA-and-ESA-Telescopes-Help-Solve-Mystery-of-Ultra-Compact-Burned-Out-Galaxies.

dust between August 2012 and January 2013, possibly the result of two large asteroids colliding.[180]

NASA's Hubble Space Telescope provided significant findings throughout 2014. Hubble and Spitzer were used to discover and characterize four unusually bright galaxies as they appeared more than 13 billion years ago, with luminosities that are about 10–20 times greater than anything seen previously. Hubble first detected the galaxies with its sharp imaging, while Spitzer allowed researchers to estimate the stellar masses by measuring the total luminosity of the galaxies.[181] The pair were also used to find one of the youngest galaxies in the universe, Abell2744_Y1 within the Abell 2744 cluster of galaxies, using gravitational lensing where specific foreground galaxy clusters are used to amplify the faint light from far more distant background objects.[182] Hubble also enabled astronomers to precisely measure the rotation rate of the galaxy Large Magellanic Cloud (LMC) for the first time. By recording its stars' slight movements over a 7-year period, it was determined that the LMC galaxy completes a rotation every 250 million years.[183] Hubble data also showed that the largest known galaxy cluster in the distant universe, ACT-CL J0102-4915 (or El Gordo), is roughly 43 % more massive than earlier estimates based on X-ray and dynamical studies.[184] Another first for the Hubble telescope occurred when it viewed a binary star system that later produced an unusual and weaker Type Iax supernova explosion of a white dwarf, SN 2012Z, providing clues about the similarities between normal Type Ia supernovas and their Type Iax progenitors.[185] And a companion star to a rare Type IIb supernova occurring in the galaxy M81 enabled astronomers to put constraints on the properties of the surviving star in estimating its luminosity and mass, which provides

[180] "NASA's Spitzer Telescope Witnesses Asteroid Smashup." 28 Aug. 2014. NASA 9 Mar. 2015. http://www.spitzer.caltech.edu/news/1696-ssc2014-06-NASA-s-Spitzer-Telescope-Witnesses-Asteroid-Smashup.

[181] "NASA Great Observatories Team Up to Discover Ultra-Bright Young Galaxies." 7 Jan. 2014. Hubblesite 9 Mar. 2015. http://hubblesite.org/newscenter/archive/releases/cosmology/2014/2014/05/full/results/100/.

[182] "Hubble and Spitzer Space Telescopes Find One of the Youngest Galaxies in the Universe." 7 Feb. 2014. Hubblesite 9 Mar. 2015. http://hubblesite.org/newscenter/archive/releases/galaxy/2014/2014/17/results/100/.

[183] "Hubble Watches Stars' Clockwork Motion in Nearby Galaxy." 18 Feb. 2014. Hubblesite 9 Mar. 2015. http://hubblesite.org/newscenter/archive/releases/galaxy/2014/2014/11/full/results/100/.

[184] "Hubble Finds That Monster 'El Gordo' Galaxy Cluster Is Bigger Than Thought." 3 Apr. 2014. Hubblesite 9 Mar. 2015. http://hubblesite.org/newscenter/archive/releases/galaxy/2014/2014/22/full/results/100/.

[185] "NASA's Hubble Finds Supernova Star System Linked to Potential 'Zombie Star'." 6 Aug. 2014. Hubblesite 9 Mar. 2015. http://hubblesite.org/newscenter/archive/releases/galaxy/2014/2014/32/full/results/100/.

insight into the conditions that preceded the explosion.[186] Hubble also helped to find the smallest known galaxy, with 140 million stars grouped in a diameter of about 300 light-years, that has a supermassive black hole that is five times the mass of the black hole at the centre of our 100,000 light-year diameter Milky Way.[187]

2.4 Satellite Applications

2.4.1 Space-Based Communications

The industry continued to expand its technology development programmes with additional investments in larger spacecraft with enhanced power and transponder capacity.[188]

On radio frequency spectrum allocation, the ITU's biennial World Radiocommunication Seminar 2014 (WRS-14) was held on 8–12 December 2014. The seminar provided deeper insight into the revised ITU radio regulations following the World Radiocommunication Conference 2012, and addressed the application of the provisions of those regulations with respect to the use of radio-frequency spectrum and satellite orbits. The seminar aimed at providing its 400 participants from over 90 countries with technical and regulatory background information to aid in preparation for the World Radiocommunication Conference (WRC-15) to be held on 2–27 November 2015 in Geneva, Switzerland.[189] Prior to the WRS-14 seminar, representatives from the top satellite operators expressed concern over an ITU demand study for C-band spectrum; they believed the study exaggerated the demand by 10 or even more than 100 times, which might result in skewed estimates of how much spectrum will be required for terrestrial wireless broadband when distributed at the WRC-15.[190] Nevertheless, on 9 October, it was confirmed that the European Commission had decided to authorise terrestrial broadband operators to use a portion of the C-band spectrum normally reserved exclusively for satellite

[186] "Hubble Finds Companion Star Hidden for 21 Years in a Supernova's Glare." 9 Sept. 2014. Hubblesite 9 Mar. 2015. http://hubblesite.org/newscenter/archive/releases/galaxy/2014/2014/38/full/results/100/.

[187] Hubble Helps Find Smallest Known Galaxy with a Supermassive Black Hole." 17 Sept. 2014. Hubblesite 9 Mar. 2015. http://hubblesite.org/newscenter/archive/releases/galaxy/2014/2014/41/full/results/100/.

[188] See generally "State of the Satellite Industry Report—September 2014." 10 Sept. 2014. SIA 9 Apr. 2015. http://www.sia.org/wp-content/uploads/2014/09/SSIR-September-2014-Update.pdf.

[189] "ITU World Radiocommunication Seminar focuses on regulatory aspects of the use of the radio-frequency spectrum and satellite orbits." 8 Dec. 2014. ITU 6 Apr. 2015. http://www.itu.int/net/pressoffice/press_releases/2014/CM11.aspx.

[190] De Selding, Peter B. "News from Satellite 2014 | Satellite Operators On Guard Against Ground Attack at 2015 Spectrum Conclave." 11 Mar. 2014. SpaceNews 9 Apr. 2015. http://spacenews.com/39819news-from-satellite-2014-satellite-operators-on-guard-against-ground/.

use; but that it would not extend to frequency used by satellites in the upper part of the band.[191] The European Commission Decision 2008/411/EC[192] on the harmonisation of the 3400–3800 MHz frequency band for terrestrial systems capable of providing electronic communications services in the Community was modified through the Implementing Decision 2014/276/EU[193] of 2 May 2014.[194]

As had occurred in the World Radiocommunication Conference of 2007 (WRC-07), the satellite sector will have to compete with the terrestrial wireless broadband industry over the use of C-band radio spectrum and of Ka-band spectrum at the World Radiocommunication Conference in 2015 (WRC-15). In addition to the loss of spectrum, the satellite sector is concerned about the risk of having Ka-band frequencies shared with the terrestrial wireless industry as it is bound to cause frequency interference at a time when satellite operators are heavily investing in Ka-band systems that allow higher bandwidth transfer in satellite communications for use in high-throughput satellite systems.[195] Moreover, a disinterested opinion by the 48-nation European Conference of Postal and Telecommunications Administrations (CEPT) remarked that sharing radio frequency without interference does not appear to be an option (at least for Wi-Fi and radar Earth observation signals), since the prevention of interference would need radio wireless broadband devices (Radio LANS) to operate at ten times lower levels than are being used in adjacent bands. Moreover, the strength of satellite signals are already depleted by the time they reach Earth, making them even more vulnerable to Wi-Fi networks. If allowed to be shared, Europe's Sentinel 1 radar satellite system, and Canada's Radarsat Constellation Mission are just two examples of what is at stake for the space sector.[196]

---

[191] De Selding, Peter B. "European Commission Backs Reallocation of C-band Spectrum to Terrestrial Broadband." 9 Oct. 2014. SpaceNews 9 Apr. 2015. http://spacenews.com/42136european-commission-backs-reallocation-of-c-band-spectrum-to/.

[192] Commission of the European Communities. on the harmonisation of the 3400–3800 MHz frequency band for terrestrial systems capable of providing electronic communications services in the Community. European Commission Decision C (2008) 1873 of 21 May 2008. 2008/411/EC. Brussels: European Union. http://eur-lex.europa.eu/LexUriServ/LexUriServ.do?uri=OJ:L:2008:144:0077:0081:EN:PDF.

[193] Commission of the European Communities on amending Decision 2008/411/EC on the harmonisation of the 3400–3800 MHz frequency band for terrestrial systems capable of providing electronic communications services in the Community. European Commission Implementing Decision C (2014) 2798 of 2 May 2014. 2014/276/EU. Brussels: European Union. http://eur-lex.europa.eu/legal-content/EN/TXT/PDF/?uri=CELEX:32014D0276&from=EN.

[194] Viola, Roberto. "Keynote speech on: Interference and telecommunication services." 9 Oct. 2014. Secure World Foundation 3 Sept. 2015. http://swfound.org/media/178605/Roberto%20Viola%20Keynote.pdf.

[195] SpaceNews Editor. "Editorial | Facing Down the Next Spectrum Challenge." 13 Jan. 2014. SpaceNews 9 Apr. 2015. http://spacenews.com/39077editorial-facing-down-the-next-spectrum-challenge/.

[196] De Selding, Peter B. "European Government Group Says Radar Satellites and Wi-Fi Cannot Coexist." 29 Jan. 2014. SpaceNews 9 Apr. 2015. http://spacenews.com/39322european-government-group-says-radar-satellites-and-wi-fi-cannot/.

2.4.2 *Space-Based Positioning, Navigation and Timing Systems*

The development of GNSS systems continued during this period, with the relevant actors increasing their efforts to complete their full satellite constellations.

In 2014, Russia's Glonass GNSS constellation was still in the midst of fleet replenishment, following the successful launches of a GLONASS-M satellite on 23 March and on 14 June on a Soyuz launcher. One of Russia's next-generation model GLONASS-K satellites was also launched on 30 November on a Soyuz launcher. The -K series will succeed the -M predecessor, and will most likely incorporate significant technical improvements, including a new more accurate timing device and a non-pressurized structure, bringing operational performance close to U.S. and European standards. The constellation requires 24 operational satellites to provide complete global navigation coverage, with 3 sets of 8 satellites operating on three orbital planes.[197]

At the beginning of 2014 Europe's four Galileo GNSS in-orbit validation (IOV) satellites were reported to have exceeded their expected level of accuracy, marking a substantial milestone for the European Commission's flagship navigation system.[198] Nevertheless, an unexplained brief power drop in one of the IOV satellites early on 27 May required the satellite to be shut down for the following month.[199] As a precaution, ESA reduced the broadcast power on all four satellites, as its ground team continued to investigate the cause of the anomaly.[200] On 22 August the first pair of Galileo's fully operational satellite models were launched from the Guiana Space Center in French Guiana aboard a Europeanized Soyuz launcher; yet shortly thereafter it was determined that the Soyuz' Fregat upper stage had injected the satellites in the wrong orbital perigee and inclination.[201] The cause of the Fregat malfunction was determined to be due to the installation position of its hydrazine fuel line, which put it in too close proximity to a supercold helium line, which caused the hydrazine to freeze long enough to upset the Fregat stage's orientation

---

[197] "Glonass System." Glonass.it 12 May 2012. http://www.glonass.it/eng/glonass-story.aspx.

[198] De Selding, Peter B. "ESA Says It Won't Be Penalized for Galileo Delays." 17 Jan. 2014. SpaceNews 19 May 2015. http://spacenews.com/39141esa-says-it-wont-be-penalized-for-galileo-delays/.

[199] De Selding, Peter B. "Sudden Power Loss Leaves a Galileo Satellite in Safe Mode." 3 July 2014. SpaceNews 19 May 2015. http://spacenews.com/41130sudden-power-loss-leaves-a-galileo-satellite-in-safe-mode/.

[200] De Selding, Peter B. "ESA Proceeding with Galileo Launches Despite In-orbit Satellite Issues." 20 Aug. 2014. SpaceNews 19 May 2015. http://spacenews.com/41616esa-proceeding-with-galileo-launches-despite-in-orbit-satellite-issues/.

[201] De Selding, Peter B. "Galileo Launch, Initially Hailed as Success, Is a Failure." 23 Aug. 2014. SpaceNews 19 May 2015. http://spacenews.com/41650galileo-launch-initially-hailed-as-success-is-a-failure/.

and premature payload release.[202] Correcting both satellites' orbital positions without exhausting operational fuel reserves initially appeared futile; however, following a series of 11 firings to raise one of the satellites to a higher orbit in the following months, it was deemed to have sufficient fuel to operate for 12 years, and could be reinserted into the planned broader Galileo constellation.[203] By late December, ESA had begun to repeat the process for the second wayward satellite, and expected it to be also suitable for inclusion into the nominal 30-satellite Galileo navigation constellation.[204]

China's development of its Beidou navigation system has been accelerated with its 21 May announcement that it will begin launching the third generation of its navigation satellites in 2015. China began fielding its regional satellite navigation capability offering services including positioning, navigation, time and text messaging to users in the Asia-Pacific region in December 2012.[205] It now expects to complete its third phase, deploying its entire 35 satellite Beidou GNSS constellation by 2017 instead of 2020.[206]

In light of China's expansion in the Asia-Pacific maritime region, and North Korea's missile launch activity, Japan's Committee on the National Space Policy outlined a draft proposal which includes increasing its regional Quasi-Zenith Satellite System (QZSS) to a total of four satellites, to increase Japan's overall surveillance systems by supplementing its surveillance satellites with positioning information.[207] Following the first QZSS satellite launched in September 2010, the program slowed due to unwillingness by the private sector to manage QZSS as a business without substantial government guarantees. On 29 March 2013, Mitsubishi Electric Co. was contracted to construct three additional QZSS satellites, i.e. one GEO satellite and two spacecraft following highly elliptical orbits, to

---

[202] De Selding, Peter B. "65th International Astronautical Congress | Fregat Plumbing Blamed for Soyuz Failure that Stranded Galileo Satellites in Wrong Orbit." 30 Sept. 2014. SpaceNews 19 May 2015. http://spacenews.com/4204365th-international-astronautical-congress-%e2%80%8bfregat-plumbing-blamed-for/.

[203] De Selding, Peter B. "Galileo Satellite Left in Bad Orbit Reaches Operating Position." 5 Dec. 2014. SpaceNews 19 May 2015. http://spacenews.com/42861galileo-satellite-left-in-bad-orbit-reaches-operating-position/.

[204] De Selding, Peter B. "Galileo Satellite Left in Bad Orbit Reaches Operating Position." 5 Dec. 2014. SpaceNews 19 May 2015. http://spacenews.com/42861galileo-satellite-left-in-bad-orbit-reaches-operating-position/.

[205] Associated Press. "China Satellite Navigation Starts Services to Asia." 27 Dec. 2012. SpaceNews 18 May 2014. http://www.spacenews.com/article/china-satellite-navigation-starts-services-to-asia.

[206] Gibbons, Glen. "China Plans to Complete BeiDou Ahead of Schedule." 21 May 2014. InsideGNSS 21 May 2015. http://www.insidegnss.com/node/4040.

[207] GPS World staff. "QZSS May Expand to Meet Japan's Surveillance Needs." 22 Aug. 2014. GPS World 20 May 2015. http://gpsworld.com/qzss-may-expand-to-meet-japans-surveillance-needs/.

complete the QZSS space architecture by 2017.[208] The QZSS system is designed to augment the regional accuracy of the GPS signal. The system scheme will provide Japanese authorities with a more accurate, secure and independent service. The future deployment of four spacecraft will allow for 24 h regional coverage, while deployment of the full seven satellite constellation will dramatically decrease Japan's dependence on GPS for regional coverage.[209]

In 2014 India continued to achieve several milestones in its regional navigation system launching its second regional navigation satellite, IRNSS-1B, into orbit on 4 April, followed by the launch of IRNSS-1C on 16 October.[210] The regional system will consist of a ground segment, along with 3 IRNSS satellites in geostationary orbit and 4 satellites in inclined geosynchronous orbits. The remaining IRNSS satellites are to be launched in 2015.[211]

2.5 Technology Developments

The new developments in space-related technologies merit discussion, as they indicate current trends in space technology and reveal the focus of relevant policy decisions. The following chapter provides an overview of recent activities, spanning all major space faring powers and institutions, both established and emerging ones.

2.5.1 Propulsion

Significant advancements have occurred within Europe and the U.S. in the field of propulsion. Newly developed rockets and other methods of propulsion are being explored with favourable results.

In Europe, following the December 2014 Ministerial Council meeting, ESA Member States agreed on funding the development of the Ariane 6 next generation launcher, in addition to an enhanced Vega launcher. The Ariane 6 main stage will

---

[208] De Selding, Peter B. "Melco To Build Three QZSS Navigation Satellites." 3 Apr. 2013. SpaceNews 19 May 2014. http://www.spacenews.com/article/civil-space/34676melco-to-build-three-qzss-navigation-satellites.

[209] Kallender-Umezu, Paul. "Japan Commits To Deploying Satellite Navigation System by 2020." Space News 17 Oct. 2011: 14.

[210] Javaraman, K.S. "India Launches Third Satellite for Regional Navigation Constellation." 16 Oct. 2014. SpaceNews 20 May 2015. http://spacenews.com/42206india-launches-third-satellite-for-regional-navigation-constellation/.

[211] "India's first ever dedicated navigation satellite launched." 2 July 2013. DNAindia.com 30 May 2014. http://www.dnaindia.com/scitech/report-indias-first-ever-dedicated-navigation-satellite-launched-1855830.

be based on the Ariane 5 liquid oxygen/liquid hydrogen Vulcan engine, while its P120 solid rocket boosters are of a similar type to the next evolution of the four-stage Vega launcher series, Vega-C. Vega is a single body launcher with three solid propulsion stages and an additional liquid propulsion upper module used for attitude and orbit control, and satellite release.[212] The Ariane 6 will also have a cryogenic liquid oxygen/liquid hydrogen upper stage propelled by a Vinci engine, which will be adapted from the Ariane 5ME upper stage.[213]

On 5 January 2014, ISRO's Geo-synchronous Satellite Launch Vehicle (GSLV-D5) powered by an indigenous cryogenic engine in its third stage was successfully launched.[214] And ISRO successfully test launched its newest and largest GSLV variant, the GSLV Mk.III on 18 December 2014. The GSLV Mk.III is a two-stage rocket, with its first stage powered by two French Vikas engines that burn unsymmetrical dimethylhydrazine (UDMH) propellant oxidised by dinitrogen tetroxide (N2O4). Its intended C25 second stage engine was not tested, and was replaced instead by a CE30 engine which normally burns liquid hydrogen and liquid oxygen, but for demonstration purposes was loaded with liquid nitrogen to simulate the full second stage.[215]

In the U.S., Aerojet Rocketdyne completed testing its J-2X upper-stage rocket engine on 26 June 2014.[216] Nine liquid oxygen/liquid hydrogen fuelled J-2X engines had been built, i.e. seven for development tests and two for certification tests.[217] The engine, previously envisioned as the upper stage engine for the Ares 5 rocket to be used in the cancelled Constellation programme, is now the upper stage of the heavy-lift U.S. Space Launch System. For now, following the completion of development testing, the engine will be shelved for several years before it is needed for the Mars effort. For precursor missions to the Moon, the all-cryogenic J-2X is somewhat over-powered, with the ability to lift a 130 metric ton SLS to LEO, whereas missions to the Moon simply need to lift a 105 ton SLS. Until NASA tests for a Mars mission, it will likely rely on the use of three or four RL-10 engines

---

[212] Launch Vehicles—Vega. 30 Oct. 2012. ESA 8 Mar. 2013. http://www.esa.int/Our_Activities/Launchers/Launch_vehicles/Vega.

[213] Ariane 6. 3 Dec. 2014. ESA 13 May 2015. http://www.esa.int/Our_Activities/Launchers/Launch_vehicles/Ariane_6.

[214] Ram, Arun. "Isro successfully launches indigenous cryogenic engine-powered GSLV-D5." 5 Jan. 2014. The Times of India 12 May 2015. http://timesofindia.indiatimes.com/india/Isro-successfully-launches-indigenous-cryogenic-engine-powered-GSLV-D5/articleshow/28437867.cms.

[215] Graham, William. "India debuts GSLV Mk.III with prototype crew capsule." 17 Dec. 2014. NASA spaceflight.com 12 May 2015. http://www.nasaspaceflight.com/2014/12/india-gslv-mk-iii-prototype-crew-capsule/.

[216] "Aerojet Rocketdyne Completes J-2X Testing." 26 June 2014. Aerojet Rocketdyne 12 May 2015. https://www.rocket.com/article/aerojet-rocketdyne-completes-j-2x-testing.

[217] SpaceNews Staff. "J-2X Engine Test Fired at NASA Stennis Space Center." SpaceNews 1 Aug. 2011: 9.

that could produce the dialled-back amount of force needed for lunar precursor missions.[218]

The 28 October 2014 failure of Orbital Science's third Commercial Resupply Services (CRS) mission to the ISS brought into question the future of the Antares launcher—particularly in regard to its first stage AJ-26 engines. Supplied by Aerojet Rocketdyne as refurbished versions of the NK-33 engines, the AJ-26 has experienced several failures in recent times, including a failed test at the NASA Stennis Space Center in Mississippi, USA in May 2014, and another failed test in June 2011 due to a fuel leak in the engine.[219] The latest failure, attributed to the engine's turbo-pump machinery,[220] accelerated Orbital's plans to replace the engine with a derivative of the RD-180 engine from Russian manufacturer NPO Energomash, and on 17 December 2014 it was confirmed that the RD-181 would begin to be delivered starting from 2015, with a launch already scheduled for some time in 2016.[221]

With relations between the U.S. and Russia deteriorating following Russia's annexation of Crimea, subsequent sanctions by the U.S., and the resulting Russian ban in using its RD-180 engine for U.S. military purposes; the U.S. government has begun seeking a replacement for the RD-180 engine and its derivatives.[222] On 16 September 2014, the United Launch Alliance (ULA) selected Blue Origin to develop a new main engine for a future rocket that would combine elements of the Atlas 5 and the Delta 4 launchers. Blue Origin has been developing its BE-4 engine for more than 3 years, giving it a head start against other developers, and it is meant to be relatively inexpensive to develop and build through the use of liquefied natural gas as its propellant. Liquefied natural gas was preferred over kerosene as the engine's propellant due to its cost and engine reusability; similar to methane, the fuel is cheaper than rocket-grade kerosene and leaves less burn residue, making the engine easier to refurbish and reuse.[223] However, as this fuel source is less dense than the kerosene fuel that is used in the RD-180, larger fuel tanks than are available

---

[218] "NASA's J-2X Engine To Be Mothballed After Testing." 4 Oct. 2013. Aviation Week 9 Apr. 2014. http://aviationweek.com/awin/nasa-s-j-2x-engine-be-mothballed-after-testing.

[219] Foust, Jeff. "Antares Failure Raises Questions About Vehicle's Future." 31 Oct. 2014. SpaceNews 9 Apr. 2015. http://spacenews.com/42395antares-failure-raises-questions-about-vehicles-future/.

[220] Foust, Jeff. "Turbopump in AJ-26 Engine Implicated in Antares Failure." 10 Nov. 2014. SpaceNews 9 Apr. 2015. http://spacenews.com/42502turbopump-in-aj-26-engine-implicated-in-antares-failure/.

[221] De Selding, Peter B. "Orbital Sciences Orders RD-181 Engines for Antares Rocket." 17 Dec. 2014. SpaceNews 9 Apr. 2015. http://spacenews.com/orbital-sciences-orders-rd-181-engines-for-antares-rocket/.

[222] Gruss, Mike, and Warren Ferster. "Senate Defense Bill Provides $100M To Develop New U.S. Rocket Engine." 23 May 2014. SpaceNews 9 Apr. 2015. http://spacenews.com/40676senate-defense-bill-provides-100m-to-develop-new-us-rocket-engine/.

[223] Ferster, Warren. "ULA To Invest in Blue Origin Engine as RD-180 Replacement." 17 Sept. 2014. SpaceNews 9 Apr. 2015. http://spacenews.com/41901ula-to-invest-in-blue-origin-engine-as-rd-180-replacement/.

on the Atlas 5 launcher would be needed, which would call for substantial design changes if it is to replace the Atlas 5 main stage RD-180 engine. The BE-4 engine will be capable of providing 550,000 lbs of thrust and is a variant of the BE-3 engine that is used on Blue Origin's New Shepard experimental suborbital rocket that will launch and land vertically.

In the meantime, following the request by the U.S. Air Force for information on RD-180 replacement options, Aerojet Rocketdyne responded on 29 September 2014 by announcing it is developing a kerosene-fuelled main engine propulsion concept designated AR-1 as a potential replacement for the Atlas 5 RD-180 main stage engine. The announcement also clarified that the AR-1 would require minimal changes to the Atlas 5 launcher, including its ground system and launch infrastructure. The AR-1 is a liquid-oxygen/kerosene-fuelled engine, and will be capable of generating 500,000 lbs of thrust and could be ready to fly by 2019 for an investment of less than $1 billion, making it a low-risk transition engine for multiple launch vehicles.

On 20 November 2014, XCOR Aerospace announced the completion of another milestone in the development of its liquid oxygen and liquid hydrogen (LH2) programme. Being developed for the ULA, XCOR successfully performed hot fire testing of the XR-5H25 engine's regeneratively cooled thrust chamber, with both liquid oxygen and liquid hydrogen propellants supplied in pump-fed mode.[224] Building on their earlier success with liquid oxygen and kerosene pumps, the group has overcome the technical challenges presented with the extreme low temperature and small molecule size of liquid hydrogen to develop and successfully operate the engine's liquid hydrogen pump at full design flow rate and pressure conditions—opening the way for integrated testing of the LH2 demonstrator engine.[225] The LH2 engine program is intended to produce a flight-ready cryogenic upper-stage engine in the 25,000 lb thrust class with growth potential of 50,000 lb of thrust or more, which should result in much lower cost and more capable commercial and government launch capabilities. It is intended as a potential successor to the Delta and Atlas series upper stage engines that are currently used.

Space Exploration Technologies' (SpaceX) Falcon 9 launcher is the only currently operating launch vehicle that has engine out capability, able to lose 2 of its 9 Merlin engines in its first stage and still complete its mission.[226] The engine's 150 vacuum thrust-to-weight ratio, and enhanced design for improved manufacturability, makes it one of the most efficient booster engines ever built. Integrated on the upgraded Falcon 9 v1.1 launcher, the engines replaced the previous Merlin 1C

---

224 Press Release. "XCOR Aerospace Announces Latest Milestone in ULA Program." 20 Nov. 2014. XCOR Aerospace 12 May 2015. http://www.xcor.com/press/2014/14-11-20_XCOR_announces_ULA_milestone.html.

225 "XCOR Aerospace and United Launch Alliance Announce Important Milestone in Liquid Hydrogen Engine Program." 23 Sept. 2013. XCOR Aerospace 14 Apr. 2014. http://www.xcor.com/press/2013/13-09-23_XCOR_ULA_announce_hydrogen_engine_milestone.html.

226 "Merlin Engines." 29 July 2013. SpaceX 10 Apr. 2014. http://www.spacex.com/news/2013/03/26/merlin-engines.

engines and require a 60 % increase in the size of the rocket's propellant tanks; the Falcon 9 v1.1 had its debut launch on 29 September 2013.[227] By 22 October 2014, eighty Merlin 1D engines had been launched, exceeding the propulsion heritage of the RS-68/68A engine (41 flown) on the Delta and the RD-180 engine (55 flown) on Atlas variants.[228]

Reaction Engines is developing the Synergistic Air-Breathing Rocket Engine (Sabre) to power a planned single-stage-to-orbit (SSTO) spaceship, Skylon, as a part chemical rocket and part jet engine. Sabre will have the ability to use oxygen in airspace rather than from external liquid-oxygen tanks, and will eliminate the need for expendable boosters. When travelling at speeds of up to Mach 5.2, the super-heated air travelling through the engine is rapidly cooled to −150 °C, and then channelled through the engine's turbo-compressor, and into the thrust chambers, to be mixed with liquid hydrogen and ignited to produce thrust for the spacecraft.[229] ESA and the British government have invested a combined $92 million in the project, however completion of the engine will require an additional investment of $3.6 billion before it could be ready for flight tests. The Skylon itself would require a $14 billion investment. The Skylon could begin test flights in 2019.[230]

In the U.S., Ball Aerospace and Technologies and other members of the Green Propellant Infusion Mission (GPIM) project are developing a green propellant demonstrator for NASA's Technology Demonstration Mission (TDM). To be used on a Ball BCP-100 spacecraft designed for easy integration of "ride-share" payloads, this mission will be the first time the U.S. will use a spacecraft to test green propellant technology.[231] The propellant is a hydroxyl ammonium nitrate fuel/oxidizer blend known as AF-M315E, which offers nearly 50 % better performance than traditional hydrazine fuel, in addition to reducing the environmental impact during propellant loading, potentially increasing payload capacity, enhancing spacecraft manoeuvrability, and extending mission durations. As milestone progress is continuing to be made, following supplier Aerojet Rocketdyne's successful test of its 1 Newton (1 N) thruster in July 2014, the completed small satellite is expected to launch sometime in 2016.[232]

---

[227] "Upgraded Falcon 9 Mission Overview." 14 October 2013. SpaceX 9 Apr. 2014. http://www.spacex.com/news/2013/10/14/upgraded-falcon-9-mission-overview.

[228] "SpaceX Completes 100th Merlin 1D Engine." 22 Oct. 2014. SpaceX 12 May 2015. http://www.spacex.com/news/2014/10/16/spacex-completes-100th-merlin-1d-engine.

[229] "The Next Space Shuttle: Hybrid Engines Make Runway-To-Orbit Missions A Reality." 10 Sept. 2013. Popular Science 15 Apr. 2014. http://www.popsci.com/technology/article/2013-08/runway-orbit-and-back.

[230] Passary, Sumit. "Sabre Jet can Fly Mach 5 and Bring You Anywhere in the World in Just Four Hours." 18 Dec. 2014. TechTimes 13 May 2015. http://www.techtimes.com/articles/22463/20141218/sabre-jet-can-fly-mach-5-and-bring-you-anywhere-in-the-world-in-just-four-hours.htm.

[231] "Green Propellant Team Propels Itself Through Preliminary Design Review." 20 Sept. 2013. NASA 15 Apr. 2014. http://www.nasa.gov/mission_pages/tdm/green/green-propellant-preliminary-review.html.

[232] "GPIM Spacecraft to Validate Use of "Green" Propellant." 19 Aug. 2014. NASA 13 May 2015. https://www.nasa.gov/content/gpim-spacecraft-to-validate-use-of-green-propellant/.

On 28 March 2014 Boeing was selected by the U.S. Defense Advanced Research Projects Agency (DARPA) to build and demonstrate a low-cost, airborne satellite launching system that is capable of relocating and launching from any major runway around the globe. DARPA's Airborne Launch Assist Space Access (ALASA) programme started in 2012, when it awarded design contracts to Boeing, Lockheed Martin, and Virgin Galactic, in addition to related technology development contracts to three other companies.[233] Boeing's cost-effective design moved the rockets engines forward, allowing both the first and second stages to be powered by the same engines, thereby enabling a reduction in weight and system complexity. Upon being released from a F-15E fighter aircraft at an altitude of 12 km, the rocket would be capable of lifting a microsat weighing up to 45 kg satellite to low Earth orbit, at a cost of $1 million per launch; amounting to a 66 % launch cost reduction when compared to Orbital Science's Pegasus XL air-launched rocket system. A demonstration launch of the ALASA system is expected sometime in 2015.[234]

2.5.2 *Telecoms*

In April 2014 the information security firm IOActive released a technical white paper on the vulnerabilities of satellite communications security.[235] The report found multiple high-risk vulnerabilities in satellite ground terminal equipment from Harris Corp., Hughes, Thuraya, Cobham, JRC, and Iridium; these flaws included hardcoded credentials, undocumented protocols, insecure protocols, backdoors, and weak password reset features.[236] Those vulnerabilities have implications for many critical services e.g., such as revealing the location of units and soldiers in military communications, or by being open to a malicious hacker to block or disrupt emergency communications to aircraft and ships.[237]

[233] Gruss, Mike. "DARPA Picks Boeing To Demonstrate Airborne Launcher Concept." 25 Mar. 2014. SpaceNews 9 Apr. 2015. http://spacenews.com/39967darpa-picks-boeing-to-demonstrate-airborne-launcher-concept/.

[234] Gruss, Mike. "Boeing Targets 66 Percent Launch Cost Reduction with ALASA." 28 Mar. 2014. SpaceNews 9 Apr. 2015. http://spacenews.com/40023boeing-targets-66-percent-launch-cost-reduction-with-alasa/.

[235] Santamarta, Ruben. "A Wake-up Call for SATCOM Security." 17 Apr. 2014. IOActive 22 Aug. 2015. http://www.ioactive.com/pdfs/IOActive_SATCOM_Security_WhitePaper.pdf.

[236] Higgins, Kelly Jackson. "Satellite Communications Wide Open To Hackers." 17 Apr. 2014. Information Week 22 Aug. 2015. http://www.darkreading.com/vulnerabilities---threats/satellite-communications-wide-open-to-hackers/d/d-id/1204539.

[237] Lever, Rob. "Satellite telecom vulnerable to hackers, researchers find." 17 Apr. 2014. Phys.org 22 Aug. 2015. http://phys.org/news/2014-04-satellite-telecom-vulnerable-hackers.html.

2.5.3 Spacecraft Operations and Design

In Europe, the European Space Agency's (ESA) Rosetta comet chasing spacecraft was awoken from a 31 month hibernation period on 20 January 2014, while on course to rendezvous with the 67/P Churyumov-Gerasimenko comet by mid-2014.[238] Rosetta arrived within a 100 km distance from the comet on 6 August, and proceeded to manoeuvre itself in triangle-shaped paths to stay mainly between the Sun and the comet to view its illuminated terrain. The spacecraft berthed its own lander, Philae, which has its own suite of instruments, including a drill capable of penetrating as much as 23 cm beneath the surface. Upon reaching within 20 km of the comet, Rosetta deployed Philae on 12 November; yet as both the lander's two harpoons and its Active Descent System failed to function as planned, the lander bounced by an elevation of 1 km before alighting in a rocky and less illuminated region of the comet. From there, its solar panels receive only 90 min of sunlight every 12 h, rather than the anticipated 6 and 7 h of sunlight per 12-h day expected.[239] Nevertheless, on 15 November, following attempts to better orient Philae's solar array by lifting the lander by 4 cm and rotating it about 35°, the lander was put into hibernation. Yet before expending Philae's 72 h battery life, the lander managed to complete all the measurements planned for the final block of experiments on the comet's surface, and relayed the data to Earth via Rosetta, marking the unprecedented success of the Rosetta mission in exploring the surface of a comet.[240] While Philae's hibernation is likely to be permanent, there is a chance that it can be reawakened as the 67/P Churyumov-Gerasimenko comet nears the Sun.[241]

NASA's Orion Multi-Purpose Crew Vehicle (MPCV) conducted its first mission, Exploration Flight Test-1 (EFT-1), on 5 December 2014. Prior to the penultimate mission, the Orion's parachutes were tested on 25 June 2014, to test how they would respond at higher speeds in the final stages of the spacecraft's descent, and in the case where one of the three main parachutes were to expand quicker than intended. Another drop in August was to test the combined failure of one drogue parachute, which opens ahead of the three main parachutes to slow Orion down

[238] De Selding, Peter B. "Rosetta Probe Revived after 31-month Slumber for Comet Rendezvous." 20 Jan. 2014. SpaceNews 14 May 2015. http://spacenews.com/39185rosetta-probe-revived-after-31-month-slumber-for-comet-rendezvous/.

[239] De Selding, Peter B. "Rough Landing Putting Comet Probe's Science Team to the Test." 13 Nov. 2014. SpaceNews 14 May 2015. http://spacenews.com/42541rough-landing-putting-comet-probes-science-team-to-the-test/.

[240] De Selding, Peter B. "Philae Mission May be Done." 15 Nov. 2014. SpaceNews 14 May 2015. http://spacenews.com/42565philae-mission-may-be-done/.

[241] De Selding, Peter B. "Philae Managers Say Recovery Possible as Comet Approaches Sun." 18 Nov. 2014. SpaceNews 14 May 2015. http://spacenews.com/42586philae-managers-say-recovery-possible-as-comet-approaches-sun/.

before the spacecraft's final landing phase, and one main parachute, as well as new parachute design features.[242] On its December flight test, the Orion crew module reached an apex orbit of nearly 5800 km above Earth, conducting two revolutions before entering Earth's atmosphere to test the spacecraft's heat shield when re-entering the atmosphere at speeds verging on those that would occur during a return from deep space. The key purpose of this launch was to test whether Orion's heat shield can withstand the forces of atmospheric re-entry at about 32,000 km/h, with temperatures reaching up to 2200 °C.[243] On landing in the Pacific Ocean, 1015 km off the coast of California, the spacecraft had recorded only a minor problem with the system that inflates the five airbags that are to keep the module upright after splashdown, as only three of those airbags inflated fully. Orion is now expected to fly again in 2018 on the first launch of the Space Launch System heavy-lift rocket, followed by the first crewed Orion launch on the SLS in 2021.[244]

Boeing is developing its own Commercial Space Transportation-100 (CST-100) crew capsule, designed to send astronauts to the ISS as early as 2015.[245] On 17 October 2014 Boeing reached its final milestone, having undergone an extensive battery of tests on its mission flight software, including launch, docking, on-orbit, and re-entry and landing manoeuvres; mission simulations to advance communications and mission operations planning; wind tunnel testing of both the CST-100 and the Atlas V launch vehicle; and test-firings of the spacecraft's launch abort engines and thrusters used for manoeuvring in space.[246] Similar to the SpaceX Dragon capsule, the CST-100 has a pusher-style abort system, wherein four such engines would propel CST-100 and its crew to safety in the event of a launch problem.[247] Measuring 4.5 m across at its widest point, this seven-seat gumdrop-shaped capsule will be reusable for up to 10 flights.[248]

In November 2014, SpaceX completed an additional two milestones in the development of its Dragon 2.0 crew capsule, designed to send astronauts to the ISS, including the Dragon Primary Structure Qualification, and the Delta Crew

---

[242] Leone, Dan. "NASA Tests Orion Parachutes Ahead of First Space Launch." 30 June 2014. SpaceNews 9 Apr. 2015. http://spacenews.com/41075nasa-tests-orion-parachutes-ahead-of-first-space-launch/.

[243] Leone, Dan. "NASA Proposes Orion Test Flight in 2014." 8 Nov. 2011. SpaceNews 12 Mar. 2013. http://www.spacenews.com/article/nasa-proposes-orion-test-flight-2014.

[244] Foust, Jeff. "Orion Aces First Flight Test." 5 Dec. 2014. SpaceNews 9 Apr. 2015. http://spacenews.com/orion-aces-first-flight-test/.

[245] SpaceNews Staff. "Boeing Picks Atlas 5 for CST-100 Test Flights." SpaceNews 8 Aug. 2011: 3.

[246] "Boeing Finishes Commercial Crew Space Act Agreement for CST-100/Atlas V" 17 Oct. 2014. NASA 15 May 2015. http://www.nasa.gov/content/boeing-finishes-commercial-crew-space-act-agreement-for-cst-100atlas-v.

[247] "CST-100 Launch-abort Engines Complete Testing Milestone." 17 Dec. 2013. SpaceNews 15 Apr. 2014. http://www.spacenews.com/article/launch-report/38731cst-100-launch-abort-engines-complete-testing-milestone.

[248] Chow, Denise. "Boeing's CST-100 Capsule Shooting for 2015 Debut." SpaceNews 30 Apr. 2012: 14.

Vehicle Critical Design Review, along with additional critical design reviews of systems and operations. However, SpaceX still needs to conduct two abort tests for the Dragon 2.0 in 2015, as required under the CCiCap.[249] SpaceX will flight-test its uncrewed 'DragonRider' to demonstrate the ability of the Dragon spacecraft abort system to lift the spacecraft clear of a simulated launch emergency. The second flight test involves simulating an in-flight emergency abort scenario during ascent at high altitude at maximum aerodynamic pressure at about 60 s into the launch. Both abort tests are essential for demonstrating that the spacecraft will activate its SuperDraco thrusters and separate in a split second to save astronaut lives in the event of a real life emergency.[250] When crewed, the DragonRider will be able to lift up to seven astronauts to the ISS, and remain docked for at least half a year. SpaceX hopes to launch an initial crewed Dragon orbital test flight to the ISS as early as 2015.

The Sierra Nevada Corporation is developing the Dream Chaser, designed to send astronauts to the ISS on a winged, lifting-body spacecraft, with the capability of returning to Earth by landing on a conventional airstrip. The design of the reusable spacecraft is derived from NASA's HL-20 Personnel Launch System from the 1990s that had undergone years of development, analysis, and wind tunnel testing, along with related synergy with the retired U.S. space shuttles. Capable of holding a crew of up to seven astronauts, the spacecraft would launch atop an Atlas 5 rocket.[251] In December 2014 SNC announced the successful completion of milestone 15a, relating to the spacecraft's Reaction Control System (RCS) which provides on-orbit control for many critical manoeuvres such as docking to the ISS. This marked the successful completion of 12 out of 13 critical milestones.[252] The Dream Chaser has attracted interest from Germany's DLR and OHB System AG, wishing to finance a study to explore ways in which the spacecraft can be used to cover German and European requirements for the transportation of payloads and astronauts to the ISS and for deployment as a manned or unmanned space vehicle allowing German and European scientists to conduct research under weightless conditions over extended periods of time. Moreover, given its capability of reaching orbits at a substantially greater altitude than the ISS, the study will

[249] Rhian, Jason. "SpaceX completes first CCtCap milestone as pace of Commercial Crew accelerates." 22 Dec. 2014. Spaceflight Insider 15 May 2015. http://www.spaceflightinsider.com/organizations/space-exploration-technologies/spacex-completes-first-milestone-commercial-crew-transportation/.

[250] "What's Ahead for Human Rated SpaceX Dragon in 2014—Musk tells Universe Today." 30 Dec. 2013. Universe Today 15 Apr. 2014. http://www.universetoday.com/107505/whats-ahead-for-human-rated-spacex-dragon-in-2014-musk-tells-universe-today/.

[251] "NASA CCiCAP Funding for SpaceX, Boeing and SNC's Crew Vehicles." 3 Aug. 2012. NASA Spaceflight 15 Apr. 2014. http://www.nasaspaceflight.com/2012/08/nasa-ccicap-funding-spacex-boeing-sncs-crew-vehicles/.

[252] "Sierra Nevada Corporation Successfully Tests Critical Dream Chaser® Spacecraft Propulsion System, Completes CCiCap Milestone 15a." 2 Dec. 2014. PRWeb 15 May 2015. http://www.prweb.com/releases/2014/12/prweb12367203.htm.

determine the extent to which it is able to supply satellites or remove decommissioned satellites from their orbits.[253]

Blue Origin is also in the running to develop its orbital reusable launch vehicle program outside of the CCiCap programme, participating with NASA in an unfunded partnership. In October 2014, NASA and Blue Origin signed an agreement to extend their CCDev2 partnership into 2016. Three additional unfunded milestones were added to the agreement to continue the development work and partnership, involving Blue Origin's propellant tank, BE-3 engine, and pusher escape system.[254]

2.5.4 Suborbital Activities

In mid-2014, U.S. lawmakers began preparations to update the U.S. Commercial Space Launch Amendments Act of 2004. The new legislation *will inter alia* address whether the FAA can begin to write human spaceflight safety regulations.[255] Up to October 2015 suborbital spacecraft developer Virgin Galactic and similar U.S. firms have a regulatory grace period extension on developing suborbital spacecraft without Federal Aviation Administration (FAA) imposed passenger and crew safety rules. Across the Atlantic, on 15 July 2014, the British government unveiled a broad strategy document designed to create the necessary regulatory regime to permit suborbital spaceplane flights by 2018 as part of a long term goal of establishing a domestic small-satellite launch capability. The UK Civil Aviation Authority (CAA) has selected eight existing airports as potential suborbital spaceports, six of them in Scotland. The U.S. FAA and UK CAA signed an MOU on the following day for better coordination between the two regions, yet the UK-based industry has some challenges to overcome especially in the areas of U.S. ITAR restrictions, flight safety regulations, and liability exposure.[256]

In the first half of 2014, Virgin Galactic was on the verge of completing the final phases toward upgrading its FAA experimental permit to also include a commercial

---

[253] "Contract Signed with DLR for the Study Phase for the Utilization of U.S. Company Sierra Nevada Corporation's Dream Chaser® Spacecraft." 13 Nov. 2013. OHB System 16 Apr. 2014. https://www.ohb-system.de/press-releases-details/items/contract-signed-with-dlr-for-the-study-phase-for-the-utilization-of-us-company-sierra-nevada-corporations-dream-chaser-spacecraf.html.

[254] Money, Stewart. "Blue Origin, NASA Add Three More Unfunded Milestones." 14 Nov. 2014. Innerdpace.net 15 May 2015. http://innerspace.net/cotscommercial-crew/blue-origin-nasa-add-three-more-unfunded-milestones/.

[255] Leone, Dan. "Hill Staffers: Commercial Space Launch Bill Is Coming This Year." 26 May 2014. SpaceNews 9 Apr. 2015. http://spacenews.com/40694hill-staffers-commercial-space-launch-bill-is-coming-this-year/.

[256] De Selding, Peter B. "News from the Farnborough International Airshow | Britain Plots Course for Domestic Small-satellite Launcher." 18 July 2014. SpaceNews 9 Apr. 2015. http://spacenews.com/41296news-from-the-farnborough-international-airshow-britain-plots-course-for/.

launch licence for SpaceShipTwo, following a successful third rocket-powered test flight on 10 January. In the first launch, the spacecraft reached a top altitude of 21 km and top speed of Mach 1.4, in addition to testing the vehicle's reaction control system and a thermal coating on its tail boom.[257] Following Virgin Galactic's May announcement that it will change the fuel used in the vehicle's hybrid rocket motor from hydroxyl-terminated polybutadiene (a form of rubber) to a polyamide-based plastic, the company held its next flight test on 28 August. This time it was an unpowered release from the WhiteKnightTwo to conduct a "cold flow" test of the spacecraft's rocket engine, which ran liquid oxidizer through the motor without igniting it.[258] Tragically, on its next powered flight test on 31 October 2014, SpaceShipTwo experienced an 'inflight anomaly' shortly after release from the WhiteKnightTwo and ignition of its hybrid rocket engine; SpaceShipTwo was destroyed, killing one of its pilots, and injuring the other. Investigations of the accident attributed the cause of the explosion to the premature unlocking of the spacecraft's feathering mechanism that, subject to subsequent aerodynamic forces, resulted in the in-flight separation of the wings and the vehicle. Following this disaster Virgin Galactic stated that it plans to continue the construction of a second SpaceShipTwo, and that it was about 65 % complete. It expects the second SpaceShipTwo to be ready for test flights in 2015 and commercial service in 2016.[259]

In a continued effort to foster a viable market for American commercial reusable suborbital platforms that allow testing of new space technologies within Earth's atmosphere, in September 2014 NASA selected Virgin Galactic and three other companies (Masten Space Systems, Paragon Space Development Corp., and UP Aerospace Inc.) to integrate and fly technology payloads on commercial suborbital reusable platforms. The new 3-year contracts for integration and flight services build on previous contracts awarded by NASA in 2011; they also carry 2-year extension options and a minimum value of $100,000. As part of the contracts, the flights will carry a variety of payloads during five diverse flight profiles to help meet NASA's research and technology needs while also maturing cross-cutting technologies to flight readiness status for future space missions.[260]

---

[257] Ferster, Warren. "SpaceShipTwo Reaches Highest Altitude to Date in Test." 10 Jan. 2014. SpaceNews 9 Apr. 2015. http://spacenews.com/39031spaceshiptwo-reaches-highest-altitude-to-date-in-test/.

[258] Foust, Jeff. "Virgin Galactic Delays First Commercial Flights to 2015." 11 Sept. 2014. SpaceNews 9 Apr. 2015. http://spacenews.com/41837virgin-galactic-delays-first-commercial-flights-to-2015/.

[259] Foust, Jeff. "SpaceShipTwo Investigation Expands To Include Human Factors." 4 Nov. 2014. SpaceNews 9 Apr. 2015. http://spacenews.com/42445spaceshiptwo-investigation-expands-to-include-human-factors/.

[260] "NASA Selects U.S. Firms to Provide Commercial Suborbital Flight Services." 8 Sept. 2014. NASA 17 May 2015. https://www.nasa.gov/press/2014/september/nasa-selects-us-firms-to-provide-commercial-suborbital-flight-services/.

Masten is developing the Xaero; a vertical-takeoff, soft vertical-landing vehicle that uses the firm's isopropyl alcohol- and liquid oxygen-burning Cyclops-AL-3 engine. While the first experimental suborbital launcher was destroyed on 11 September 2012 during the spacecraft's 110th test-flight, when a stuck engine valve triggered the rocket's flight termination system during its descent,[261] a second slightly larger Xaero-B has continued the programme's development, and completed four successful launches on 16 September 2014.[262] Also, Paragon Space Development Corp. is developing a passenger-carrying stratospheric balloon that will be operated by its World View subsidiary.[263] The vehicle, a 40 million cubic-foot (1.1 million cubic-meter) helium balloon and a steerable parafoil, is expected to carry six passengers and two pilots on a two hour journey; and its initial demonstration tests claimed a world record for the highest parafoil flight when its parafoil was inflated at about 15 km during the balloon's descent.[264] In addition, UP Aerospace Inc. has operated expendable solid-fuelled suborbital rockets from Spaceport America, New Mexico since 2006; its latest mission conducted on 23 October 2014 carried four payloads for NASA's Flight Opportunities Programme.[265]

Finally, XCOR's Lynx suborbital spacecraft is the company's entry into the commercial reusable launch vehicle (RLV) market. The Lynx is a horizontal takeoff and horizontal landing vehicle that uses its own fully reusable rocket propulsion system to take off and land on a runway. This two-seat, piloted space transport vehicle will take humans and payloads on a half-hour suborbital flight to 100 km and then return to a landing at the takeoff runway. The piloted, two-seat spacecraft can be used to lift humans and payloads on a 30 min suborbital flight up to four times in a day. Lynx will be FAA AST-licensed, and has already passed the AST licensing process with an earlier vehicle concept.[266] XCOR announced rapid progress towards final assembly on 18 December 2014, following the bonding of both the Lynx' cockpit and carry-through spar to its fuselage. The next integration

[261] "Masten's Xaero Rocket Lost During Mojave Test Flight." 17 Sept. 2012. SpaceNews 14 Apr. 2014. http://www.spacenews.com/article/masten%E2%80%99s-xaero-rocket-lost-during-mojave-test-flight.

[262] Commercial Space Transportation Advisory Committee (COMSTAC) Systems Working Group. "DARPA XS-1 | MSS XEPHYR" 16 Sept. 2014. Masten 18 May 2015. http://www.faa.gov/about/office_org/headquarters_offices/ast/advisory_committee/meeting_news/media/2014/sep/Dennis_Poulos.pdf.

[263] SpaceNews Staff. "NASA Awards 4 Companies Suborbital Flight Contracts." 15 Sept. 2014 SpaceNews 18 May 2015. http://spacenews.com/41867nasa-awards-4-companies-suborbital-flight-contracts/.

[264] Brooks, Karen. "Balloon space flight company completes small-scale test flight." 25 June 2014. Reuters 18 May 2015. http://www.reuters.com/article/2014/06/25/us-space-ballooning-idUSKBN0F01MK20140625.

[265] "Welcome to UP Aerospace Past Missions." 23 Oct. 2014. UP Aerospace Inc. 18 May 2015. http://www.upaerospace.us.com/Past-Missions.html.

[266] "About Lynx." XCOR Aerospace 11 Apr. 2014. http://www.xcor.com/lynx/.

step involves attaching the strakes to the spacecraft.[267] With interest already created among commercial industry and research institutes, XCOR's Lynx Mark I will begin commercial flights in 2015 and will carry payloads smaller than 1 kg as a "ride share" or "secondary payload", and up to one 120 kg "primary" mission payload [integrated into the Lynx by the Czech Space Office (CSO)]. Payloads may be placed inside the Lynx pressurized cabin or exposed to the vacuum and radiation conditions of space.[268]

2.5.5 Other Technologies

Developments in general technology and science continued to advance in 2014, with implications reaching beyond the space sector.

In mid-2014, NASA's Satellite Servicing Capabilities Office (SSCO) had achieved several critical milestones involving its continuing Robotic Refuelling Mission (RRM), such as the ground-based Remote Robotic Oxidizer Transfer Test (RROxiTT) in February, which tested how robots can transfer hazardous oxidizer at flight-like pressures and flow rates through the propellant valve and into the mock tank of a satellite.[269] The RRM mission is being used to demonstrate that remotely controlled robots using current-day technology could refuel satellites that were not designed to be serviced. Now entering its second phase, following the 29 July delivery of its fourth mock satellite task board and the Visual Inspection Poseable Invertebrate Robot (VIPIR) tool, the RRM mission will demonstrate how space robots can replenish coolant in the instruments of similar legacy satellites.[270]

NASA's Robonaut 2 (R2) had its climbing legs attached to its torso by September 2014, for added mobility in conducting regular and repetitive tasks within and outside the ISS.[271] The legs were delivered to the station on 18 April

[267] Press Release. "In Pictures: XCOR announces Further Progress on XCOR Lynx Spacecraft." 18 Dec. 2014. XCOR 18 May 2015. http://www.xcor.com/press/2014/14-12-18_Lynx_development_in_pictures_carry_through_spar.html.

[268] "Czech Space Office and XCOR Aerospace Sign Payload Integrator Agreement for Suborbital Flights." 21 Nov. 2013. XCOR Aerospace 14 Apr. 2014. http://www.xcor.com/press/2013/13-11-21_czech_space_office_xcor_payload_integrator.html.

[269] Washington, Dewayne and Adrienne Alessandro. "NASA Tests New Robotic Refuelling Technologies." 5 Mar. 2014. NASA 18 May 2015. http://www.nasa.gov/content/goddard/nasa-tests-new-robotic-refueling-technologies/.

[270] Alessandro, Adrienne. "NASA's Space Station Fix-It Demo for Satellites Gets Hardware for 2.0 Update." 12 Aug. 2014. NASA 18 May 2015. http://www.nasa.gov/mission_pages/station/research/news/rrm_update/#.U-o_jCQcB9U%EF%BF%BD%EF%BF%BD.

[271] Howell, Elizabeth. "Space Station's Robonaut 2 Is Getting More Astronaut-Like By The Day." 5 Sept. 2014. UniverseToday 18 May 2015. http://www.universetoday.com/114346/space-stations-robonaut-2-is-getting-more-astronaut-like-by-the-day/.

2014.[272] With a fully extended leg span of 2.7 m, each leg has seven joints, giving it great flexibility, and an end effector with a camera, allowing it to grasp handrails and sockets in and on the station.[273] Earlier in the year, the R2 underwent tests to show how its dexterity could be used to perform telemedical procedures, through the application of an appropriate level of force and tracking of progress using R2's vision system.[274] The same technology could foreseeably also be used by doctors to conduct complex medical procedures on humans in remote locations.

Other technological advances inspired by NASA's R2 include the RoboGlove. The RoboGlove was developed by NASA in collaboration with General Motors (GM), as a wearable human grasp assist device, to help reduce the grasping force needed by an individual to operate tools for an extended time or when performing tasks having repetitive motion.[275]

DARPA's Membrane Optical Imager for Real-Time Exploitation (MOIRE) program, currently in its second phase, aims to create technologies that would enable future high-resolution orbital telescopes to provide real-time video and images of the Earth from GEO orbit. In May 2014, the final tests on MOIRE's components were completed, with both the primary mirror's composite back structure and the deployment hinges having undergone environmental tests designed to simulate the temperature range and vacuum of a space environment.[276] In the past, size and cost constraints prevented large-scale imaging satellites from being placed within GEO. A new system incorporating MOIRE optics would be roughly one-seventh the weight of a comparable traditional system, and could be tightly packed into a configuration roughly 6 m in diameter, wherein upon reaching GEO, it would unfold to create the full-size multi-lens optics reaching 20 m in diameter. If successful, the satellite could have a 40 % field of view of the earth's surface, and would be able to focus on a 10 km-by-10 km area at 1-m resolution, providing real-time video at 1 frame per second.[277]

On 20 November 2013, the U.S. Secretary of Defence completed a Memorandum of Understanding with Australia's Defence Minister to have DARPA's Space

---

[272] Hall, Loura. "A Step Up for NASA's Robonaut: Ready for Climbing Legs." 23 Apr. 2014. NASA 18 May 2015. http://www.nasa.gov/content/a-step-up-for-nasa-s-robonaut-ready-for-climbing-legs.

[273] "Robonaut 2 | Robonaut 2 Getting His Space Legs." 27 Aug. 2014. NASA 18 May 2015. http://www.nasa.gov/mission_pages/station/main/robonaut.html.

[274] Gannon, Megan. "NASA Teaches Humanoid Robonaut 2 Medical Skills for Space Emergencies (Video)." 18 Feb. 2014. Space.com 18 May 2015. http://www.space.com/24717-nasa-robonaut-telemedicine-training.html.

[275] "Robo-Glove." 1 Sept. 2014. NASA 19 May 2015. http://technology.nasa.gov/media/NP-2014-08-1136-HQ_JSC_RoboGlove_1_Web.pdf.

[276] "Final tests completed on giant surveillance telescope components." 23 May 2014. Optics.org 18 May 2015. http://optics.org/news/5/5/43.

[277] "First Folding Space Telescope Aims to "Break the Glass Ceiling" of Traditional Designs." 5 Dec. 2013. DARPA 14 Apr. 2014. http://www.darpa.mil/NewsEvents/Releases/2013/12/05.aspx.

Surveillance Telescope (SST) fully relocated to Australia by 2016. The SST will be moved from its current mountaintop location in White Sands, New Mexico, USA where the system underwent operational testing and evaluation, to Australia, where it will provide key space situational awareness from the largely unexplored southern portion of the geosynchronous belt. The relocation process began in 2014, and the telescope is expected to resume operations sometime in 2016. From its new location, the SST—with ten times more sensitivity than current state-of-the-art systems—will transmit its observations to the Space Surveillance Network (SSN).

The ISS received its first 3D printer on 23 September 2014, having been delivered aboard a Dragon supply ship that had launched on a Falcon 9 rocket 2 days earlier. The printer was installed aboard the ISS on 17 November, and over the following weeks it underwent testing to produce an initial set of more than 20 demonstration parts, including tools and test coupons that will also be printed by a duplicate printer on Earth for comparison.[278] The 3D printer has the capability of building an estimated 30 % of the spare parts on the station, in addition to specialty tools and experiment upgrades. The main goal of the 3D Print experiment is to help jump-start an off-planet manufacturing capability, which could aid in deeper space exploration by making life in space easier and cheaper. In this pursuit, NASA has also recently funded the development of a prototype 3D printer designed to make space food products out of cheap raw materials that have a long shelf life; useful for long space journeys, such as the 500-day return trip to Mars.[279]

[278] Clark, Stephen. "3D printer activated aboard the International Space Station." 18 Nov. 2014. Spaceflight Now 10 May 2015. http://spaceflightnow.com/2014/11/18/3d-printer-activated-aboard-the-international-space-station/.

[279] "3D Printer Passes Zero-Gravity Test for Space Station Trip." 19 June 2013. Space.com 16 Apr. 2014. http://www.space.com/21630-3d-printer-space-station-tests.html.

Part II
Views and Insights

Edited by Arne Lahcen

Chapter 3
Governance in Space

Marco Aliberti and Stephen D. Krasner

3.1 Governance

Governance structures are comprised of some combination of norms, rules, adjudication procedures, and enforcement mechanisms. Norms are general principles; rules are specific stipulations with regard to specific behaviours; adjudication procedures are established to decide when rules have been broken; enforcement mechanisms impose penalties when it has been determined that a rule has been violated. For instance in international trade the general principle informing the World Trade Organization is that free trade is good; a specific rule would refer, for instance, to the circumstances that define dumping or state subsidies; the dispute settlement mechanism of the WTO adjudicates disagreements about the rules; the enforcement mechanism in the WTO is the withdrawal of concessions. Governance structures are necessary in situations where the unconstrained behaviour of actors will result in sub-optimal outcomes; situations in which without governance structures the Pareto frontier would not be reached.

There are three different configurations of interest: harmony, coordination, and cooperation. Each has different implications for the kinds of governance structures that must be put in place if actors are to avoid sub-optimal outcomes. Harmony is a situation in which each actor, acting on its own, leads to a Pareto optimal outcome. Governance structures are entirely unnecessary when there is harmony, when every

M. Aliberti (✉)
European Space Policy Institute, Schwarzenbergplatz 6, 1030 Vienna, Austria
e-mail: marco.aliberti@espi.or.at

S.D. Krasner
Graham H. Stuart Professor of International Relations, Stanford University, Stanford, CA, USA

Freeman Spogli Institution, Stanford University, Stanford, CA, USA

Hoover Institution, Stanford University, Stanford, CA, USA

C. Al-Ekabi et al. (eds.), *Yearbook on Space Policy 2014*, Yearbook on Space Policy, DOI 10.1007/978-3-7091-1899-3_3

actor can maximise its own utility by acting in ways that do not need to take any account of the behaviour of others.[1]

Coordination problems arise when a sub-optimal outcome occurs if each individual actor fails to take account of the behaviour of others. Coordination problems require some level of governance. Minimal governance structures are needed when the choice of a focal point or rule has no distributional consequences. Actors must coordinate their behaviour around one of a number of possible focal points to reach the Pareto frontier (the set of points along which no one actor can be made better off without making other actors worse off), but they are indifferent about which one is chosen. Once a focal point has been chosen there is no incentive for any one actor to defect, because defection would leave the actor worse off. Every actor is just as happy with the focal point chosen as they would have been with any other. Enforcement mechanisms are not necessary because if an actor violates the agreed upon rule that actor would be worse off. For instance, the universal practice for traffic lights is that red means stop and green means go, but these same values could have initially been assigned differently without any impact on the utility of any actor. What is critical is that all drivers agree on the same meaning for each colour. (This is not to say a driver will never run a red light; but no drivers are trying to make green stop and red go).

It may be more difficult to establish governance structures if the solution to a coordination problem has distributional consequences. If actors have different preferences over which focal point or rule is chosen then coordination problems may be more difficult to solve. Every actor is better off coordinating, but they have different preferences over which coordination point or points should be chosen. Battle of the sexes is the most well-known example of this payoff matrix. A couple prefers to go to the same entertainment venue but one prefers opera and the other prefers basketball. They are both better off going to the same event than going their separate ways, but they do care which event is chosen. Here first mover advantage or bargaining power, not just agreement on the same outcome, matters. Even though coordination problems have distributional consequences, enforcement mechanisms are generally not necessary. Once a focal point has been chosen no one actor has a short-term incentive to defect. However, an actor might defect if that actor believed that it would be possible to choose new focal point, one that would be more favourable to that actor's interests over the long term.

Cooperation as opposed to coordination problems require more ambitious governance structures. Cooperation problems arise when individual and collective incentives cannot be aligned without adequate adjudication and enforcement mechanisms. Shared principles and rules are not enough. Cooperation problems occur because some actors will be better off in the short run if they can violate a rule without being identified and punished. However, if all parties then violate the rule everyone is worse off. Cooperation problems arise because of market failures, situations where each player acting to maximise his or her own self-interest without regard for the behaviour of others, adopts behaviours that lead to Pareto sub-optimal outcomes. Prisoners' dilemma is the most well-known example of such a

[1] See Keohane R. (1986). *Neorealism and Its Critics*. Columbia University Press: Chapter 6.

situation. In this classic game the two prisoners both have an incentive to talk regardless of what the other player does but if they both talk their prison sentences will be higher than if they both kept silent. The problem here is that there is no way for one prisoner to guarantee that the other will not talk. In international trade, for instance, if one actor is able to adopt an optimal tariff then it will be better off and everyone else will be worse off, but if all actors sequentially adopt optimal tariffs, trade barriers will spiral upward and all actors will be worse off. The World Trade Organization is designed to deal with the temptation of each single country to defect from agreed upon principles and rules, adjudicating disagreements about the rules, and punishing defectors.

Public goods offer another example of a cooperation problem. Public goods are non-excludable and non-rivalrous. Defence, for instance, is a public good. Once security is provided no one can be excluded and the provision of security to one actor does not lessen the amount available for others. Security, however, will not be provided voluntarily. Every single actor prefers that others pay for defence while it pays nothing. Because defence is non-excludable the entity that does not pay gets security without paying anything. But if no one pays, everyone is worse off because there will be no security. To provide public goods some authoritative actor is needed (a government) that can tax individuals and pay for public goods provision.

For cooperation problems it is usually necessary not only to have a set of rules that all actors agree to, but also a mechanism for enforcing these rules. Without such a mechanism each individual actor will be tempted to defect, because defection will leave it better off, especially if other actors do not retaliate. Without mechanisms for identifying and enforcing violations of the rules each actor would have an incentive to cheat but if everyone cheats everyone will be worse off than they would have been if the rules had been adhered to. (For some market failures governance structures might not be necessary if certain very specific conditions are met. For prisoners' dilemma these conditions include repeated play and an uncertain endpoint, but such conditions are rare in the international environment where, without an elaborated regime, actors may disagree about what constitutes cheating or how it should be punished.)

The issues associated with space do not fall neatly into some one category: harmony, coordination, or cooperation. The space regime has, however, worked well up to the present because most space issues are characterised by harmony, where every state can do as it likes without affecting others, or coordination where states need to agree on a focal point, but the focal point then becomes self-enforcing. There should, however, be no expectation that this relatively benign environment will continue indefinitely.

3.2 The Durability of the Space Regime

Although the United Nations General Assembly has adopted a number of non-binding resolutions over the last several decades articulating principles for outer space, such as the resolution dealing with remote sensing in 1986, the formal international treaties related to space, the "hard law" of space, have remained unchanged for more than 40 years. The international regime for space, the rules and norms for space exploration, liability, registration, militarisation, weaponisation and other activities were originally formulated in the 1960s and 1970s. They were embodied in five international treaties related to space concluded between 1967 and 1979: the 1967 Treaty on Principles Governing the Activities of States in the Exploration and Use of Outer Space, including the Moon and Other Celestial Bodies (the "Outer Space Treaty"); the 1968 Agreement on the Rescue of Astronauts, the Return of Astronauts and the Return of Objects Launched into Outer Space (the "Rescue Agreement"); the 1972 Convention on International Liability for Damage Caused by Space Objects (the "Liability Convention"); the 1975 Convention on Registration of Objects Launched into Outer Space (the "Registration Convention"); and the 1979 Agreement Governing the Activities of States on the Moon and Other Celestial Bodies (the "Moon Treaty"). In addition, the 1963 Treaty Banning Nuclear Weapon Tests in the Atmosphere, in Outer Space, and Under Water ("Partial Test Ban Treaty") banned the testing of nuclear weapons in outer space. The Outer Space Treaty, which has been signed by more than 100 countries, is the core formal document governing outer space. The Moon Treaty, which would give authority over celestial bodies and their orbits to the international community, has essentially been rejected, having been ratified by less than twenty states, all of which remarkably are not spacefaring nations. A number of additional agreements have been signed by a limited number of states, such as the agreement governing the International Space Station (ISS), which was concluded in 1998. Forums for formal and informal interaction have also been created and a variety of guidelines and standards of behaviours introduced. On the whole, however, the formal international law governing space has remained more or less unchanged for decades.

At the same time the number of countries, the array of activities, and the kinds of entities launching vehicles into space has grown dramatically. Over the last 50 years the number of countries with the capability to put a satellite into orbit has risen from two (the United States and the USSR) to about twenty (including the members of the European space consortium). The Soviet Union launched its first satellite in 1958 followed a year later by the United States. France, China, the U.K., the European Space Agency (ESA), and Japan sent objects into space in the 1960s and 1970s; India, Israel, Iran, and North Korea since 1980. There are now more than 1000 active satellites in orbit and even more inactive ones. Space is no longer just a rich country's game. India has launched more than 50 satellites even though it was cut off from cooperation with most other countries for much of the last 40 years because of sanctions imposed following its nuclear tests. An Indian satellite orbited Mars in September 2014 joining 21 other successful Mars missions from the United States, Russia and the ESA.

The range of activities in space has exploded. The focus has moved from demonstrating national pride and power, notably the first successful manned orbital flight by Yuri Gagarin, the Soviet cosmonaut, in 1961, and the American manned moon missions from 1969 to 1972, to undertakings that affect the daily activities of billions of people. Indeed, all the major space faring nations operate earth observation and communication satellites for both civil and military purposes, as well as space science and robotic exploration missions. Moreover, the United States, Europe, Russia, and China have, or are developing their own global navigation systems.

While the most dramatic manned spaceflight, the American project to land a man on the moon, took place more than four decades ago, other endeavours involving human beings in space, most notably the ISS, have been ongoing. The Chinese government is building its own space station and has set at least a rhetorical marker to send a man (or perhaps a woman if it is clever) to the moon, a reflection of China's ongoing efforts to define itself as one of the leading powers, perhaps in the long run the leading power, in the world. Chinese taikonauts returning from a mission on the surface of an extra-terrestrial body would be a dramatic symbol of China's status on Earth.

At the beginning of the space age all activities were carried out by states. Over time commercial actors, private firms, have grown in importance. In the process, the distinction between military and non-military uses has become increasingly blurred; space technology being inherently dual-use in nature. Financing for space activities now comes not only from states but also from a wide variety of non-governmental sources including venture capitalists, very wealthy individuals, private corporations, even crowd sourcing such as Kickstarter. Private actors have aspired to a wide range of activities including communications, remote sensing, mining, 3D printing, even tourism.

This surge of activity has not been devoid of conflict, but the most striking observation about the international regime for space is how durable the principles, norms, and rules have been despite having mostly been formulated several decades ago when most of the activities that are taking place in space now were only a glimmer in the eye of a few futurists.

3.3 Antarctica and the Oceans

The two issue areas that appear to be most closely related to outer space are Antarctica and the sea and seabed. The international regimes associated with both are well-elaborated and widely recognised and understood, but neither provides a definitive resolution for outstanding issues. The international agreement for the Antarctic essentially puts on hold the single most contentious issue, sovereignty. The United Nations Convention for the Law of the Seas (UNCLOS) has still not been ratified by the United States and there are conflicts between the UN Treaty and American law with regard to the exploitation of minerals in the international sea.

There is, however, a fundamental difference between these two issue areas and outer space. Both the Antarctic and the sea and seabed are strategic environments in which the behaviour of one actor can be disrupted by others, often many others. There would be Pareto sub-optimal outcomes absent rules that coordinate behaviour. The Antarctic Treaty, for instance, simply elided around the fundamental issue of sovereign control. While clearly banning military activity and forbidding the commercial exploitation of resources, Article 5 of the Treaty states that: "Nothing contained in the present treaty shall be interpreted as: (a) a renunciation by any Contracting Party of previously asserted rights of or claims to territorial sovereignty in Antarctica..." Had one of the seven states with sovereignty claims in the Antarctic tried to physically prevent another country from engaging in, for instance, scientific activity within an area that it claimed, there could easily have been conflict. Anyone of the seven claimant states could engage in military activity in the Antarctic, and other countries like the United States and the Soviet Russia could as well. The benefits that could be derived from any such clash would surely outweigh the costs.

Likewise, for the oceans shared rules were necessary to prevent clashes that would have led to sub-optimal outcomes. Many actors, including many states, have the capacity to impede or interfere with the activities of others in the Exclusive Economic Zone (EEZ) and the open ocean. (The territorial sea, within twelve miles of land, is even easier for most states to control.) Fishing vessels can be captured. Oil rigs can be blown up. The ownership of resources extracted from below the seabed floor in the EEZ, such as oil, could be challenged in court absent some general agreement governing such activities. The United States has not ratified the UNCLOS but the rules are, with the exception of deep seabed mining, consistent with American interests. With the most powerful blue water navy in the world and the longest coast line the United States benefitted more than any other state from the creation of the exclusive economic zone.

The Law of the Seas Treaty has not, however, fully resolved all issues. Some tensions are evident. The United States and China disagree about the kinds of activities that can take place in the EEZ. The EEZ, which gives states control over commercial activities in the area between the territorial sea and generally 200 nautical miles, but not over the movement of vessels, is a complete innovation in international law. The United States has embraced this concept not only because it has the largest EEZ in the world but also because, at least according to American interpretations, the Treaty provides full freedom for naval ships and military planes within the EEZ. The Chinese, however, have challenged this interpretation, arguing that the coastal state has the right to govern certain kinds of activities, such as naval reconnaissance in the EEZ. This has already led to a number of uncomfortable clashes in the waters around China. This question, about whether or not there is total freedom of flight and navigation in the EEZ, will ultimately be settled within the shadow of power cast by the relative military strengths of the United States and China.[2]

There is also uncertainty about the exploitation of resources, primarily deep seabed nodules from beneath the international zone. The UNCLOS declares that the international zone, the area that is not part of the EEZ of any country is the

[2] See Dutton P. (ed.) (2009). *Military Activities in the EEZ: A U.S.-China Dialogue on Security and International Law in the Maritime Commons*. China Maritime Studies Institute U.S. Naval War College Newport, Rhode Island, for a discussion of these issues.

"common heritage of mankind", an expression also used in the Moon Treaty. The UNCLOS provides for an international regime that is governed by the International Seabed Authority located in Jamaica. Under the provisions of UNCLOS the International Seabed Authority is empowered to issue licenses for the exploration and exploitation of deep seabed hard mineral resources. The Authority has signed contracts with at least 15 "contractors." These contractors or companies have been given licenses to explore 150,000 square kilometres each. As part of the contract, they must relinquish half of this area after a fixed period of time. They are also required to train individuals from developing countries.[3] Successful contractors would pay royalties to the International Seabed Authority beginning at one percent of the value of exploited minerals in year six and increasing at one percent a year to a maximum of seven percent. The International Seabed Authority would distribute these revenues. Presumably most would go to developing countries.[4]

Despite recommendations from several administrations, the U.S. Senate has not ratified the UNCLOS. The United States has enacted its own law governing the exploitation of hard minerals in the international zone, the Deep Seabed Hard Mineral Resources Act. This act does not require licensing from the International Seabed Authority. It does not require payments to the Authority. If the U.S. does not ratify, and if the exploitation of deep seabed hard minerals becomes economically viable, there will be a direct clash between the United States and the Seabed Authority, which has been joined by about 160 states. Would other states be willing to attack the operations of a firm operating under American law in the open sea? Absent some more general conflict, this seems unlikely. Would another state be willing to bring suit against minerals or products produced by such a firm if these minerals or products came under its jurisdiction? Certainly possible. How such disputes will ultimately be resolved, if they occur will, like the disagreement between China and the United States over naval activities in the EEZ, be determined by national power and interests.

3.4 Common Aversions and Common Interests in Space

Strategic interaction in space is more limited and power asymmetries are greater than in either the oceans or Antarctica. This situation has generally enabled the more powerful states to secure their first best outcome through unilateral action. Consequently, the incentives to establish and maintain international regimes have been lower and, not surprisingly, the institutional arrangements that have been

---

[3] International Seabed Authority (2015). Overview. http://www.isa.org.jm/deep-seabed-minerals-contractors/overview. Accessed 4 June 2015.

[4] Perry J. (2012). Deep Seabed Mining Emerges From the Depths. King & Spalding Energy Newsletter. http://www.kslaw.com/library/newsletters/EnergyNewsletter/2012/December/article6.html. Accessed 5 June 2015.

agreed to in the basic space treaties have sanctioned the freedom of use and access to the space environment for civil, commercial and even military activities. Where restrictions have been accepted, these have not, at least to date, imposed much constraint upon the actors.

For instance, according to the Outer Space Treaty states cannot assert sovereignty over bodies in outer space. If eschewing sovereignty claims has an impact on any state it will be on the big, rich, and powerful not the poor and weak. Given, however, the present state of technology, this concession is costless. The Space Treaty also bans placing weapons of mass destruction in space. This provision, like denying sovereign claims in space, will only affect the rich and powerful but also, like the denial or sovereignty, the acceptance of this provision has at least to date been costless. Weapons of mass destruction in space would not alter nuclear deterrence, which rests on the existence of platforms, especially submarines, which cannot be tracked with high levels of confidence. A weapons race in space would be costly and would not alter the relative position of the competitors.

For many activities related to space, individual states can basically do what they want without damaging the interests of others. There is a harmony of interests. Still, there are a number of situations in which self-regarding behaviour would not be conducive to Pareto optimal outcomes. In such situations, as explained below, the creation of regimes has become consequential. Institutional arrangements were generally necessary to provide solutions to dilemmas of common aversions and prevent the collective sub-optimality resulting from individual decision-making. The regime has ensured coordination among actors by establishing a focal point from which no actor has had an incentive to defect. However, the existence of different preferences over the coordination points along the Pareto frontier has inevitably triggered distributional conflicts, the resolution of which has been generally determined by the relative power capabilities of the actors involved. By contrast, cooperation regimes, which are generated by dilemmas of common interests, have to date been more limited.

3.4.1 Allocation of Radio Spectrum

The electromagnetic spectrum is a fundamental resource for the conduct of space activities. Radio frequencies are utilised for both the uplink of commands to satellites and the downlink of data and services, including the transmission of broadcasts, positioning services, imagery and scientific data.

Since the very beginning of the space age, there has been a clear understanding that the utilisation of the spectrum resources involves a classic coordination problem.[5] Purely unilateral action would have left all actors worse off. Absent a general

[5] It was also apparent that coordination should be adjusted on a continuous basis in order to respond to relevant technological change in the use of the radio spectrum.

agreement on the frequencies' utilisation, the various signals could easily interfere with each other and cause a temporary or even permanent disruption of a costly service in addition to loss of data. This common aversion to radio interference has forced actors to eschew independent decision-making and coordinate their behaviours. Coordination has been accomplished through the International Telecommunication Union (ITU), the world's oldest international organisation. The ITU sponsors world radio communications conferences (WRC), which are held every three or four years and entrusted with the task of managing frequency allocation for both space and non-space services.

In creating a system for assigning frequencies within radio frequency bands allocated to space, however, the WRCs have been faced with questions of distribution in addition to those of coordination. Since the radio spectrum is a finite resource, states cannot be indifferent to the specific set of rules upon which the partition is based, because different rules would privilege some at the expense of others. Therefore, even though states agreed on the need to avoid a mutually undesired outcome, they disagreed on their individually optimal outcome. In such a situation, the configuration of interest can be characterised as a Battle of the Sexes payoff matrix. The core issue was to determine on which focal point along the Pareto frontier actors would converge.

Different focal points could be chosen. One possible alternative was to develop a comprehensive plan for the radio spectrum where frequencies within defined frequency bands would be assigned on the basis of sovereign equality. Another alternative was to assign frequencies within the bands on a first-come-first-served set of rules, thus through an a-posteriori scheme based on actual usage. The eventual prevalence of the latter approach was essentially determined by the stronger bargaining power of the technologically advanced nations, and the predictable outcome was the progressive monopolisation of the majority of available frequencies.

Over the years, the pressure exercised by developing countries in the ITU has, however, stimulated a degree of dynamism within the regime: by the late 1970s the concept of equitable access to the spectrum/orbit resources was acknowledged by the ITU, while by the mid-1980s an a-priori allotment scheme was eventually elaborated/applied for certain frequency bands like the Ku and X.

The elaboration of this mixed utilisation rules certainly contributed to ease the fears of developing countries and decrease the potential for more acute conflicts. Distributional conflicts, however, have been also mitigated by the introduction of technological improvements for accessing the spectrum resources as well as by the gradual expansion of the portions of the spectrum frequencies allocated to space-related uses.

In this regard, it is of interest to note that coordination problems involving distributional conflicts have also characterised the negotiations for the partitioning of the spectrum between space and non-space uses. Given that interference can also derive from space systems transmitting on the same or similar radio frequencies to terrestrial systems, there is a need to continuously determine what part of the spectrum is to be allocated to a specific use. However, any type of partition would inevitably introduce a *de-facto* prioritisation and privilege some types of activities instead of others.[6] The question here is thus what type of activity should be given priority.

When the space-related uses of the spectrum were brought into the radio regime, they initially had to give way to terrestrial uses in case of frequency interference even if the space based use had started earlier. This changed with the creation of a dedicated "space services" category in the ITU Table of Allocations, but only a small percentage of the then available frequencies was assigned for this type of activities at the 1963 WRC. The successive expansion became a reflection of the changing balance between space and non-space uses that resulted from the rapid development of satellite technology and the ever-increasing demands for accessing the spectrum resources for space services. What is interesting about the dynamics within this Battle of Sexes payoff matrix is the complicated nature in the contention over possible focal points: the divide was not between developed and developing countries; rather between space users of different countries on the one hand and terrestrial users of different countries on the other.

The allocation of the spectrum resources has thus always been subject to iterated negotiations and frequent adjustments that continue to put pressure on the WRCs, but in light of the dramatic increase in the use of the radio spectrum over the last hundred years, including most recently satellite broadcasting and cellular phones, there is no doubt that the WRC has been an amazingly successful coordination mechanism. This success has been secured by multiple elements, including the progressive expansion of the accessible spectrum and the choice to approach issues on technical rather than political grounds. The common ground was the shared aversion to signal interference, which forced convergence towards a focal point from which no actor had an incentive to defect.

There is, however, one visible major exception to this common aversion to radio interference that is not fully resolved by the present coordination regime: jamming. During the Cold War there was disagreement between the United States and the Soviet Union on the issue with the Americans asserting the right to broadcast freely and the Soviets asserting their right to jam. The relative success of both sides depended on the technologies that they could deploy and, to a lesser extent for Communist regimes, the demands of their own populations. The US and its western allies found ways to frustrate some Soviet efforts at jamming. Broadcasts from West Germany were so popular in East Germany that the GDR eventually backed away from jamming them.

The ITU Convention now clearly bans jamming. There is general agreement that jamming constitutes a harmful interference, and as such it is in violation with article 9 of the Outer Space Treaty and Article 19 of the UN Charter.[7] Enforcement

[6] Hulsroj, P. (2002). Beyond Global: the international imperative of space. Space Policy Vol. 18 (2): pp. 107–116.

[7] The article affirms "the freedom to seek, receive, and impart information and ideas through any media and regardless of frontiers".

mechanisms are, however, weak. When a complaint about jamming is brought to the ITU the complainant must identify the location of the jamming signal. Although the attribution of a jamming source to a territory is a relatively easy task, the ITU itself is not empowered to verify or reject this information and thus to take a decisive action.[8] In recent years a number of countries have engaged in jamming including Iran, Bahrain, China, Syria, Ethiopia, and North Korea, but no resolving measure has been effected.

Thus with regard to the allocation of the radio spectrum, coordination has worked up to a point. For the most part coordination has been very successful, accommodating the dramatic increase in demand brought by new technological developments worldwide. There is, however, no effective instrument to deal with the issue of jamming; and this is perhaps inevitable, since the foundation of the regime is a shared aversion to radio interference that does not require enforcement mechanisms and adjudication procedures. With the development of satellite television broadcasting and space-based internet services, the global dissemination of information is now much more compelling than it was in the era of radio when Voice of America and the BBC were committed to penetrating the communist bloc. However, without more elaborated governance structures, governments that want to control the information available to their own citizens are not likely to abandon efforts to jam signals that they regard as politically threatening.

3.4.2 *Utilisation of the Geostationary Orbit*

Like the allocation of the radio spectrum, and directly connected to it, the utilisation of the geostationary orbit provides another example of a resource management regime, in which the potential for conflict has been to date mitigated by coordination, availability and technological change.

The geosynchronous or geostationary earth orbit (GEO) provides the most valuable spaces in which to park a satellite. In GEO the satellite remains stationary in relation to the earth's surface, thus keeping a constant view of the area beneath it. For broadcasting and some other activities this is a big advantage. For one thing, satellite antennas on the ground can be pointed at a fixed location to transmit and receive data. In addition, only a limited number of satellites is needed to provide global coverage. Not surprisingly, the number of GEO satellites has grown dramatically over the last 40 years: from in five in 1972, to 138 in 1984, to 458 in 2014.

Unlike other earth orbits, GEO is a single and well-defined environment having a finite number of parking space and therefore potentially subject to an over-

[8] De Rosen M. (2014). Orbital slots and spectrum use in an era of interference. Interference and freedom of information. Secure World Foundation. http://swfound.org/media/178208/Michel%20de%20Rosen%20Discours%20ifri%209%20octobre%202014_EN%20prononce.pdf. Accessed 10 June 2015.

crowding problem. Though the number of orbital slots is quite ample, the distribution in longitude is not uniform, as certain locations tend to be more crowded than others. For instance, the orbital locations above the Americas have more satellites than those above the Pacific. In utilizing the GEO resources, coordination is necessary because satellite-launching entities must know which spaces are already occupied before launching and whether their satellite would interfere with other systems operating nearby. However, given the finite number of parking spaces and the fact that countries located at the same longitude but at different latitudes might understandably want to gain access to the same orbital locations, how should the allocation be organised?

As for the partitioning of the radio spectrum, states were clearly averse to the absence of coordination, but they were not indifferent to the specific set of rules for coordinating their actions. Different rules—an *efficient* utilisation versus an *equitable* utilisation—would affect actors differently. And also in this case, spacefaring nations—the prime movers in the creation of a regime for GEO activities—eventually got the upper hand.

Since the 1963 WRC the ITU has been tasked to manage the GEO resources also in terms of orbital slots. The decision to associate the allocation of orbital slots with that of radio frequencies responded not only to technical exigencies, but also to clear political interests. Bringing the regulation of the orbit resources into the radio regime implied abiding by the first-come-first-served rules, and thus implemented a coordination mechanism favourable to the interests of spacefaring nations.

The nature of the institutional arrangements was thus also here a reflection of the relative power of the states involved in the bargaining process. However, if acceptance of the so-called a-posteriori allocation system could be easily accomplished in the 1960s, when the prospected use of GEO was still modest, the rapid development of satellite technology and the growing perception about the scarcity of orbital locations during the next decade led developing countries to challenge the rules of the game. In the Bogota Declaration of 1976 eight equatorial states claimed that they had sovereign control over the segments of the GEO above their respective territories and thus that their national authority extended beyond the air space all the way up to the GEO ring. In putting forward their claim, the states involved pointed out that no demarcation between air and outer space was ever agreed and that GEO should not be considered part of outer space but part of their national territory.[9] This claim was rejected by all the industrialised countries, including those of the communist blocs. The equatorial states had no way to enforce their claim. The declaration, however, was effective in in stimulating progressive change within the regime.

By the late 1970s, developing countries' demand for new focal points in the utilisation of the GEO entered more strongly in the agenda of the ITU, in which the one-nation, one-vote principle gave them some leverage to modify the procedures

[9] Tronchetti F. (2009). *The exploitation of natural resources of the moon and other celestial bodies: a proposal for a legal regime*. Martinus Nijhoff Publishers: p. 176.

for the allocation and assignment system. At the 1979 WRC those countries operating as a single voting bloc forced the conference to pledge new institutional arrangement.[10] These have been accepted and progressively incorporated into the ITU regime during the 1980s, leading ultimately to the formal entitlement of each ITU member to at least one orbital location for fixed satellite services. The assignment system has thus evolved and now ensures, a degree of equal access to the orbit resources, at least in principle. [11] The change that occurred within the regime was necessary to maintain stability and prevent distributional conflicts to become more intense.

To date major conflicts over the allocation of the orbit resources have been avoided because of an ample number of parking spaces. While orbital locations on GEO are a limited resource, there has always been room for additional users. In addition, technological improvements have through the years reduced the size of the slot into which each GEO satellite must be inserted, thus mitigating the perceived scarcity of the available resources. Complementary coordination mechanisms, such as the practice of placing satellites at the end of their operational life into the GEO graveyard—a super-synchronous orbit lying several hundred km above GEO—have also limited the risk of depletion found in other resource management regimes. However, were future demands to exceed the availability of orbital locations, the current coordination regimes could no longer be viable. New institutional arrangements would be required to prevent the likely scramble for the remaining resource and the ensuing pure conflict configuration of interests.

3.4.3 Satellite Navigation

Satellite navigation offers another example of space activities that have spread fast and where there have been incentives to establish and maintain an international regime. The first thing to note, however, is that navigation has not moved toward governance through a cooperation regime that would limit the freedom of action of individual states. It might have been more cost effective to develop a single Global Navigation Satellite System (GNSS) for the entire world, instead of deploying multiple systems potentially interfering with each other. This has, however, not

[10] The resolutions adopted at the 1979 WRC affirmed the equal rights of all states in the use of the GEO resources and called for a new WRC in 1985 specifically dedicated to the planning of space services using GEO. Ibid., p. 178.

[11] See 1982 ITU Convention Article 33.: "In using frequency bands for radio services, Members States shall bear in mind that radio frequencies and any associated orbits, including the geostationary-satellite orbit, are limited natural resources and that they must be used rationally, efficiently and economically, in conformity with the provisions of the Radio Regulations, so that countries or groups of countries may have equitable access to those orbits and frequencies, taking into account the special needs of the developing countries and the geographical situation of particular countries"

happened. The level of trust among major space faring nations is too low, especially since navigation is an area where considerations of strategic independence are the main driver and where the technical interconnections between civilian and military uses are very strong. Instead, the major actors in space have each developed or started to develop their own navigation systems. In addition to the GPS and the GLONASS systems, respectively maintained by the United States and Russia, two global systems are currently under development—the Galileo system of the European Union and the Chinese Beidou/COMPASS satellite system. Moreover, two Regional Navigation Satellite Systems (RNSS) are in the process of being completed—the Indian regional navigation satellite system (IRNSS) and the Japanese Quasi Zenith Satellite System (QZSS). It is expected that by 2020 these six systems will be fully operational.

Together, the systems will make use of more than 100 satellites operating in various MEO and GEO and generating a large number of radio signals (roughly 20 different signals) for both civil and military users. With this ongoing steep increase in positioning, navigation and timing (PNT) services, the complexities of the environment in which providers operate have inevitably grown; and so has the need to respond to the ensuing dilemmas of common aversions and interests. Dilemmas of common aversion are once again manifest in the need to ensure interference-free operations and signal integrity for each individual navigation service. In this regard, the ITU has defined a range of frequencies (mainly in the L band) to be used by the different navigation constellations. These are, however, quite narrow compared to the radio-frequency spectrum dedicated for communications. Distributional conflicts have thus been particularly intense, with prolonged disagreements between providers on the use of particular frequencies.

Dilemmas of common interest mainly involve compatibility and interoperability intended to enhance the utility and performance of services of each GNSS/RNSS provider. Specifically, interoperability refers to the ability of GNSS/RNSS to be used together to provide better capabilities (and reliability) at user level than those achieved by relying solely on the open signals of one system.[12] From a national security perspective compatibility is essential to ensure that military uses are not interfered with—clearly a shared goal, but one fraught with potential for disagreement. Interoperability is a common interest both in terms of the provision to citizens of a public good and in terms of enhancing military utility. However, mutual dependencies become a concern particularly for military uses when they start to rely on access to more than the own system.

Whereas the ITU provides a general framework for addressing compatibility issues stemming from the lack of sufficient spectral separation and unavoidable signal overlap, dedicated agreements between providers concerned have proved necessary to define mutually acceptable solutions to the whole range of issues.

[12] International Committee on Global Navigation Satellite Systems. (2009). Providers Forum Working Principles of Compatibility and Interoperability and their Further Definition. ICG/WGA/REC2/SEP2009

Several bilateral GNSS/RNSS cooperation agreements are in force. The United States and the European Union, for instance, signed an agreement in June 2004, which provides strong compatibility and interoperability between the signals emitted by GPS and Galileo at the non-military user level. Maximum compatibility and interoperability has been also achieved between the QZSS and GPS, and between Galileo and the QZSS. In some other cases (for instance between the Beidou and the Galileo systems), compatibility issues have not been fully resolved.

Especially for the leading GNSS providers, such as the US and Russia, bilateral negotiations represent the first best institutional venue, as it allows to maximise their bargaining power. However, the possibility to ensure compatibility and interoperability among providers generally exceeds the pure bilateral dimension, because of the presence of multiple systems using the same bands. When negotiating a compatibility agreement, providers must thus take also other systems into account. In such a context, the incentives to establish a framework for interaction and discussions among multi GNSS providers have progressively grown.

The International Committee on GNSS (ICG) was established in 2005 under the umbrella of the UN as an informal body to "promote voluntary cooperation on matters of mutual interest related to civil satellite-based PNT and value-added services".[13] The ICG has set-up four Working Groups (WGs) to deal with key issues regarding multi-GNSS: compatibility and interoperability (WG A), enhancement of performance of GNSS services (WG B), information dissemination and capacity building (WG C), reference frames, timing and applications (WG D). In addition, a Providers Forum has been established as a mechanism "to continue discussions on important issues that require focused inputs from system providers".[14]

All in all, these mechanisms have played a crucial role in bringing together all the different GNSS providers, in reaching consensus on general principles and agreeing on what needs to be done to ensure that as a minimum, all signals from navigation satellite systems are compatible and, to the maximum extent possible, also interoperable. It is obvious, however, that such mechanisms do not, per se, provide the focal point ensuring an optimal functioning of the GNSS/RNSSs. To do so, providers would also have to adopt a shared set of management and exploitation rules.[15]

[13] United Nations Office for Outer Space Affairs. (2015). International Committee on Global Navigation Satellite Systems (ICG). http://www.oosa.unvienna.org/oosa/SAP/gnss/icg.html. Accessed 10 June 2015. The stated mission of the ICG is to "encourage coordination among providers of global navigation satellite systems (GNSS), regional systems, and augmentations in order to ensure greater compatibility, interoperability, and transparency, and to promote the introduction and utilization of these services and their future enhancements, including in developing countries, through assistance, if necessary, with the integration into their infrastructures. The ICG also serves to assist GNSS users with their development plans and applications, by encouraging coordination and serving as a focal point for information exchange".

[14] United Nations Office for Outer Space Affairs. (2007). International Committee on Global Navigation Satellite Systems (ICG). http://www.unoosa.org/pdf/publications/icg_book01E.pdf. Accessed 10 June 2015.

[15] See Gibbon G. (2011). GNSS Interoperability. Not So Easy, After All. Inside GNSS. Vol. 6 (1): pp. 28–31.

In the absence of cooperation mechanisms it might not be possible to achieve the goal of using multiple navigation systems interchangeably, seamlessly and in an operationally optimum fashion. At the same time, formal cooperation would considerably reduce important prerogatives of the different providers, each of which has already laid out its own set of managing and exploitation rules. Formal cooperation mechanisms would thus to some extent run counter to some of the strategic imperatives which led to the establishment of independent systems in the first place. Not surprisingly, the possibility to construct a fully-fledged system of systems managed within a more authoritative cooperation scheme appears some way off.

3.4.4 Remote Sensing

In contrast to satellite communications and navigation, remote sensing is an issue-area in which the incentives to establish and maintain international regimes have been limited by the capability of the more powerful actors to obtain their first best outcome through unilateral action. As a result, most of the rules and decision-making processes that have emerged in the field of remote sensing are not constraining.

Apart from the general mechanisms established to minimise the chances of collisions and radio interferences, states have not confronted situations of potential market failure that might require eschewing independent decision-making. The more powerful states have been able to maximise they own utility by simply pursuing self-regarding behaviour. Therefore, there has been no need to coordinate policies by agreeing on a common set of principles and rules.

Indeed, remote sensing is an issue-area that has been for long characterised by disagreement about basic principles - the freedom of observation versus the need to ensure prior consent. While the advanced industrialised countries preferred rules to allow their satellites to sense wherever they pleased without prior approval from the sensed states, several socialist and developing countries championed a restrictive regime based on the principle of national sovereignty that would require prior approval before sensing takes place. In the end, a number of possible regimes were precluded. For instance, the proposal, advanced during the 1970s, to create an international regime in which the gathering and dissemination of all earth observation data would have been managed by an international organisation was simply rejected by industrialised countries.

The richer and more powerful states have prevailed not because all states agree on the rules implied by the principles of free use of outer space but because weaker states have not had the power to prevent remote sensing from taking place, or induce other states to endorse their view. Unlike the oceans or even the Antarctic where many actors might have the physical ability to disrupt activities that they might view as violations of international norms or even their own self-interest, in space only a small number of countries have the leverage to stop activities with which they might disagree. Moreover, even those with the power to stop such activities, such as China, would be deterred by the likelihood of retaliation from states whose activities they might impede.

In 1986 the United Nations General Assembly adopted, after 15 years of contentious negotiation, a resolution entitled Principles Relating to Remote Sensing of the Earth from Space. In an attempt to balance between different interests, the resolution endorses the principle that remote sensing shall not be conducted in a manner that is detrimental to the legitimate rights and interests of the sensed state. Principle II of the resolution also states that:

> Remote sensing activities shall be carried out for the benefit and in the interests of all countries, irrespective of their degree of economic, social or scientific and technological development, and taking into particular consideration the needs of the developing countries.

However there is no mandate for prior consent (nor prior consultation) in this resolution, which on the contrary acknowledges the freedom of observation from space.[16] States engaged in remote sensing agree to make information available on a non-discriminatory basis but there is no obligation to make information available at all to other countries. This general assembly resolution is soft law, not an international treaty, but even as soft law while it bows to principles advocated by the developing countries it does not require space faring states to actually do anything that they would not otherwise choose to do.

Policies and practice of governments confirm this interpretation. For one thing, access to data has been allowed on a selective basis (think the high-resolution satellite imagery). In addition, even where the principles have found application, this has been the result of unconstrained decision-making. Unlike the freedom of observation, which has become a well-established principle in international customary law, the principles of cooperation and non-discriminative access to information have not reached complete legal relevance: they remain procedural, not substantive rules.[17] At the same time, it is undeniable that the 1986 principles have provided a basis for convergent expectations and common initiatives.

The principles relating to environmental harm (Principle X and XI) have for instance provided the basis for the creation of the International Charter on "Space and Major Disaster". The Charter is a mechanism of coordination subscribed on a voluntary basis through which members agree to provide data from their satellites to states affected by natural or man-made disasters. Since its implementation in 2000, more than 400 activations in the interest of more than 100 countries have occurred. The Charter is an example of harmony or perhaps a minimal level of coordination without distributional consequences. Each country acting on its own, whether requesting or providing information, is better off.

---

[16] This is to be found in Principle IV, XI and XIII, which refer to a sensed state, not a state to be sensed.

[17] Marchisio S. (2004). The 1986 United Nations Principles on Remote Sensing: A critical assessment. In: *Scritti in onore di Gaetano Arangio- Ruiz*. Napoli: pp. 1311–1340.

With the proliferation of remote sensing missions and data, additional coordination mechanisms have through the years emerged in certain application-areas of remote sensing, such as global climate change and environmental monitoring. Here, the convergence in behaviour has become consequential because jointly accessible outcomes have been viewed more preferable than those that could be reached unilaterally. For instance, the creation of the Committee on Earth Observation Satellites (CEOS) stemmed from the recognition that no single program, agency or nation, could satisfy all the observational requirements which are necessary for improved understanding of the Earth System.[18] The objective of CEOS is to optimise the benefits of remote sensing through harmonisation of earth observation programs and maximum utilisation of their data. To do so it fosters the development of complementary and interoperable data products, information services and applications, which are accessed by the members of the Committee (14 as of 2015) on a reciprocity basis. In addition, since 2005 CEOS has been acting as the space arm of GEO, the intergovernmental Group on Earth Observations. GEO joins 97 member state and 87 participating organisations with the goal of integrating space-, air-, cryosphere-, oceans-, and land-based observation systems into a Global Earth Observation System of systems (GEOSS). Information derived from these observations is intended to improve understanding and support decision-making on issues related to sustainable development, resource and energy management, climate change forecasting, etc.[19]

While the initiatives undertaken by CEOS and GEO are certainly bound to grow in importance, they cannot be seen as the prelude for an elaborated cooperation regime acquiring control over the collection and distribution of earth observation data. Rather, these initiatives simply help to maximise actors' utility as well as that of the system as a whole through voluntary coordination, which does not however impose constraints nor create real dependencies among parties. All countries were better off, once they had agreed on some simple rules for providing and requesting information, although providing and requesting countries operate with different sets of rules.

3.4.5 Orbital Congestion: Collision Avoidance and Debris Mitigation

With the dramatic increase in space activities, the need to deal with the orbital congestion problem, and the danger of collision that it poses to the safety of satellite operations, has become pressing.

---

[18] Committee on Earth Observation Satellites. (2008). *The Earth Observation Handbook—Climate Change Special Edition 2008*. ESA Communication Production Office: pp. 21–25.

[19] Group on Earth Observations. GEOSS. http://www.earthobservations.org/geoss.php. Accessed 10 June 2015.

There are currently more than 1200 active satellite in Earth orbit and some 22,000 other tracked objects measuring more than 10 cm in diameter. In addition, there are an estimated 500,000 additional pieces of debris measuring 1–10 cm. Space debris arises from several sources, including upper stages of launch vehicles, payloads carriers and non-operational satellites. Deliberate destructions of satellites, most notably ASAT tests, have also contributed to increase the debris population dramatically.

As the number of objects in space continues to increase, so do the possibilities for collisions. In 2009 an American and a Russian satellite did collide, generating a large amount of debris that now pose collision risk to other satellites along their orbit. Even paint chips, which are traveling at thousands of miles an hour, might cause serious damage and possibly cause the loss of critical functions or even the termination of a mission. Future debris-on-debris and debris-on-active-satellite collisions may progressively generate a "collisional cascading" process that could dramatically increase risks and cost of space operations or even discourage investments in space activities.[20]

Against this background, coordination mechanisms have been primarily established with regard to the provision of Space Situational Awareness services (Earth orbit monitoring, conjunction assessments and information distribution) and the elaboration of internationally agreed debris mitigation standards and guidelines. The coordination regime for SSA sharing is, however, manifestly fragile, as it is essentially maintained by the United States. Several countries possess orbital tracking capabilities, but the most accurate and comprehensive information is provided by the Joint Space Operations Center (JSpOC) of the United States military's Strategic Command. Using the Space Surveillance Network (which includes both civilian and military arrays around the world and in orbit), the Joint Space Center can make about 400,000 observations a day. The Center maintains a catalogue of space objects, which is in part publicly available, to provide a number of services, most notably conjunction assessment warnings. If it detects a possible collision it will notify the affected satellite owner-operator along with the probability of collisions. These operators can then, if they have the capability, instruct their satellites to take evasive action. (Not all satellites have this capability, also due to the astonishing growth in the number of small satellites). Since the Iridium-Cosmos collision, the United States has been informing all countries of potential collisions with hundreds of warnings each year. There is no formal treaty, but maintaining a regime for public safety in space is in the United States' own best interest. Debris pose a threats to all satellites in orbit without discrimination, but considering that the United States is the country most heavily reliant on space capabilities, potential collisions resulting in creation of new debris would impact the environment in which the United States is the primary operator.

[20] See Weeden B. (2014). Space Traffic Management: Preventing a Real Life Gravity. Hearing of the Committee on Science, Space, and Technology. U.S. House of Representatives. http://science.edgeboss.net/sst2014/documents/5.9.14_charter.pdf. Accessed 10 June 2015.

This, however, cannot be taken to mean that the regime represents the most effective solution. There are relevant shortcomings, most notably with regard to the accuracy and reliability of the service,[21] which could be more effectively addressed if the JSpOC had given access to the tracking and trajectory data of its High Accuracy Catalog, instead of performing internal assessments for all operational satellites of the entire world.[22] This would have enabled satellite operators to perform their own conjunction assessments, relieve the United States of some burdens and increase the flexibility and responsiveness of this service. However, due to national security considerations, the United States has preferred withholding its most accurate SSA data and information, as well as the methods used for processing them; [23] a decision that has come with a cost for accuracy of this service.

Dissatisfaction over the current situation has already led to the emergence of complementary coordination mechanisms, the most important being the Space Data Association (SDA). Formed in 2009 by three leading commercial satellite operators (Inmarsat, Intelsat, and SES), SDA is a non-profit entity that provides participating operators with warnings of potential collisions. Its Space Data Center, located on the Isle of Man, gathers the information provided by each member satellite operator on the positions and other aspects of its satellites, process them and then delivers operators with operational data critical to safe and efficient operations. As at 2015, 27 satellite operators responsible for more than 360 operational satellites in Earth orbit are SDA members. It is likely that this and similar initiatives (like the International Scientific Optical Network) could in the medium term lead to a more elaborated mechanism where multiple data-providers coordinate to provide a more accurate and reliable public safety service for all operators in space.

Sharing information about space debris and warning of possible collisions is not the only mechanism established to cope with the congestion problem. Efforts have been also directed to the development of international standards and regulatory measures for reducing the creation of new orbital debris during the planning, manufacture and operation of space activities.

Voluntary guidelines for debris mitigation were first published in 2002 by the Inter-Agency Space Debris Coordination Committee (IADC), an international forum of governmental bodies established in 1993 for the coordination of activities

---

[21] Not all the locations are known with accuracy. In several cases the calculated orbital locations are incorrect, because the Center does not receive and process information about upcoming maneuvers of third satellite operators.

[22] Some of the more accurate data contained in the High Accuracy Catalog can be accessed through the SSA Sharing Program which is subscribed on a bilateral basis.

[23] The fragility of the system also stems from the lack of a clear common, mutually agreed approach to processing the information. Given that conjunction assessments are almost entirely based on predictive techniques, the methodology behind them is key to the accuracy or reliability of the information received. At international level, however, there are no agreed models for obtaining and processing information, or methods of interpretation and practical use of the information. See United Nations Committee on the Peaceful Uses of Outer Space (2014). Draft Report of the Working Group on the Long-term Sustainability of Outer Space Activities. United Nations General Assembly. A/AC.105/C.1/L.343.

related to the issues of natural and man-made debris in space.[24] These guidelines include the requirement to design space systems in such a fashion as to reduce the amount of debris released during their operation, the proper disposal of spacecraft and rocket stages after the end of their missions, and the development of methods to minimise the risk for post-mission break-ups resulting from stored energy.[25]

After five years of deliberations, most of these guidelines were included in a resolution proposed by the UN COPUOS and endorsed by the UNGA in December 2007. The resolution presents general recommendations in the form of seven principles to be implemented at national level through regulatory mechanisms.[26]

Even though they are voluntary and legally non-binding rules and procedures, it has taken a long time to achieve consensus on the specific content, because of the distributional consequences involved: different standards would have affected the utility of the different actors differently. As the guidelines have been eventually modelled on practice and standards of the major spacefaring nations, which were prime movers in the creation of the IADC, it is evident that the system serves their interests, as opposed to more intrusive regimes.

More importantly, the effective implementation of the guidelines is proving more difficult to achieve than their negotiations. This is because the agreed to focal points are not self-enforcing. On the contrary, there are clear incentives to defect, as any systematic application of these measures would impose a considerable burden on operators. In addition, the IADC guidelines simply indicate what shall be done, but not how to do it. Without institutional mechanisms for generating and monitoring information, actors may disagree about what constitute cheating and elude the adoption of specific and stringent domestic procedures and requirements for mitigating debris. Thus the guidelines remain far from being fully and effectively implemented by everyone. Like controlling global warming, the protection of the earth orbit environment pose a clear problem of collective goods, and it would thus require a cooperation rather than a simple coordination regime based on voluntary actions and self-regulation. The development of some enforcement mechanisms to define, monitor and inhibit cheating (e.g. cap and trade, debris tax, etc.) remains however unaddressed, and so does the possibility to take the more ambitious collective action that is required for the active removal of space debris.

[24] The stated purposes of the IADC are to exchange information on space debris research activities between member space agencies, facilitate opportunities for cooperation in space debris research, to review the progress of ongoing cooperative activities, and to identify debris mitigation options.

[25] See United Nations Office for Outer Space Affairs. (2007). IADC Debris Mitigation Guidelines. http://www.unoosa.org/pdf/spacelaw/sd/IADC-2002-01-IADC-Space_Debris-Guidelines-Revision1.pdf. Accessed 10 June 2015.

[26] The International Organization for Standardization has also endorsed a series of space debris mitigation requirements based on the IADC guidelines.

3.5 An Uncertain Future

The most telling sentence in the transmittal letter for the National Security Space Strategy issued by the United States in 2011 is the following: "Space, a domain that no nation owns but on which all rely, is becoming increasingly congested, contested, and competitive".[27] The document notes that there are 22,000 larger objects (the figure is now closer to 23,000 with a diameter of 10 cm or more) in space of which 1100 are active satellites. There are now almost 60 states or government consortia operating in space. The number of transponders for satellite communications will increase to 9000 by 2015, increasing the chance of radio frequency interference.[28]

The document outlines a basic American strategy for dealing with these challenges which is to bring other states that are willing to ally with the United States into a coalition that would be able to set rules and standards for activities in space. Partnering with like-minded states is given more prominence than working with international organisations, where one nation one vote rules often lead to clashes between developing and industrialised countries. The operative language in the document reads as follows:

> The United States will lead in building coalitions of like-minded space-faring nations and, where appropriate, work with international institutions to do so. With our allies, we will explore the development of combined space doctrine with principles, goals, and objectives that, in particular, endorse and enable the collaborative sharing of space capabilities in crisis and conflict. We will seek to expand mutually beneficial agreements with key partners to utilise existing and planned capabilities that can augment U.S. national security space capabilities.[29]

Rule setting through coalitions of the willing is the only way that the United States can square the circle. The United States wants, on the one hand, cooperation in space from which all countries might benefit, and, on the other, continued dominance in space to assure American national security. The tension between these two objectives is apparent in a number of different issue areas.

Private entities interested in private profits are playing an ever larger role in space. These companies are based in the OECD world. They are dependent for their success on scientific, legal, financial, organisational, and technological infrastructures that are available for the most part only in the developed world. SpaceX might be able to operate from India, Brazil, or even China but there is no chance that it will have major subsidiaries in Nigeria, Bolivia, or Burma in the near future. For the United States private entities will play a significant role in the future in space.

---

[27] US Department of Defense. (2011). National Security Space Strategy. Unclassified Summary. http://www.defense.gov/home/features/2011/0111_nsss/docs/NationalSecuritySpaceStrategyUnclassifiedSummary_Jan2011.pdf. Accessed 10 June 2015.

[28] Ibid., p. 2.

[29] Ibid., p. 9.

Private entities, however, do not necessarily operate in the common interests of mankind. They operate to make a profit for their own shareholders.

Satellite tracking and collision avoidance is now provided on a global basis by the United States military. The global surveillance system is expensive and technologically sophisticated. But the United States military is not the only entity engaging in such activity. In the future it is likely that states that distrust the U.S. will develop their own capabilities just as other entities have developed navigation systems of their own rather than relying solely on the American GPS system.

Despite dramatic changes in the global distribution of power and tensions between the United States and other countries, between industrialised and developing countries, between spacefaring nations and new entrants, the present regimes will not unravel any time soon. This is not because the fundamental principles associated with outer space have been embraced by nearly everyone. It is not because specific rules can be, with confidence, derived from general principles. It is not because all of the specific issues have been resolved. Rather it is because the extant principles, norms, rules, and decision-making procedures that have evolved in space have not inhibited states, and now private actors from pursuing their space related objectives. Many issues associated with space are characterised by harmony, states and other entities can do what they want without harming others. In those areas where unconstrained activities could damage the interests of others such as the use of radio frequencies and geostationary orbit, market failures have been resolved by relatively simple coordination regimes. These regimes are much less institutionally complex than cooperation regimes because coordination regimes are self-enforcing. Once a focal point or rule has been adopted, states do not have an interest in defecting because defection would make them worse off.

This, however, can neither be taken to mean that the present regime for space has eliminated potential for conflict nor that it can continue indefinitely. The currently agreed focal points do have distributional consequences privileging some actors at the expense of others. Dissatisfaction with the current coordination outcome could, for instance, trigger new controversies in the allocation of the spectrum/orbit resources and even cause departures from regime-governed behaviour in the absence of new institutional arrangements. Arguably, the more visible tension between the relatively unrestrained utilisation of radio frequencies and associated orbits sanctioned by the first-come-first-served rule and the ever-increasing scarcity of these resources could eventually generate the typical "dilemma of the commons" for which only a cooperative allocation mechanism (a comprehensive planning in the ITU parlance) would guarantee stability.

The general progression towards cooperation problems is even more apparent with respect to other governance issues. The debris problem offers one clear example. It involves a classic prisoner's dilemma payoff matrix in which each individual actor is better off if not contributing to the mitigation of debris (or in the future participating to active removal) but in which mutual defection will leave all actors worse off. Indeed, the equilibrium outcome resulting from each actor being a free rider would inevitably deteriorate the environment in which everyone operates and swiftly increase the threats to the safety of space operations. In order to ensure that conditions in outer space will be suitable for continued use, it would thus be necessary not only to have a set of rules that all actors agree to, but also a mechanism for enforcing these rules and monitor their compliance. Without such a mechanism each individual actor will be tempted to defect, and the sustainability of space activities ultimately endangered.

Similarly, other critical issues like the potential weaponisation of space could in the near future require the establishment of institutionalised mechanisms of cooperation. Current rules are assumed to not prohibit the placement of conventional weapons in space. Yet, it is hardly in anybody's interest to do so, unless others do so first. Only a cooperation regime, with proper enforcement mechanisms, would do. The guidelines and codes of conduct to which the international space community has recently turned, although laudable, are not true deterrents on placing arms in space. There must be clearly defined rules that specify what is legal, what constitutes cheating as well as mechanisms of verification and monitoring that make defection counter-productive. Yet, the high degree of institutionalisation that is required to ensure cooperation makes the viability and stability of such a regime questionable. Constructing international regimes to solve cooperation problems is inherently complex, but that does not mean that the international space community can continue to shirk from the challenge.

Chapter 4
Space Governance and Globalisation

Jacques Blamont

4.1 Status of Space at the Beginning of the Twenty-First Century

Space is now established as a steady and stable, if evolutive, activity. It is estimated that 1 150 satellites will be launched from 2013 to 2022, against 810 during the previous decade (2003–2012), with two thirds as government-paid missions. Let us say 100 per year, providing total revenue of 248 Gigadollars.[1] If the 10 countries possessing a space industry will receive 90 % of the market of government satellites, 35 emerging countries will represent a market of 2 Gigadollars per year. About 25 geostationary commercial satellites are launched per year (24 in 2013, with new players: Azerbaijan, Bolivia, Qatar). The other commercial satellites are devoted to observation from helio-synchronous orbits. In 2013, they were at least 8 of them launched. Including minisats, the 78 successful launches of 2013 have placed in orbit 207 payloads (among which 82 cubesats from Austria, Ecuador, Estonia, Peru, etc.).

Launches in 2013 (with 78 successes):

[1] Euroconsult. *Satellites to be built and launched*, July 2014.

J. Blamont (✉)
CNES, 2 Place Maurice Quentin, 75039 Paris, France
e-mail: Jacques.blamont@cnes.fr

C. Al-Ekabi et al. (eds.), *Yearbook on Space Policy 2014*, Yearbook on Space Policy, DOI 10.1007/978-3-7091-1899-3_4

Russia	32
USA	20
China	15
Europe	7
Japan	3
India	3
South Korea	1
Total	**81**

Non GTO commercial satellites generate 1 Gigadollar recipes per year. During the next decade, 350 commercial satellites will be launched, 83 % of which in GTO. For human space flight launches, the year 2013 has seen 13 missions, including 12 to ISS and 1 to the Chinese Tiangong. Built on borrowed Russian technology, the Chinese man-in-space programme accelerates towards, after a station in LEO, missions to the Moon. For interplanetary missions, the year 2013 has seen six launches. In conclusion, diversity of applications and therefore of missions, has become the characteristic of space, with an increasing number of players.

As a consequence, space hardware has diversified its origin: American launchers rely on Russian engines, astronauts of various nations, including the United States, are ferried to the International Space Station by Russian rockets, India utilises Ariane motors, the Chinese human in space mission is built on redesigned Russian equipment ... The launch is today considered as an ordinary merchandise without strategic connotation. Satellites of all nations are placed in orbit by any rocket provided by any other nation (with some exceptions). U.S. military missions are launched by ILS Protons from Baïkonour.

Agencies have to manage this diversity of missions and players. Let us examine first its impact on the status of the most prominent of the Space Agencies, NASA. Since major applications of space are now driven by outside Agencies, by operators and industries, we will concentrate the analysis on man-in-space and space science missions.

4.2 The U.S. Space Policy

4.2.1 Governance by Politicians[2]

NASA was created in the midst of the Cold War as a multi-purpose agency to pursue goals in robotic and human spaceflight and to build on aeronautics technology development. The largest of NASA's many missions over the past half-century has been the pursuit of human spaceflight, and approximately half of the agency's current budget is devoted to this pursuit.[3]

[2] This section relies heavily on a number of NRC reports.

[3] NRC Report. *NASA's Strategic Direction*, 2012.

Human spaceflight goals have generally been established by presidential policy, subject to congressional authorisation and appropriations. The goal of human spaceflight has itself changed over the years. During the 1960s the ultimate purpose was geopolitical—to compete against the Soviet Union and demonstrate U.S. technological prowess on an international stage. During the 1970s the purpose was to reduce the cost of launching spacecraft to orbit and to develop routine operations for humans in space. By the 1980s, the goal had become to develop a space station with Western allies. By the 1990s this goal had evolved to include engagement with post-Cold War Russia. Since the early 1970s, human spaceflight has been confined to low Earth orbit.

By the dawn of the new millennium, there was no dominant rationale for the space programme, and in particular for the human spaceflight programme. The Colombia accident in February 2003 was the catalyst for a new space exploration policy (the Vision) that attempted to bring policy clarity. In 2004, after many years of uncertainty about the future of the space shuttle and the International Space Station, President George W. Bush announced a "Vision for Space Exploration" (NASA 2004).

President George W. Bush did not propose adding significant funding to NASA's budget. His plan was to terminate the space shuttle programme in 2010 after completing construction of the ISS and to end U.S. involvement in the ISS in the 2015–2016 timeframe. The space shuttle and ISS funds would be redirected to achieving the Moon/Mars goals to return to the Moon by 2020 and someday to go to Mars

In 2005, a Republican-controlled Congress passed the 2005 NASA Authorization Act, which supported President Bush's Moon/Mars programme. Three years later, a Democratic-controlled Congress passed the 2008 NASA Authorization Act, which was similar to the 2005 act. At that point in time, Congress and the White House and Democrats and Republicans were all in general agreement about the future of the human spaceflight programme. NASA pursued the presidential and congressional policies by initiating the Constellation programme, in order to build capabilities to send people back to the Moon and to Mars, including new launch vehicles and spacecraft.

In January 2009, President Barack Obama convened a special committee, chaired by Norman Augustine, to look at the human spaceflight programme and offer options. In its report, *Seeking a Human Space-flight Program Worthy of a Great Nation* (Executive Office of the President 2009), the committee concluded that in order for NASA to pursue a mission of sending humans beyond low Earth orbit (LEO), NASA required additional funding of $3 billion more per year.

As a consequence, in February 2010, the White House proposed terminating the Constellation programme and replacing it with a NASA effort to develop technologies and capabilities for human exploration beyond LEO. Neither specific destination nor timeframe for human expeditions beyond LEO was included.

Meanwhile, the President decided that instead of NASA developing a replacement capability for the space shuttle, NASA would build on its Commercial Orbital Transportation Services (COTS) partnership agreements with U.S. industry, initiated in 2006, with the expectation of the emergence of a private human spaceflight market.

Congress wanted a destination and a timetable for sending astronauts beyond LEO. In April 2010, the President announced his goals of sending astronauts to an asteroid by 2025 and to orbit Mars in the 2030s.

The totality of the decisions—to proceed with President Bush's plan to terminate the space shuttle, but also to end the Constellation programme that was developing a replacement U.S. crew transportation capability—resulted in programmatic disruptions. These decisions also resulted in an indefinite extension of the number of years the United States would need to depend on Russia to take NASA astronauts to and from the ISS. In addition, the decisions to rely on the commercial sector to build a new U.S. crew space transportation System, and to replace the Moon with an unspecified asteroid as the next destination for human spaceflight were met with congressional scepticism.

A number of influential members of Congress insisted that the government—NASA—build a new crew transportation System regardless of any commercial crew aspirations. Congress wanted a new large rocket reminiscent of the Saturn V used for the Apollo programme to enable trips beyond LEO, whatever the destination, and to accelerate as much as possible restoring U.S. ability to launch people into space rather than relying on Russia.

In October 2010, Congress and the White House reached a compromise in the 2010 NASA Authorization Act. In essence, the agreement was for NASA to do both what the White House and Congress wanted. NASA would proceed with the White House plan for commercial crew transport as well as Congress's plan for a NASA-developed Space Launch System (SLS), and an Orion spacecraft that would take humans beyond LEO.

The budget outlook for NASA, meanwhile, worsened. The President had planned to add $6 billion to NASA's budget over 5 years when he announced his new plan in the FY2011 budget request. A year later, with Republicans regaining control of the House and deficit reduction becoming the dominant political theme, NASA was hoping for level running at best. Today, the same NASA that was deemed by the Augustine committee to be unable to afford the Constellation programme now must fund Constellation's replacement and also fund commercial crew transport. NASA still must find funds for a habitation and support module to enable long-duration trips beyond LEO.

Because of the overall impact of the Space exploration programme on the budget of other missions, this chain of events shows that all decisions on the civil space programme depend only on the consideration of internal politics. The outside world is totally ignored ...

NASA is at a transitional point in its history and is facing a set of circumstances that it has not faced in combination before. The agency's budget, although level-funded in constant-year dollars, is under considerable stress, servicing increasingly expensive missions and a large, aging infrastructure established at the height of the Apollo programme. Other than the long-range goal of sending humans to Mars, there is no strong, compelling national vision for the human spaceflight programme, which is arguably the centrepiece of NASA's spectrum of mission areas. The lack of national consensus on NASA's most publicly visible mission, along with out-year budget uncertainty, has resulted in the lack of strategic focus necessary for national agencies operating in today's budgetary reality. As a result, NASA's distribution of resources may be out of sync with what it can achieve relative to what it has been asked to do.

There is no evidence that a current stated goal for NASA's human spaceflight programme-namely, to visit an asteroid by 2025—has been widely accepted as a compelling destination by NASA's own workforce, by the nation as a whole, or by the international community. This lack of national and international consensus on the asteroid-first mission scenario undermines NASA's ability to establish a comprehensive, consistent strategic direction that can guide programme planning and budget allocation.

The asteroid mission is ostensibly the first step toward an eventual human mission to Mars. This goal has been studied extensively by NASA and received rhetorical support from numerous U.S. Presidents, and has been echoed by some international space officials, but it has never received sufficient funding to advance beyond the rhetoric stage.

There also is no national consensus on what would constitute an appropriate mix of NASA's capability-driven and mission-driven programmes. While a capabilities-driven approach may be the most reasonable approach given budget realities, such an approach still has to be informed by a clear, consistent, and constant path to the objective.

Earth and space science goals are ostensibly established in the decadal survey process led by the NRC, a process that has been highly successful at developing priorities in Earth and space sciences and leading to their eventual implementation. This process has been under strain in recent years. Some projects, such as the James Webb Space Telescope, have run over budget, and the administration has rejected the proposed planetary science programme and also postponed work on a key element of the astronomy community's decadal survey.

However, NASA has clearly demonstrated the success of the strategic planning process for Earth and space science that is founded on the NRC's decadal surveys The decadal survey process has matured into a robust method for developing a set of goals and objectives for various programmes, that are based on a community consensus on an achievable suite of science programmes in pursuit of high-priority, compelling science questions. However, even the best strategic plan is vulnerable to severe changes in the assumptions that underlie its development, whether those changes are applied internally or externally. As an example, the recent set of surveys on astronomy and astrophysics (NRC 2010) and planetary science (NRC 2011) were based on budget projections provided to the relevant decadal committees, and now these projections exceed the current budget as well as current budget projections. Rising costs associated with increasingly complex missions, declining science budgets, international partnerships that fell apart, and mission cost overruns have strained science budgets to their breaking point. As a result, key decadal priorities in astrophysics, planetary science, and Earth science will not be pursued for many years, or not at all. The carefully crafted strategic planning process, with its priority setting and consensus building, which has led in the past to the United

States leading the world with science missions such as the Curiosity rover on the surface of Mars and the Hubble Space Telescope, is now in jeopardy because it no longer may lead to a tangible programme outcome.

The funding for NASA's total budget has been remarkably level in constant-year dollars for more than a decade. However, there has been instability at the programmatic level and the out-year projections in the President's budget are unreliable, which makes it difficult for programme managers to plan activities that require multi-year planning. Put another way, although the budget may have been level over time, NASA experienced substantial programme instability over the same period. Numerous times the agency initiated new programmes with the *expectation* that budgets would increase to support them (a basic requirement for optimising any development programme's budget), only to have no increases emerge.

NASA cannot execute a robust, balanced aeronautics and space programme given the current budget constraints. For example, major components needed for future human exploration (including important life sciences experiments on the ISS) are not currently in the budget; high-priority science missions (including robotic planetary exploration missions that are precursors to human exploration) identified in the most recent NRC decadal survey are unfunded; and aeronautics now accounts for only about 3 % of the total NASA budget. In addition, individual NASA centres are finding it necessary to selectively reduce their infrastructure or find alternative ways to support it (e.g., through external collaborations).

The ten centres kept operational by NASA as an Apollo legacy need 2 billion for urgent repairs and maintenance of facilities which are not budgeted.

In today's multipolar and increasingly interconnected world, the global competitiveness of the U.S. aerospace industry has been significantly hindered by what are generally acknowledged to be outdated and oftentimes counterproductive International Traffic in Arms Regulations.[4] Further, which countries the United States chooses to partner with in space is influenced by foreign policy considerations. National security considerations regularly prevail over desires for international scientific collaborations.

The United States can no longer base the foreign policy aspects of space activities on the assumption of being the first choice for global partnerships based on its unchallengeable dominance. They are no longer in a position to control the space technology that is available and is increasingly being developed in other nations. As a result, the United States is no longer the only, or in some cases even the best, option for countries interested in space partnerships. Rather, as with other aspects of globalisation, U.S. leadership in space activities must be based on the competitiveness of its capabilities; the attractiveness of its ingenuity, entrepreneurialism, and willingness to take risks; and its recognition of mutual interdependencies.

Confronted with this erosion of the so-called American leadership in space, the U.S. has failed to recognise the threat and to evolve a strategy.

[4] NRC Report. *America's future on Space*, 2009, p. 11.

Future U.S. leadership in space requires a foundation of sustained technology advances that can enable the development of more capable, reliable, and lower-cost spacecraft and launch vehicles to achieve space programme goals. A strong advanced technology development foundation is needed also to enhance technology readiness of new missions, mitigate their technological risks, improve the quality of cost estimates, and thereby contribute to better overall mission cost management.

The responsibility to provide for this advanced technology base for civil space activities rests with NASA, in partnership with universities, other government agencies, and industry. The "customers" for the products of technology are NASA, NOAA, industry, and military space programmes in which multiple-use technology is applicable. Because of budget pressures and institutional priorities, however, NASA has largely abandoned its role in supporting the broad portfolio of civil space applications, and the space technology base has thus been allowed to erode and is now deficient. The former NASA advanced technology development programme no longer exists. Most of what remained was moved to the Constellation Programme and has become oriented specifically to risk reduction supporting the ongoing internal development programme.

For the United States to be a leader in space, as required by the 1958 National Aeronautics and Space Act, it must be a country with bold ideas, science and engineering excellence, and the ability to convince others to work with it in the pursuit of common goals. Leadership depends on the perception of others that whoever is in the lead knows the way forward, is capable of forging the trail, and is determined to succeed despite inevitable setbacks.

Leadership is more nuanced today than during the Cold War rivalry with the Soviet Union over which country would achieve the next space "first." Countries that once depended on partnerships with the United States to execute their space programmes now have other choices, including going it alone. If the United States is to continue to maintain international leadership in space, it must have a steady, bold, scientifically justifiable space programme in which other countries want to participate, and, moreover, it must behave as a reliable partner.

Despite decades of U.S. leadership and technical accomplishment, many of these elements are missing today. Abrupt changes in the goals the United States is pursuing for human spaceflight, coupled with concerns about U.S. unreliability in key international partnerships, can erode this country's leadership position. The thrilling Mars Curiosity mission may be a testament to U.S. leadership in robotic space exploration today, but the sudden and dramatic proposed cut to the Mars exploration budget and withdrawal from the ExoMars programme with Europe cast doubt on the future. Human spaceflight capabilities historically have served as a symbol of a country's leadership in space. This multi-year period when the United States cannot launch humans into space, requiring reliance on Russia for access to the ISS, further undermines any claim to leadership despite the programmatic success of the development of the ISS, which is, in fact, led by the United States.

In conclusion, the challenge launched by the new players in space against U.S. leadership has not been met by the successive administrations with the tool provided by the federal institution NASA, which has been left without the building of a national consensus by a common effort of the two branches of the government; there is today no national consensus on strategic goals and objectives for NASA. Absent such a consensus, NASA cannot reasonably be expected to develop enduring strategic priorities for the purpose of resource allocation and planning. This is very different of the approach taken by DOD for military space, where adequate budget have been maintained, and the U.S. predominance in the military uses of space remains intact.

What is the explanation for this failure? Globalisation is not the cause since we have seen that considerations of internal U.S. politics are dominant in the decision-making process. The main reason comes from space itself: everybody is convinced that the man-in-space activities are both useless and extremely expensive. Their justification is only prestige, and it is a very weak motivation when huge sums are required. Going to the Moon? Done that, been there. Going to Mars? Technically impossible today. Then we go down to a job motivation, and then the fight starts between politicians and what is called in the U.S. "pork barrel" with unpredictable results.

Today it is common to declare among space Agencies managers that all future human spaceflight or large-scale Earth and space science projects will be international. Many U.S. leaders also assume that the United States will take the lead in such projects. However, American leadership in international space cooperation requires meeting several conditions. First, the United States has to have a programme that other countries want to participate in, and this is not always the case. Second, the United States has to be willing to concede substantial responsibility to its partners. In the past, the approach of the United States to international partnership has too often been perceived as being based on a programme conceived, planned, and directed by NASA. Third, other nations must be able to see something to gain—in other words, a reason to partner with the United States. Finally, the United States has yet to demonstrate its reliability and attractiveness as an international partner.

The capabilities and aspirations of other nations in space have changed dramatically since the early days of the space race between the Soviet Union and the United States. It is said in the U.S. that the most important success of the ISS was its international character and the role of the United States as the managing partner in a global enterprise. For the U.S. again, this success calls for adopting such a model for future ventures. If the United States does seek to pursue a human mission to Mars, such a mission will undoubtedly require the efforts and financial support of many nations.

4.2.2 Recourse Outside the National Space Agency

4.2.2.1 Recourse to Private Initiative

The policy of diminished support of the federal institutions for the implementation of space programmes contrasts sharply with the recourse to private initiative and resources for filling the gap.

Beginning with the Commercial Orbital Transportation Services (COTS) programme initiated by then-Administrator Michael Griffin, NASA has invested about $925 million of taxpayer seed money to help create a private human spaceflight industry. That money has attracted significant private investment, and the agency is optimistic that this can generate not only one or more new ways to get humans and cargo to the International Space Station (ISS) for less money than traditional government procurements, but also cheap commercial launches—NASA's commercial approach—to carrying astronauts to low Earth orbit could lower the cost of space access to the point that the business cases for new LEO applications can close.

The history of space launch is littered with paper rockets, concept vehicles that never got off the ground or never reached full-up operations. Conestoga, Aquila, Roton, K-l and Gryphon are among the also-rans of commercial spaceflight. Usually there wasn't enough money to keep them going long enough to iron out the inevitable technical problems and turn a profit.

That old equation has changed with the arrival of billionaires like Allen, Musk and Amazon.com founder Jeff Bezos, who is using some of his dot-com money to endow the secretive Blue Origin vertical-take-off-and-landing space launch vehicles. The addition of serious U.S. government funding for commercial access to the ISS also has spurred companies that must answer to shareholders to enter the fray.

The COTS programme is already paying off since Musk's Space Exploration Technologies Inc. (Space X) has succeeded getting its Dragon capsule off the ground with its Falcon 9 rocket and berthed at the ISS, carrying food and other relatively low-value cargo. Orbital Sciences Corp., NASA's other COTS partner, has also reached the station this year with its Cygnus cargo vehicle riding the company's new Antares liquid-fuelled launcher.

That could be just the beginning. With the space shuttle fleet retired, NASA is paying Russia more than $60 million a seat to train and fly astronauts to the ISS in Soyuz capsules. Starting with $50 million in federal economic stimulus package funding in fiscal 2009, NASA has been seeding development of commercial space taxis under its Commercial Crew Development (CCDev) and Commercial Crew Integrated Capability (CCICap) initiatives.

The latter has attracted several different proposals for full-scale systems that would ferry ISS crews to orbit beginning as early as 2017. Funding commercial crew is contentious in Washington, and the first human flights already have been delayed a year because Congress does not want NASA to put all of its eggs in the commercial crew basket.

At the heart of the commercial new thrust is a change in the basic approach to space, which translates into a real gain in price. NASA-backed commercial crew vehicles have already created a space-launch market which did not exist before.

4.2.2.2 Recourse to Foreign Hardware

The adoption of Russian hardware or launch systems has not had any influence on governance. The development of the Chinese space station from improved Russian equipment has not meant that Russia would introduce itself into the Chinese decision making process and the apparition of Soyuz on the slate of Arianespace launchers has had no influence on European space policy.

Amid heightened tensions with Russia, the Defense Department has been making preliminary preparations for life without the RD-180, which powers the first stage of United Launch Alliance's Atlas 5 rocket, one of two the company uses to launch military and other government satellites.

The request for information, posted to the Federal Business Opportunities website Aug. 20, is a step in that direction.

Companies are being asked to respond to questions. Among them: "What solution would you recommend to replace the capability currently provided by the RD-180 engine?"

In the request for information, the Air Force says it is open to a variety of options including an RD-180 facsimile, a new design and alternative configurations featuring multiple engines, and even a brand new rocket.

The Air Force is also trying to decide on the best acquisition approach. Options include a traditional acquisition or a shared investment as part of a public-private partnership.

The RD-180 is built by RSC Energomash of Russia and sold to Denver-based ULA by RD Amross, a joint venture of Energomash and United Technologies Corp. When that arrangement was set up during the 1990s, the idea was to eventually set up a U.S. production line for the hardware. However, the U.S. production line never materialised due in large part to cost considerations.

Congress, meanwhile, is preparing bills that would fund a full-scale engine development programme starting next year; the White House is advocating a more deliberate approach that begins with an examination of applicable technologies.

A 2015 defense authorisation bill drafted in the U.S. Senate would prohibit U.S. space companies from entering new contracts or renewing existing contracts for launch vehicle supplies from Russia.

4.3 The International Links in the Exploration Programmes

4.3.1 The Different Modes

Space exploration consists in extending the human presence far above the Earth surface and beyond the limits of our planet's gravitational field, either directly (human missions), or indirectly (robotic missions). These ventures are restricted to the places that could be visited by humans.

Because of its price and also because no interest other than science involved, exploration could be conceived as a domain where international programmes (compare with CERN) could be implemented. It is not the case. What is lacking is model of governance for space projects led by many nations together. There are three modes for such ventures: discussion, coordination and cooperation.

For the first mode, **discussion**, a good forum is offered by ISECG (International Space Exploration Coordination Group), a group where 14 Space Agencies are represented in order to exchange ideas on planetary exploration and draw tentative roadmaps (latest issue, August 2013).

The ISECG Global Exploration Roadmap creates a framework for interagency discussions in three areas: (1) common goals and objectives, (2) a long-range human exploration strategy, and (3) coordination of exploration preparatory activities. By understanding the elements common to their exploration goals and objectives, and by collaborating to examine potential long-range scenarios, Agencies continue to inform near-term decisions affecting their exploration preparation and to create opportunities for the partnerships which will realise future missions.

It is to be noted that China, a major Space power, has only become an active member of ISECG in 2013.

If ISECG succeeds in organising exchange of information, it does not provide a frame for progressing beyond that stage.

The second modes **Coordination** is more talked about than real:

- The ISS, essentially driven by NASA, has no relationship at all with the future Chinese station, essentially because of the ITAR philosophy which prohibits any relations between the western countries and China.
- Uncoordinated missions to the Moon and Mars are sent by newcomers in the space race, such as Japan, India and China, all trying not really to explore but to acquire the status of a technological power of first magnitude. The motivation for these various efforts is prestige, in the best vein of the twentieth century's ideology. There is pseudo coordination between these missions decided and managed unilaterally and independently in the frame of ILEWG and IMEWG (international Lunar or Martian Exploration Working Group).

The third mode, **cooperation**, is vastly more difficult. Cooperation is usually initiated and managed by a major Space Power (NASA or in the Soviet case Interkosmos) with a governance strictly imposed by the major partner to the others.

For NASA, "international cooperation" has been conceived as the acceptance by other partners of a programme conceived, planned and directed by NASA. As a matter of principle, the U.S. does not want to relinquish responsibility for a critical element in a mission or in a programme to an outsider (always minor) partner. Same domination is observed by Russia with its partners (China on the Grunt Mars-mission in 2012 as a last example).

4.3.2 The Human Station in LEO

We have an example of a successful international cooperation in space with the ISS (International Space Station). Conceived, led, managed and directed by NASA, the ISS was built on a series of bilateral agreements with the governments of Russia, Japan, Canada and ESA. All major decisions were made unilaterally by the U.S.; examples are the retirement of the shuttle in 2010, the unilateral withdrawal of NASA from the programme in 2015, and the reluctant acceptance by NASA of an end-date in 2020 (very likely to be not the last word). This situation is justified by the relative volume of financial contributions to the programme by each of the partners: 100 billion dollars from NASA compared to much less from ESA, Japan and Canada. . .

As intensive utilisation of the ISS developers, essentially as a platform on which to prepare for future mission out of the Earth's gravity, a better equilibrium among partners is to be hoped for. There is some feeling that at least for deep space, there will be an evolution.

Large multinational exploration missions will require agencies to accept and manage interdependency at different levels: architecture, mission, infrastructure, and systems. The nature of human exploration beyond low-Earth orbit will necessitate acceptance of, and commitment to, a level of interdependency that is beyond our current experience and that will increase interoperability across the architecture.

Efforts to promote future interoperability of space systems are critically important. Pursuing interoperability initiatives, such as international standards and common interfaces, will ensure different systems and nations can work together in exploring the solar system. A doctrine has to be elaborated.

Agencies participating in the Interagency Operations Advisory Group (IOAG), Space Frequency Coordination Group (SFCG), and the Consultative Committee for Space Data Systems (CCSDS) have collaborated on establishing data communications and mission operations architectures, coordinating spectrums for space communications, and technical standards for cross support which take advantage of current and anticipated state-of-the-art technologies.

Some of that work is already underway. These teams have developed service catalogues and technical standards which respond to the anticipated needs of future exploration missions. These services and standards will enable highly internet worked mission operations and facilitate the integration of new partners into complex human space exploration missions. More information about these international organisations is available at: www.ioag.org, www.sfcgonline.org, and www.ccsds.org.

In addition to the efforts in communications and operations, on-board systems standards are equally important. For example, initiatives such as the International Docking System Standard and On-board Data Interface Standards are essential for fostering on board interoperability. Work on such standards continues. By applying them, vehicles developed by many nations can conduct missions such as those shown in the ISECG Mission Scenario and increase robustness in space exploration endeavours.

If we find a number of areas where cooperation is tried, the major decisions are still kept far away from the "minor partners". The U.S. has shown, when in 2004 the Vision was adopted by President George W. Bush, that American leadership with non-American participation "needed" but in fact kept marginal, is still very much of the order. Same attitude from President Obama deciding a human mission to an asteroid. We can expect the same attitude in the future for a return to the Moon and eventual missions to Mars, even if, as we have seen, American leadership is not what it has been.

4.3.3 The Chinese Space Station

China is seeking "collaboration" in the exploration and utilisation of outer space. Its future space station (CSS) is being advertised as an international collaborative project on an unprecedented scale for China. The key policy problem is finding a model that is effective for fair and rational cooperation, which is defined in Chinese white papers as mutually beneficial, transparent, reciprocal and sharing the costs, while striking a balance with partners over ownership, intellectual property and utilisation rights. "Mutual benefit" is echoed in several documents pertaining to cooperation in outer space.

The 1996 U.N. Declaration on International Cooperation says that in all aspects of international cooperation states must cooperate on an equitable and mutually acceptable basis. Understanding what China believes is a "mutually acceptable basis" requires further examination of current practices of international cooperation.

Chinese intentions become more apparent when viewed through the theory of international modernisation, which suggests an interaction between national transformation and the international environment resulting in a dynamic process that involves multinational activities. For Chinese space activities, this includes fostering multilateral bodies such as the Asia-Pacific Space Cooperation Organization (APSCO), bilateral cooperation with several Latin American and African countries, and increased participation in U.N. space-related organisation. In the event that another country lags far behind in science and technology, China must figure out what can be exchanged for training and cooperation—perhaps unfettered access to natural resource rights, business contract preferences and related agreements, or port calls for the People's Liberation Army navy from developing coastal countries like Venezuela.

The model for cooperation on CSS has not yet been determined. China should pursue a hub-and-spoke model similar to the International Space Station, control the majority share, establish a chain of command and be in a position similar to the U.S. on ISS.

We do not know today if the cooperation on CSS will be limited to regional, non-ISS nations. If CSS is not truly an international platform but rather an Asia-centred space station, this model would be similar to the Soviet Union accommodating visitors from the Interkosmos organisation on Salyut. In 1976, a number of socialist countries signed an accord with the Soviet Union to cooperate in outer space. Shortly after its formation, the Soviet Union introduced the idea of having a guest visit the Salyut station from one of the Interkosmos nations. In 1978, a Czechoslovak became the first non-American or Soviet to fly into space.

The Soviet Interkosmos missions of the 1970s and 1980s should serve as an effective model for CSS missions. The visitors could come from APSCO, a Chinese-determined consortium of strategically important nations, and possibly a U.N. selection. CSS participants would be expected to provide their own experiments and focus research on their home countries. Chinese experiments would remain separate from the guests. If China defines sovereignty more like on ISS, where supreme authority rests in the hands of the owner of a given element, and independent nations operate their given areas but exercise jurisdiction with partners in mind, then cooperation will be much more open. These levels of openness, however, will not further Chinese strategic goals. Chinese rule over major elements should only be yielded to a nation that is commensurate to the partners' technical contributions on ISS.

China should ideally pursue technical and operational interfaces with ISS while both are in orbit. Cooperation may not be on the level of the shuttle-Mir or even Apollo-Soyuz, but any increased interaction with the ISS partners, even just multilateral meetings, will make later "cooperation" on CSS more possible.

The most realistic depiction of a Cooperative Framework on CSS in the 2020s does not include the major space powers on the same space station. CSS will remain a China-led regional platform for cooperation during its time in orbit.

4.3.4 Beyond LEO

Today space exploration is stuck in uncertainties concerning two major problematic areas:

- What to do with the International Space Station (ISS) and with the future CSS?
- Will astronauts be sent to the Moon, to Mars, . . . or elsewhere?

The United States exercise a leadership in this domain, but the vagaries of their policy (stoppage of Apollo, development, then withdrawal of the Shuttle, inception, followed by the cancellation of the return to the Moon initiated by President G.W. Bush) show the absence of the long term strategy required by ambitious aspirations. Newcomers in the space race, such as Japan, India and, above others, China, want all to acquire the status of a technological power by following the tracks paved by the Big Two in the 1960s.

Meanwhile, European thinking, planning and implementing in the domain of space exploration is in a limbo. Ministerial declarations have stated that Europe wants to participate in space exploration, without proposing any precise direction, and therefore no real action has been undertaken. The planning effort initiated in the framework of ESA's Aurora programme has been reduced to the single programme ExoMars, for which no continuation is expected.

However, Europe has to face the fact that space exploration, including human spaceflight, **will happen** under American or Chinese leadership; therefore Europe cannot be absent without a major loss of status and cannot continue to say that it will passively follow the lead of other space-faring nations. Abstention would be interpreted everywhere as decline in all domains.

In this general vacuum, we can only advance a few ideas.

Meanwhile, China's manned space programme is ascendant, with its own ambitions for a 2025 landing on the Moon. An American congressman a few years ago suggested naming the first lunar base after Neil Armstrong. Another lawmaker, Tom Feeney, Republican of Florida, responded "What makes you think the Chinese would give U.S. permission to name their base after one of our astronauts?"

Let us take for granted that the United States is not going to abandon its policy of permanent occupancy of space by American astronauts, and will keep the objective of returning men to the Moon. My contention is that these objectives cannot bear any significance without an international cooperation of a new nature.

We are not in the twentieth century anymore. Shortly after the 2004 speech of President Georges W. Bush launching the Vision, on the following February 16th, the Director General of ESA, Jean-Jacques Dordain and the former EU Commissioner, Philippe Busquin, issued a joint reaction to the Bush-Initiative with the spirit that a coherent Exploration Space policy would not make sense if not seen in the wider global context: *"Unlike the days of the Cold War, getting to the moon and Mars is not about proving one's superiority over a political enemy. It is about all of us around the world, working together for a common goal."*

However we have seen that this approach will not be followed in the next 10 years in the governance of the space stations which will remain apart, *separate but equal*.

In order to follow the idea that Man has embarked upon a second space age, it is essential to build space activities on a paradigm of international cooperation, which means for the Space Agencies to depart from the mode of *multibilateral* agreements and to adopt *multilateral* agreements, following many models already created in Europe as CERN or ESA. Although many difficulties have arisen that are indigenous to this type of operation, the success of CERN and ESA shows that such structures might be adopted to conduct international science on the Moon.

The new multilateralism would rely on a mix of national leadership and cooperation, building a sense of shared responsibilities under the guidance of a steering group including major nations.

This approach is a return to the philosophy that prevailed at the end of World War II. At that time, the American Administration conceived and imposed on the Allies an ambitious policy of cooperation among nations: the Organization of the United Nations, the World Bank, the International Monetary Fund and, in the supreme example of American leadership, the Marshall Plan. These initiatives shaped history; and the spirit that inspired them was still alive in 1959 when the newly born NASA offered to foreign scientists a place on its satellites. This offer, and its implementation in practice, was crucial to the development of space activities in England, France and Italy.

The aim of the space exploration programme would not be "going to the Moon" but "going to the Moon together", with socio-political objectives as explained in the IAA 1st Cosmic Study published in 1990 "The case for an International Lunar Base", in the following terms:

> *Many people fervently wish for ways to eliminate regional conflicts, the possibilities of global nuclear conflict, and the investment of world wealth in weapons and military activities. It is argued that one or more major international space programmes could divert the military nations from global struggles and focus them on growth into space.* ***The International Lunar Base*** *(ILB) is seen as one way in which influential world leaders and the citizens of many nations could be induced to work together on an outward directed project and thus learn to work together to reduce international tensions.*

The ILB would be a new Apollo, but an Apollo for all mankind, the Apollo of the twenty-first century, recapturing and focusing the public interest in a new utilisation mode of the symbolic power of space.

What would be the roadmap for such a change of paradigm? Among the space faring nations, two outstanding players, the United States and China, would have to convince all other nations or groups of nations as Canada, Europe, India, Japan and Russia to join the team. A possible process could start by an ISECG study of the international aspects of the ILB, which would be accompanied by a NASA study of the possible technical and operational concepts.

The exact form of the international cooperation proposed to establish an ILB would require considerable exploration. The creation of a multinational development agency comes quickly to mind. However, such an approach might lead to rigidity in design and operation. One danger of a global programme will be the creation of support organisations on Earth that become primarily concerned with their own survival. In order to avoid the reaction of reject which would certainly be caused by the idea of a new bureaucracy, a simple steering group including all the participants, which hopefully would be the major space faring nations, would be considered to run the project through established national and international institutions. However, other options are open.

Three mechanisms for nurturing global cooperation in building and managing a lunar base can be categorised: internationalism, requiring total cooperation among a consortium of nations, national enterprise, involving one nation which takes the major role but invites other nations to contribute and participate, and private venture, calling for a consortium of private interests to supply the high-technology management system.

An example of an international undertaking is the European Center for Nuclear Research (CERN) in Geneva. Eleven member nations combined resources in the 1950s to build an accelerator laboratory that competes well with the high energy physics programmes of other larger nations. Although many difficulties have arisen that are indigenous to this type of operation, they have been resolved successfully. CERN serves as a model that might be adapted to conduct international science on the Moon.

The International Program of Ocean Drilling (IPOD) provides a model of national enterprise fostering international participation. Since 1968, the Deep Sea Drilling Project has been supported by the United States through the National Science Foundation. Scientific planning was conducted under the auspices of an advisory group of scientists from all over the world. The international phase of Ocean Drilling (IPOD) began in 1975 when the United States extended membership in the Federal Republic of Germany, Japan, United Kingdom, France and the Soviet Union. The IPOD member nations participated in the scientific planning of the project. They took part in the field work, and they collaborated in the post-cruise scientific studies. Along similar lines a lunar base conducting scientific research could be initiated by one lead nation which in turn solicits contributions from other nations towards building facilities and planning activities.

If the economic potential for a lunar base is to be stressed, the International Telecommunications Satellite Consortium (INTELSAT) yields a successful model. INTELSAT is a user-based management System made up of participants from many countries. The participants work to coordinate the operations of international communication satellites. The INTELSAT concept could be modified appropriately to form an organisation that would bring into the management of a lunar base those states and other interests which manifest the greatest motivations for ensuring the successful implementation of that managerial system.

Until a specific lunar base programme begins to materialise, it is neither appropriate nor possible to determine which scheme will be best for promoting international cooperation. Certainly, a fourth model consisting of several lunar bases operated independently by competing nations as is the practice in Antarctica could evolve if for example launch costs are reduced by orders-of-magnitude.

Concerning the substance of the project, many options should be considered. For example, the international cooperation might focus on construction of staging facilities in low Earth orbit and large fleets of space ferries. Then nations could form sub-teams to build their own facilities on the Moon, in a small region or a several sites. Very likely an ILB programme will occur in the same time frame as one or more national or perhaps even small private programmes.

Which propulsion systems should be used, which vehicles, which equipment, which telecommunication networks, all these questions would have to be negotiated once the steering group would start functioning.

In the summer of 1950, sipping cocktails with friends at the home of Jim Van Allen in Springfield, Maryland, Lloyd Bjerkner suggested that the progress in technology and specially in rocketry would warrant the organisation in the near future of an International Geophysical Year. Clever lobbying endorsed the support of the scientific international Unions, IUGG, IAU, IURS, ICSU, and with the help of UNESCO, a special committee was formed to promote the idea, the *Comité Spé cial pour l'Année Géophysique Internationale,* or CSAGI. In September 1954, these Unions met in Rome and were followed by CSAGI which endorsed an idea floating in the mind of some visionaries, the use of artificial satellites for scientific research. During the following winter, under the impulse of CSAGI, all the scientific institutions went to lobby their authorities in the Soviet and in the United States, and as a result the governments of these two countries announced simultaneously at the end of July 1955, that they would include the launch of scientific satellites during the IGY. Space was born, out of the efforts of a few well organised scientists.

Such a surge by the space community is required today in order to rationalise and expand mankind's reach to the Moon as a unified drive.

The way would then be paved for the next step, the Man-to-Mars venture which requires deep innovation not so much in technology than in governance and management. The ILB would constitute a model.

4.4 A MUST: The Refoundation of European Space Institutions

The success of Europe in creating institutions regrouping many states around joint scientific, technical or political programme (ESA, CERN, EU itself) indicates that it could offer to the world a model of space governance. But in order to do so, a major reorganisation has to be performed.

The space institutions in Europe were created in the years 1962–1975. There was no EU, no euro, no integrated industry. And therefore these institutions have become obsolete. The consequences are huge: budgetary stagnation, quarrels on the programme of launchers, absence of a military space capacity and, above all, complete lack of vision.

Futile efforts have been spent to coerce ESA into the frame of a dependence of the European Commission. They failed because they always tried to force ESA's industrial practices to obey the Commission's rules. This approach leads nowhere.

A different approach would be to recognise two facts:

1. Space is essentially political, and the decision-making process should be placed at the political level of the European institutions.

2. Since Europe is "Europe of Nations", the implement of the programme should be pursued through a subsidiarity policy compatible with the present practice of the *various* Agencies.

I. The European institutions (Council and Parliament) would *elaborate* a European Policy for Space (ESP).

An ESP Bureau would be created and attached to the Vice-President of the Council, in parallel to the Defense European Policy. Its structure would be light (plus or minus ten persons, high level experts).

The functions of the Bureau would be to elaborate the ESP and to lobby Council, Parliament and Member States.

II. A mixed structure would *manage the realisation* of ESP.

One Commissioner (Member of the European Commission) would be in charge of the European space, either as a full or partial job. He would:

- Propose, through ESP, the elements of the elaborated programmes, to the Council and to Parliament for acceptance.
- Execute the space programme.
- Be responsible for the budget.

III. The European Space Federation

The Commissioner responsible for space would dispose of a Space Federation for his action: existing Space Agencies (ESA, CNES, . . .) would be loosely federated (EFS or European Federation for Space) under a unique directorate. The Director General of the Federation would be nominated by the Commission and would act in the frame of a multiannual programme defined by Council and Parliament.

Each Member State would keep the possibility of using its capacities for a national programme.

Each Agency included in the Federation would keep its personnel under its present status and continue to apply their present rules and practices. The status of the technical Centres would stay unchanged, but their basic policies would be coordinated at the Director's level.

Centres would compete for their part on each programme.

Applying a principle of subsidiarity, the Member States would keep the following responsibilities:

- Scientific research.
- Support of industry and national RD.
- Management of Technical Centres.
- National projects (if any).
- Security and Defense national projects.

Adopting such a governance system for Space, Europe would pave the way to world governance for exploration, from management of Space Station in LEO to the International Lunar Base to the permanent human settlement on Mars.

Chapter 5
The Space Regionalisation and Global Space Governance

Xavier L.W. Liao

5.1 Space Regionalisation and Global Space Regime Complex

Before probing to understanding the relevance between space regionalisation and global space governance, several basic concepts need to be clarified, such as *regions*, *regionalism*, *regionalisation* and *regional regime*. *Regions* generally refers to a middle level, but also to an intermediary level between that of global and national. They must be *composed of geographically clustered sets of such units, and these clusters must be embedded in a larger system, which has a structure of its own.*[1] *Regionalism* and *regionalisation* are respectively noted as the 'institutionalised practices' and 'a process that engages actors'.[2] A *regional regime* which is established par the countries in a predefined geographical region or a cluster of neighboring states who commonly claim their sameness can be '*a set of implicit or explicit principles, norms, rules, and decision-making procedures around which actors expectations converge in a given area of international relations.*[3] It is with these general definitions, we study the relevance between space regionalisation and the currently imperfect global space governance.

The development of space governance is framed by its competition-cooperation-compromise hybrid nature which is different from other type of issue-area

[1] Buzan B, Wæver O (2003) Regions and Powers—The Structure of International Security. Cambridge University Press, Cambridge.

[2] Katzenstein PJ (2006) East Asia—Beyond Japan. In Katzenstein PJ Shiraishi T (eds) Beyond Japan—The dynamics of East Asia. Cornell University Press, London.

[3] Krasner SD (1982) Structural Causes and regime consequences: regimes as intervening variables. International Organization 36 (2): 185–205.

X.L.W. Liao (✉)
Ghent Institute for International Studies (GIIS), Ghent, Belgium
e-mail: LiWen.Liao@UGent.be

C. Al-Ekabi et al. (eds.), *Yearbook on Space Policy 2014*, Yearbook on Space Policy, DOI 10.1007/978-3-7091-1899-3_5

governance dealing with merely competition-driven issues (*i.e.* security), cooperation-oriented issues (*i.e.* environmental protection), or compromise-seeking issues (*i.e.* natural resource management) which actors' competition and cooperation ought to be compromised. Outer space is commonly noted as global public goods to benefit all mankind because space technologies and their boundary-free applications have become ubiquitously valuable to handle various issue areas of safety, economic and social development, and ecological sustainability. Hence, the model of governing space related affairs throughout international cooperation and international regimes became commonplace. Nevertheless, since space remains crucial for many states to assure their national security and to gain economic interests, the internationalisation of space altruism needs habitually compelled to satisfy general interests and respective state's self-interest concerns. In this context, actors from either intra- or extra-regional dimension of a predefined regional space community take initiative to lead or to support respective *space regionalism* in pursuit of the goals of aligning regional astropolitics, harmonising different intra-regional space systems, or developing or consolidating the foundation of regional space governance architecture. With these motivations and institutionalisation practices, the formal or informal regional space regimes stood out with their conductor role in the middle to connect global and national space communities, accommodate and transmit the dynamics interflows between global and national astropolitics within the currently imperfect global space governance architecture.

To observe the role of regional space regimes alongside of their respective regionalisation processes, and moreover to analyse the impacts these regionalisation made on the architecture of global space governance, it seems essential to identify why and how the space regionalisation starts and their further expansion and fragmentation. As there is neither a single space hegemon power nor a unique global space regime which governs the space affairs alone, the start-up of space regionalisation with a group of neighboring countries in a predefined geographical area does not depend on a single actor, but on various factors from global, regional and national dimensions. These factors jointly trigger and boost the space regionalisation processes in different regions and subsequently make impacts on the consolidation of global space governance architecture. We note that the regionalisation processes are sparked and fed by a mixture of numerous inputs from the intra- and extra-regional dimensions. In the intra-regional dimension, (1) the dynamics of regional astropolitics, (2) the quest for regional space capacity, and (3) the necessity for regional space governance; are the major inputs, which jointly simulate the processes of space regionalisation. In the extra-regional dimension, the inputs are generated from (1) extra-regional space powers' stimulus, (2) inspiration from other regionalisation development (*mirror effect*), and (3) global regimes' endorsements. Not an input alone but a mix of them results in the activation and the continued existence of the space regionalisation processes in different regions.

5.1.1 Intra-regional Inputs for Space Regionalisation

First, we locate the inputs issued from the dynamics of regional astropolitics. Regional spacefaring countries are one of the designers to start the space regionalisation with their cooperative supremacy. By leading a space regionalisation, the regional space power aims to demonstrate its regional space leadership and uphold the regional power-balance against the other neighboring space powers. On the other hand, the power-driven space regionalisation can also gain strategic allies or business partners for the space power to extend the geographical coverage of its exclusive space system and enlarge the market of its space technology products and services. The regional space power determines what would be the centralities for their own cooperation networks. In the case there are more than one regional space powers in a given region, for example Egypt and Saudi Arabia in the Arab League, or Japan, China and India in Asia, once a regional space leading country starts up its proper space regionalisation process, the other follows in form of duplicating the same action to counter it. Each of leading regional space power established distinct regional space regimes and offer vital cooperation projects as incentives to satisfy its respective allies though remain rational regarding the loyal 'fair return' from their protégés.

Second, the collective quest for developing common regional space capacity or a specific or exclusive regional space system (*e.g.* for satellite TV and radio broadcasting, disaster mitigation, navigation safety, and Earth Observation) can activate and fuel space regionalisation. The regionalisation is therefore undertaken with actors' functional or cost-benefit logic. By knowing the fact that developing space capacity and upholding it is an extremely expensive and highly risky business, there is no country, even not the US, that can handle it alone. Pooling different material or immaterial resources to develop regional space capacity doubtlessly becomes the optimal and legitimate strategy for collective and individual prosperity and benefits. Since the space 'democratisation' after the Cold War, emergent industrial countries and developing continents have various ways to continue or to start to develop their own space capacity and they are all keen to enjoy the utilities of space technology applications for military, civil or dual-use. The European countries have explored a regional space centrality since the 1960s that was obviously based on the both aforementioned strategic and functional logics to fulfil both regional and national interests. The regional space institutions concentrate crucial space technology expertise and agglomerate the research investments and industrial facilities for increase the space interoperability among the intra-regional partners. The EU and ESA help the member states to increase national space capacity, share financial burdens and reduce the risks of marketing failure with partners. Additionally, the European space regionalisation also created advantages for strengthening Europe's global political and economic actor's position for increasing the bargaining forces. The space regionalisation was also expected to reduce Europe's dependency on the American space capacity in order to gain more flexibility for choosing other partners or markets. In the Arab world, it was the cultural sameness which motivated the request for developing Arab space technology alignment in order to serve regional communication, culture, education and others within the confines of the Arab League Charter. In the South East Asia, the Association South-East Asian Nations has established an ASEAN Charter on disaster management and emergency rescue which relies heavily on a regional network of satellite related services for its effectiveness. In Latin-America and in Africa, the space regionalisation initiatives mostly stress indigenous regional space capacity building to improve people's lives and the economy. The various space regionalisation processes were all motivated by a collective quest for specific functional purposes of states' cost-benefit logic.

Third, similar to the development of regional telegraph and radio governance in the past, the necessity to connect national space systems which cover satellite telecommunication, weather forecasting and environmental monitoring, satellite TV and radio broadcasting and navigation services urge the space regionalisation for the harmonisation, standardisation and interoperability among neighboring states' space systems. Throughout the regionalisation process, states negotiate common measures, such as regulations, standards, tariffs, and interference avoidance rules for heterogeneous national space systems within a given geographical region. Especially nowadays, the growing commercialisation of space technology for its design, manufacture, launch and operations and its application for telecommunication, TV and radio broadcasting, remote sensing and navigation are increasingly taking more ground, the quest of establishing regional common conducts rules and operational standards become more and more important. The necessity for institutionalise such regional space governance architecture is doubtless uncontested. Yet, whether the design as well as the perfection path for building any regional space regimes depends on whether the desired regime meets its member states' strategic calculation and functional concerns. This often made the managerial manoeuvre of a given space regionalisation more complicate and complex. The creation of aforementioned Arab Satellite Communications Organization, the Asia-Pacific Broadcasting Union (ABU) that serves as regional platform for national TV and radio broadcasters, and the Regional African Satellite Communications Organization (RASCOM) were the classic examples of the space regionalisation for the needs of governing regional space related activities and for the development of interoperability of different space systems of a given region.

5.1.2 Extra-regional Stimuli for Space Regionalisation

A part from the intra-regional inputs, the inputs from the extra-regional dimension can also play a vital role in sparking and to fuelling a continued space regionalisation. These extra-regional inputs can be analysed using three layers: (1) the stimuli from extra-regional space powers, (2) the inspiration other regionalisation from other regionalisation (*mirror effect*), and (3) the endorsement from global space related regimes. It is important to state that never a single one of these inputs but always a mix of them results in the activation and the continued existence of these space regionalisation processes in different regions.

The stimuli from extra-regional space powers, namely from the US, Russia and nowadays China, India or others, are centripetal forces that efficiently congregate a new regional space centrality. These space powers, with their crucial technology *know-how* and financial supports, push to institutionalise a regional space centrality is either to enhance their ties with the extent allies, make new friends or attract new followers from non-spacefaring countries in a given region. This outreach toward the regional level is supposed to increase the respective space power's political and strategic influences on both regional and global astropolitics. It is also commercially interesting for the space powers to conquer foreign regional clusters more efficiently. As for the choice where to do such space power stretch exercises, it depends on the each space power's geopolitical concerns and strategic interests. Furthermore, while sponsoring a given space regionalisation, the space powers do not provide full space capacity assistance and do not offer it for free neither. The attractive incentives for the accommodating countries for having and keeping the deals are often accompanied with strict conditions, such as the US International Traffic in Arms Regulations (ITAR). The ITAR has unilateral power to decide whether a piece of technology can be sold to the US allies or interested states or companies, but it can also sanction the contractor if a contracted project is leaked to a third party. Consequently, the interested parties, like European states and China were somehow pushed to seek their independency or at least non-dependency from the US, and therefore wanted to create their own regional space clusters. Finally, whether a targeted region has political desires and adequate capacity to host and develop a given space regionalisation sponsored by extra-regional space powers has no co-relationship to the efforts provided by the space powers. The Soviet Union have incorporated the Eastern European socialist states into a closer regional space community. These days, Africa, Latin America, ASEAN and Central Asia just became the new power playground for the US, Russia and China to bid for allies or followers. Non spacefaring states of a given regions often undertake their 'regime shopping' in accordance to their functional interests and preference. The stimuli from the space powers are valuable to help the regionalisation of space capacity building. Yet, it can never be the only factor which leads such processes to their final goal.

The successful story of other space regionalisation, namely that of the European Space Agency (ESA), becomes an important model for many space regionalisation initiatives to follow. The ESA was created in 1964 by merging the former European Space Research Organisation (ESRO) and European Launcher Development Organisation (ELDO). It became the regional platform for European spacefaring countries primarily to reduce the intra-regional competitions by redistributing the regional space capacity building tasks to the member states with the chartered 'fair return' clause. Secondly, the ESA regionalisation model provides a cost-risk sharing system wherein the European spacefaring states can diminish the burdens of their respective national investments and reduce the technical and market failures while advancing such costly and uncertain products and market development. The ESA also plays an important role in bridging the capacity gap between European spacefaring and non-spacefaring states. Thirdly, the ESA model integrated Europe's regional space economy to counter global competitors, such as the US, the Soviet Union, and Japan at the primary time and finally made Europe to become one of global space technological, economic, and space diplomacy actors. The successful story of the ESA therefore generated a 'mirror effect' for other regions to duplicate it for their own space regionalisation. We can point out a few, such as the Japan-led APRSAF, the China-led APSCO, the debates about creating an African Space Agency (ASA) that could merge the extant African Leadership Conference on Space Science and Technology for Sustainable Development (ALC), the Regional African Satellite Communications Organization (RASCOM) and the African Resource Management System–Constellation (ARMS-C), an ASEAN Space Agency.

Finally, with a mindset of organisational or managerial logic, the global space regimes, such as UNCOPUOS, International Telecommunication Union (ITU) often encourage the space regionalisation model in order to implement their global policy with tailor-made programs or projects adapted to respective regional needs and particular working methods. These global-led regionalisation offers vital immaterial endorsement (*e.g.* legitimacy or connection with international space community networks) and material support (financial means, administrative assistance and so on) for the start-up of a space regionalisation in pursuit the space capacity building goal, namely in the developing and under-developed regions. For example, to improve the UN development goals, the UN sponsors (1) organising an intra-regional space forum/meetings as an effective mechanism to provide space education and awareness trainings and programs, (2) boosting national and regional space-driven economies and industrialisation to improve citizens' living quality and safety, create connectivity between urban and rural or remote areas, prevent natural or manmade disasters, facilitate emergency assistance (The UN *International Cooperation Declaration*, 1996), and (3) the regionalisation of the space related global regime. The UN has been sponsoring the Space Conference of the Americas (CEA) since the 1990s to enhance social and economic development in the Latin America and Caribbean (LAC) region. The UN Office for Outer Space Affairs (UNOOSA) set up two regional centres in Africa—the African Centre for Space Science and Technology (CRASTE-LF in French), and the African Regional Centre for Space Science and Technology Education (ARCSSTE-E in English). There are also other UN Centres for Space science and technology education in Latin America and the Caribbean, Western Asia, Asia-Pacific, and in China. The ITU organises the bottom-up model regional workshops, for example in the Commonwealth of Independent States (CIS), to disseminate relevant information and collect reflections regarding the issue of how space communication systems are in demand for the economic and social development of the CIS states and what are the major technical needs and hurdles to overcome throughout the regional collective and coordinated efforts. The international treaties also trigger the creation of regional regimes to more effectively implement their regulations or global policy in the regions. The 2005 ASEAN Disaster Charter, which technically relies on remote sensing technology, satellite communication and satellite navigation system to effectively organize regional disaster management and emergency responding actions, was in fact one of the extensive regional actions following the 2005 Hyogo

Declaration and Hyogo Framework for Action 2005–2015 that deals with global disaster management and emergency response issues. The global endorsements are expected to bridge the space capacity gaps between spacefaring and non-spacefaring countries. However, the choices of the host geographical areas and the proportions of these global endorsements would unsurprisingly depend on the dynamic of global and regional astropolitics, and the strategic preferences and functional calculations of the major donor countries and host states.

To conclude this section, we have discussed that the processes of space regionalisation within a geographical area by a group of nations are often designed to calibrate regional or global astropolitics; (2) to bridge the national space capacity gap in the name of developing a regional space community or centrality; and to (3) establish and then consolidate the regional space governance architecture for the good functioning and the non-interference between different space systems within a given region. The end of space regionalisation is expected to satisfy one of or all the strategic, functional or managerial concerns of states toward the concerned geographical region or the group of neighboring nations. To start-up and nourish these space regionalisation processes, different factors jointly trigger and boost the space regionalisation which subsequently made impacts on the consolidation of global space governance architecture. We have identified that the regionalisation processes are sparked and nourished by a mixture of numerous inputs from the intra- and extra-regional dimensions. In the intra-regional dimension, (1) the dynamics of regional astropolitics, (2) the quest for regional space capacity, and (3) the necessity for regional space governance; are the major inputs, which jointly simulate the processes of space regionalisation. In the extra-regional dimension, the inputs are generated from (1) extra-regional space powers' stimulus, (2) inspiration from other regionalisation development (*mirror effect*), and (3) global regimes' endorsements. Not an input alone but a mix of them results in the activation and the continued existence of the space regionalisation processes in different regions.

5.2 Impacts of Space Regionalisation on Global Space Regime Complex

The global space governance is considered as a *global space regime complex*[4] in which the homo-astro affairs are co-managed by *an array of partially overlapping and non-hierarchical institutions*.[5] From the beginning of the Space Age, the global space affairs have been designed to be co-managed by different regimes, such as the UN Committee for Peaceful Use of Outer Space (UNCOPUOS) for the issues

[4] Liao L-WX (2014) Global space governance in the 21st century: A regime complex perspective. Astrosociology Insights 3(2): 3–5.

[5] Raustiala K, Victor DG (2004) The Regime Complex for Plant Genetic Resources. International Organization 55 (2004): 277–309.

related to the peaceful use of outer space, the space related treaties and multilateral agreements, such as the Outer Space treaty (OST) for regulating actors' behaviours, and the international institutions, *i.e.* the International Telecommunication Union (ITU) and the World Meteorological Organization (WMO). This reflects an organisational logic of labour division which gave birth to the architecture of global governance. Moreover, the continuous astropolitical power competition and the complex space interdependency among all actors also continuously urge states to adapt strategic or functional logics when taking part of the development of global space governance. Countries continuously probe to reshape the imperfect architecture of global space governance for their advantages by changing the functions or the organisational design of the extant space related regimes, or simply create new regimes side-by-side with the existing regimes to co-manage the more and more overlapping space relevant issues. Subsequently, the architecture of global space governance successively expands its magnitude yet at the same time became more and more fragmented, thus, a regime complex. The trend of current regional space institutionalisation becomes actually also a part of the global space *regime complex*. We note in any case three discernible impacts the current space regionalisation trends have made on the extant imperfect global space regime complex. (1) The regional space regimes become a new middle ground or new conductor for global and regional space powers to calibrate global or regional astropolitics. This adds one more governing layer in the current global space governance architecture to manage. It also expands the current global space regime complex toward a larger dimension. (2) The regional space regimes are valuable as building blocks to consolidate the foundation of global space governance. The space regimes are valuable as building blocks to consolidate the foundation of global space governance, because the labour division and the decentralisation of global space governance toward the regional level—space regionalisation, better the particular needs and accommodation capacity of individual regions. This trend of democratisation is vital to consolidate the foundation of global space governance. Finally, (3) a new form of general also space related inter-regional regionalisation, such as EU/ESA, BRICS models, appeared and caused the extant space regionalisation to generate more fragmentation in the global space regime complex. Further observations regarding the development of this inter-regional space regionalisation trend are required.

5.2.1 Space Regionalisation Calibrates Global and Regional Astropolitics

Deploying and developing regional space regimes to be another astropolitical playground has been and is increasingly solicited by the space powers (the developing nations in particular) as a preparatory agora to amplifying their voices in pursue of advantageous space deals at the global level. In short, space regionalisation becomes a strategic project for states to gain their leverage to calibrate regional or global astropolitics in the name of regional coalition. Various examples that have occurred as from the early space age and are still continuously appearing now, illustrate this strategic logic. The 1976 Bogotá Declaration proposed by six Equatorial developing countries, three in America and the others are in Africa, claimed that their national sovereignty should include the geostationary orbit above their terrestrial territory. This strategic regional or inter-regional coalition was precisely established at the moment when the world was reviewing the OST for negotiating a draft of the 'Moon Treaty' as well as when the new negotiating round would start to discuss the issues of radio spectrum and orbital slot allocation at the 1977 ITU Assembly. The regional or inter-regional coalition was noted legitimate due to its collective strength though it ended in failure due to other factors. Another similar case was a regional Convention on the Transfer and Use of Data on the Remote Sensing of the Earth from Outer Space that was proposed in 1978 by a group of socialist countries when the UN was negotiating the Principles related to Remote Sensing, ultimately concluded in 1986. The regional convention asked for space faring countries' equitable uses and dissemination of the data on the sensed states.

For the spacefaring counties, the regional space power-led space regionalisation can polarise the intra-regional division to the global level, therefore calibrate global astropolitical alignment as well. For example, the competition between the Japan-led APRSAF and the China-led APSCO simultaneously aligns their respective intra-regional allies in the Asia-Pacific region and urges the extra-regional actors to interact rationally and diplomatically with this regional competition. The European Union together its quasi executive space agency the ESA proactively explorer possibilities of inter-regional space diplomacy to influence the direction of global space governance construction, such as the ICoC, which has already triggered a new wave in the global astropolitical alignments. These various strategic space regionalisation processes, whatsoever aim to calibrate regional or global astropolitics, have created more fragmentation in the extent imperfect architecture of global space governance.

5.2.2 *Regional Regimes Consolidate Global Space Governance Foundation*

The global space regimes, particularly in the narrow sense related to international law, used to be positioned higher than the regional space regimes though it has not always been the reality. However, due to this customary mindset, a managerial logic is often explored by the states that regional space regimes are also valuable as an outreach instrument for the global regimes to attain their policy goals, implement the global norms and regulations because of its physical proximity to the targeted countries and regions. These regional outreach mechanisms are considered to bridge the top-down gap between the global, regional and national levels and can facilitate the bottom-up feedback flow in such vertical global-regional-national policy-regulation lifting. The regional space training and education centres of the UN Office of Outer Space Affairs (UNOOSA), UN-sponsored regional space forums, the ITU-supported regional workshops are typical space regionalisation based on this organisational logic of a division of labour. The global space related regimes also expect the regional multilateral space institutions to concert the intra-regional capacity divergence in order to better outline their own regional concerns *vis-à-vis* global expectations. Therefore, the global regimes are able to make a tailor-made improvement strategy and solutions for each targeted region. This kind of space regionalisation which decentralises the global mandates toward the regions and localises the general formula of space building program to individual regional situations is mostly welcomed by countries or regions for sharing burden of space capacity building, forging cooperation to consolidate regional and global space governance architecture, nourishing solidarity and inclusivity of the global space community. With this model of space regionalisation, the regional needs can be easily taken into account and the local voices can be properly heard and understood. This leads the democratisation of the previous 'space club' toward a global space community, in another word, consolidates the foundation of the global space governance.

The challenges for the managerial top-down space regionalisation are the dynamics generated from the regional astropolitics as well as the alleged 'regime shopping'[6] phenomena that intra- or extra regional actors are in favour of a certain regional platforms or issue-area linkages that suit their respective national interests, astropolitical strategy or their expected regional imaginary. These factors derive the development path of top-down space regionalisation away from its ideal plan of using regional mechanisms as intermediary tools. The top-down space regionalisation also meets obstacles while dealing with competitive-driven or sovereignty-stressed issues related to national security or trade interests. These issues trigger resource contention and exclusivity of the regional space community and then stimulate the creations of more overlapping and competitive regional mechanisms led by competitive intra- or extra-regional space powers, such as the cases we have seen in the Asia-Pacific region and among the nations of the Arab League. In this regard, the space regionalisation scatters the extent architecture of the regime complex of global space governance.

5.2.3 A New Hip of Inter-regional Space Related Regimes Interplay Deserves More Observation

The proactive space diplomacy of the European Union starting from 2008 stimulated a new hip of inter-regional space related regimes' interplay, because the

[6] Keohane, R, Victor DG (2011) The Regime Complex for Climate Change. Perspectives on Politics 9.1 (2011): 7–23.

EU-led initiative for a draft non-binding International Code of Conduct for space activities (ICoC) probes to create a new multilateral non-binding or voluntary space regime apart from the usual international platform—the United Nations (UN), to deal with space governance matters. The ICoC was proposed as a voluntary pragmatic mid-way solutions *vis-à-vis* the growing safety concerns for astronauts and the space assets of public and private owners (*i.e.* satellites, the ISS and space-base telescopes) to reduce harmful impacts caused by space debris. The ICoC was also expected to complete a missing link between the current reality that complex human space activities become overall interconnected with various global concerns and the imperfect global space governance architecture in which a myriad of outdated, incomplete, disproportional, dysfunctional or even contradictory space treaties and international institutions exist. Further to the EU's self-profiling as a regional entity to influence global homo-astro issues, Russia and China, not to be outdone, quickly proposed a joint draft 'Treaty on the prevention of the placement of Weapons in outer space, the threat or use of force against outer space objects' (PPWT) to the Conference for Disarmament in February 2008 as a counter-proposal. They counter-blamed that the US and Israel repeatedly voted against UN resolutions preventing an arms race in outer space, hence creating a new legal-binding regime would be prominent to outlaw the weaponisation of space. China and Russia have explored the BRICS group to amply their critics on the EU ICoC. In the 2014 BRICS Fortaleza Declaration, the international regime for us or the 'informal intergovernmental organisation' (IIGO) for some others stated that there is a lack of inclusiveness and consensus-base in the EU-led international negotiation for a draft of non-binding ICoC, and proclaimed their support for the China-Russia co-sponsored draft PPWT.[7] The BRICS group, primarily questioned as a geopolitical fad and nowadays viewed as an IIGO of inter-regional coalition composed with different regional leading or emerging economic powers, namely Brazil, Russia, India, China, and lately South Africa, seems to be continuously demonstrating their weight against the existing Western-dominated global regime.[8] For example in 2014, the BRICS decided to create the New Development Bank (NDB) with an initial capital of US$ 100 billion to finance development projects for BRICS and non-BRICS countries. The space related statement in the 2014 BRICS Fortaleza Declaration also seems a prelude that the BRICS countries will probe their inter-regional space diplomacy to expand their global leverage. The inter-regional space related regime interplay surely adds more inertia to fragment the extant global space regime complex.

Finally, it is worth noting that developing the inter-regional space regime or IIGOs to increase country's influence on particular global astropolitical or space governance agenda is not new. As mentioned in the previous section, the 1976

[7] Vabulas F (2014) The BRICS and the future of 'Informal' IGOs'. ISN Paper. International Relations and Security Network.

[8] Brütsch C, Papa M (2012) Deconstructing the BRICs: Bargaining coalition, imagined community or geopolitical fad? CPR Paper. University of Cambridge: Centre for Rising Powers, Cambridge.

Bogotá Declaration was one example of such inter-regional coalition which consisted of six Equatorial developing countries from Latin America and Africa. China has approached the Group 77 in 2011 to jointly urge for more North-South and South-South cooperation to enhance and facilitate the transfer of space related technology among nations. This coalition equally advocated that while the ITU member states manage the problems regarding the attribution of satellite radio spectrum and orbital slot, the equality among nations and the interests of the developing countries should be more taken into account. It seems clear that the inter-regional platform might not become a major mechanism to shift the global astropolitics or the extant global space governance to another order. However, the increase of inter-regional space relevant regime interplay can generate further fragmentation in the extent global space regime complex, therefore deserves further observation.

5.3 Concluding Remarks

In our discussions, we have analysed that the regionalisation processes are sparked and nourished by a mixture of numerous inputs from the intra- and extra-regional dimensions. In the intra-regional dimension, (1) the dynamics of regional astropolitics, (2) the quest for regional space capacity, and (3) the necessity for regional space governance; are the major inputs, which jointly simulate the processes of space regionalisation. In the extra-regional dimension, the inputs are generated from (1) the extra-regional space powers' stimulus, (2) the inspiration from other regionalisation development (*mirror effect*), and (3) the global regimes' endorsements. It is important to know that they appeared distinctly in our analysis yet are mostly mixed when actors initiate and try to survive the space regionalisation. Furthermore, we note at least three conceivable impacts that the current space regionalisation has made to the extant imperfect global space regime complex. (1) The regional space regimes become a new middle ground or new conductor for global and regional space powers to calibrate global or regional astropolitics. It adds one more governing layer in the current global space governance architecture and expands this global space regime complex toward a larger dimension to deal with. (2) The space regimes are valuable as building blocks to consolidate the foundation of global space governance, because the decentralisation and the localisation of global space regimes toward the regional level suit better the particular needs and the accommodation capacity of a given regions. This 'democratisation' is vital to consolidate the foundation of global space governance. Finally, (3) a new hip of inter-regional space regime creations, such as EU/ESA, BRICS and so on, appeared to lead the extant space regionalisation to the further fragmentation in the global space regime complex though any salient impact seems not noticed at the level of global space governance. This, nevertheless, requires further observation.

Chapter 6
How Governance Models Affect Geopolitics: The Asian Case Study

Kazuto Suzuki

6.1 Introduction

Governing space is quite different from ordinary international governance. In the world of traditional governance, primary responsibility lies in the hands of territorial sovereign states. States have jurisdictions and control over certain parts of the world. It is also a foundation of international governance that states will not [normally] extend their exercise of power beyond their territorial jurisdiction—at least not under normal conditions, which excludes for instance, acts of self-defence. Although there are some exceptional cases, for example, international trade regimes or the International Criminal Court where governance mechanisms transcend territorial sovereignty, the foundation of international governance has not changed for centuries.

However, governance in space is quite different from traditional international governance. There is no territorial jurisdiction in space. No rule or regulation can be implemented and executed because of the absence of coercive force in orbit. Therefore, the rules and regulations on space governance have to aim at constraining and restricting the behaviour of actors in space. Current legal frameworks for space activities, such as the Outer Space Treaty or the Debris Mitigation Guidelines, are designed to put onto states the responsibility for space activities, including those activities by private entities. In other words, governance of space is governance without government.[1] Each state entering space is responsible for the governance of space. In so doing, states must impose self-restraint on themselves to follow international norms and paradigms.

[1] James N. Rosenau and Ernst-Otto Czempiel (eds.) Governance without Government: Order and Change in World Politics, Cambridge University Press, 1992.

K. Suzuki (✉)
Public Policy School, Hokkaido University, North 9, West 7, Sapporo 060-0809, Japan
e-mail: kazutos@juris.hokudai.ac.jp

C. Al-Ekabi et al. (eds.), *Yearbook on Space Policy 2014*, Yearbook on Space Policy, DOI 10.1007/978-3-7091-1899-3_6

The main objective of space governance is to protect and maintain the sustainability of the space environment for exploiting and exploring space. Deterioration of the space environment, which means the explosive growth of space debris, will increase the risk of collision and malfunctioning of space infrastructure. Intentional destruction and disruption of space objects, such as the anti-satellite (ASAT) test by China in 2007, not only creates vast amount of debris but also increases the sense of insecurity and anxiety about activities in space. Such intentional acts threaten not only spacefaring states that are capable of developing space infrastructure, but also space-exploiting states that own and operate space assets for the benefit of their societies. The increase of space-exploiting states makes space a "Global Commons" where everyone benefits from the "commons", and therefore, the sustainability of space has become a global governance issue for the benefit of all mankind.

The other side of the governance of space actually takes place here on Earth in a traditional way. Space has become an integral part of the daily lives of many, including both developed and developing states. In particular, the role of space for building socio-economic capability through tele-education, tele-medicine and market access in rural areas of developing states, where terrestrial infrastructure is too costly to build, is extremely important. The demand for access to space technology is growing more than ever.

However, at the same time, space technology is a dual-use technology, which can be used for military purposes.[2] According to the Outer Space Treaty, it is prohibited to use the Moon and other celestial bodies for military purposes and placing weapons of mass destruction in orbit.[3] However, other uses, such as military telecommunications, reconnaissance for military purposes, or military weather monitoring, are not prohibited. Today, many states, including those states without space technological and industrial capacity, are turning to space technologies to enhance their military capabilities since this will multiply forces capabilities.[4] Even states such as North Korea and Iran, who are prohibited from developing technologies to deliver nuclear warheads, are investing their scarce resources in such space technologies in the name of "peaceful uses of space".

Therefore, controlling proliferation of space technologies, primarily launcher/missile technologies but increasingly also satellite technology, is a global geopolitical concern. The governance of space is not only about the post-modern non-territorial governance in outer space, but also about very traditional territorial governance of space technologies, which eventually lead to the governance of geopolitical issues.

---

[2] Jane C. Hu, "The Battle for Space: Russia, China, and the United States are developing "dual use" technologies. One use is peaceful. The other isn't", Slate, 23 December 2014.
http://www.slate.com/articles/health_and_science/space_20/2014/12/space_weapon_law_u_s_china_and_russia_developing_dangerous_dual_use_spacecraft.single.html.

[3] The text of Outer Space Treaty can be found at http://www.unoosa.org/oosa/SpaceLaw/outerspt.html.

[4] Robert C. Harding, Space Policy in Developing Countries: The Search for Security and Development on the Final Frontier, Routledge, Abingdon, 2013.

6.2 Governance Model in Space

Although a growing number of states are engaged in developing space technologies, there are still huge gaps between those states that have technological and industrial capabilities and those that do not. Developing space technological and industrial capabilities requires not only large scale investment but also a wide range of industrial capabilities. Space technology is no longer monopolised by a few states as it was during the Cold War period, but still the gap between the "haves" and "have not's" remains large.

In other words, those states that obtained and indigenised space technology have a geopolitical advantage over those that have not. The "haves" are able to control the distribution and transmission of knowledge, industrial know-how and technological skill, and are also able to provide technical assistance and human and capital resources. In this regard, there are three different ways of governing geopolitical relationships for space technology transfer.

6.2.1 Cooperative Leadership

The management of technology transfer from "haves" to "have-nots" can take many forms. One of these is to do it cooperatively. It is often the case that those states that have advanced capabilities transfer technologies through international institutions that treat all participating states equally. Those international institutions are, basically, created on the initiative of capable states that are seeking to share funding for space programmes in return for providing opportunities for non-capable states to participate in those programmes.

One of the most successful examples is the European Space Agency (ESA). ESA was initially created to develop, collect and share scientific findings from European Space Research Organization (ESRO) and satellite launching capability from the European Launcher Development Organization (ELDO) in the early 1960s. These organisations, due to the changing environment and improvement of efficiency, integrated into ESA in the mid 1970s with two distinctive institutional features. One is the concept of *juste retour*, or fair return, which implies that the member states will receive contracts equal to their financial contributions to ESA.[5] In this way, while those states that have advanced capabilities will take the lead, non-capable states whose contributions are smaller, can also participate in the programme and acquire technologies that can improve their technological capabilities. Although the concept of *juste retour* is criticised for its inefficiency, this concept was widely appreciated by both "haves" and "have-nots".

[5] For general description of *juste retour*, please see the section on "Industrial Policy and Geographical Distribution" in ESA website: http://www.esa.int/About_Us/Business_with_ESA/How_to_do/Industrial_policy_and_geographical_distribution

The second concept of cooperative leadership embedded in ESA's institutional arrangement is optional participation. The member states of ESA can participate in programmes when it contributes to their objectives, but it is at the member states discretion as to whether or not to participate in the programmes. In this way, non-capable states will have the possibility to avoid being controlled by dominant states and can maintain control over their objectives. With these two institutional arrangements, ESA provides leadership for those states that are capable of pursuing their objectives by involving other states, while other member states can benefit from participating in those programmes.[6]

6.2.2 Hegemonic Leadership

Although the ESA is a successful case of cooperative leadership in the governance of space technology transfer, most of the cases of leadership in this domain can be characterised by hegemonic leadership. Those states that have technological advantages often use them as leverage to change other states' attitude or behaviour. In many cases, the transfer of space technology or access to space assets and information has been traded for return from other states. For example, the United States refused to provide a launching service for the Franco-German telecommunication satellite, Symphonie, in the mid-1960s while it decided to provide Thor-Delta liquid-propellant launcher technology to Japan.[7] These decisions were, on the one hand, designed to discourage European states from developing commercial satellite technology that might jeopardise the monopoly of satellite telecommunications by the United States, while encouraging Japan to switch its solid-propellant launcher development programme that could be used for the development of ballistic missiles. Both decisions were made to change the attitudes and behaviour of European states and Japan by refusing and providing space technologies, respectively.

Hegemonic leadership can also be seen when establishing international organisations for the application of space technologies. Organisations such as the International Telecommunication Satellite Organization (better known as Intelsat before its privatisation) or Intersputnik were typical case of hegemonic leadership. Those states, the United States and the Soviet Union, took leadership with institutional arrangements to control and dominate these organisations. The follower states joined these organisations to receive services while not being able to participate in the management of the organisations. Hegemonic leadership may face some difficulties in maintaining its leadership when the leader states are not able to

[6] Roger Bonnet and Vittorio Manno, International Cooperation in Space: The Example of the European Space Agency, Harvard University Press, 1994.

[7] L. Sebesta, The Availability of European Launchers and Europe's Decision 'To Go It Alone', ESA History Study Reports, HSR-18, October 1996.

provide enough funding or services to other states, and when other states became capable of providing their own services without international organisations.

6.3 Space Governance in Asia: Competition of Different Governance Models

International governance in Asia was dominated by the Cold War framework after the World War II. Many states achieved independence after the defeat of Japan which, during the war, itself took over occupation from colonial powers. However, the establishment of the People's Republic of China and the Korean War from 1950 to 1953 introduced sharp confrontation between East and West. Because of that, the United States played a very important role in providing security to Western states whereas the breakout of confrontation between China and the Soviet Union in the 1960s created a different situation from that in Europe. The historical reconciliation between the United States and China in the 1970s made the international governance structure in Asia more complicated.

In general, the governance structure in Asia can be characterised as a "hub and spokes" model.[8] The United States plays the role of hub and it stretches spokes to individual states such as Japan, South Korea, Philippines, Thailand, Taiwan and Australia. But on the other hand, China also plays the role of hub and influences North Korea, Cambodia, Malaysia and other states in the region. The key issue here is that these states heavily depend on the relationship with their "hubs" and coordination among the states in the region is quite limited. Although there are attempts such as Asia-Pacific Economic Cooperation (APEC) and the Association of South East Asian Nations (ASEAN), those frameworks remain limited to economic cooperation and have barely achieved the level of international governance on a regional level.

Although the geopolitical structure in Asia is based on the "hubs and spokes" centred on the United States and China, space governance has a different centres. Japan, as the first state that launched a satellite in Asia, has been playing an important role in providing a framework for cooperation while China, the second state to launch a satellite a few months after Japan, is increasingly influencing the region.

[8] Victor D. Cha, "Powerplay Origins of the U.S. Alliance System in Asia", International Security, Volume 34, Number 3, Winter 2009/10, pp. 158–196.

6.3.1 Chinese Approach: Hegemonic Leadership

In 1992, China, Pakistan and Thailand signed a Memorandum of Understanding (MoU) for establishing an international organisation for cooperation in space technology and applications. This cooperation created an international space cooperation agency called AP-MCSTA (Asia Pacific Multilateral Cooperation in Space Technology and Applications) which attracted many developing states in Asia, including South Korea, Iran, Indonesia, Mongolia, Bangladesh, and so on.[9] The objective of AP-MCSTA was to promote multilateral cooperation in space applications, but it was clear that China, driven by the Politburo and the Ministry of Foreign Affairs, was the most technologically advanced member and was keen to take hegemonic leadership.

AP-MCSTA offered programmes for remote sensing data analysis on environmental protection, natural resource exploitation as well as disaster monitoring and prevention. This education and training contributed to the promotion of capacity building for the Asia-Pacific Region. The organisation held 7 international conferences, attended mainly by engineers and scientists with some policy makers, for exchanging information and setting up workshops for various application programmes. Among those workshops, cooperation in Small Multi-Mission Satellites (SMMS) was the most important one. SMMS was a programme for China, Iran, the Republic of Korea, Mongolia, Pakistan, Thailand and Bangladesh for developing microsatellites for communications and remote sensing. China has been developing smallsat technology since the early 1990s by sending students to the University of Surrey, and several universities have played incubator for such technologies brought back by those students. Consequently, the SMMS project functioned as a technology transfer mechanism from China to other members.

This Chinese leadership through AP-MCSTA has further developed by creating APSCO (the Asia Pacific Space Cooperation Organization). In 2005, China, Bangladesh, Indonesia, Iran, Mongolia, Pakistan, Peru and Thailand signed the APSCO Convention and Turkey signed in 2006. APSCO transformed the AP-MCSTA into a treaty-based cooperation framework with more explicit rules on programme finance, data use and SMMS cooperation. The Convention of APSCO has some resemblance to the ESA Convention, which suggests that APSCO aims to be an ESA-like international institution for the development of space technology.[10] However, it is undeniable that China has taken hegemonic leadership by forming the cooperation scheme without its competitors such as Japan or the United States. Also, the APSCO programmes, if successful, would provide more launching opportunities for the Chinese Long March rockets, which are covered by the American ITAR (International Trade in Arms Regulation) restrictions and are an obstacle for China to attract foreign payloads.

---

[9] For the history of AP-MCSTA and APSCO see http://www.apsco.int/history.aspx.

[10] Convention of APSCO can be found at http://www.apsco.int/UploadFile/2009924/F1TKF8A2009924.pdf.

Through AP-MCSTA/APSCO, China has used its technological superiority to lead space activities in this region, particularly for states that "have not" experience and expertise in space technology. In other words, China has exercised its hegemonic leadership to organise and control the transfer of technology through these frameworks.

6.3.2 *Japanese Approach: Cooperative Leadership*

For many years, Japan was the leading state in this region and the Japan Aerospace Exploration Agency (JAXA) and its supervisory ministry, the Ministry of Education, Culture, Sports, Science and Technology (MEXT), took the initiative to create an international governance mechanism in Asia. They created the Asia-Pacific Regional Space Agency Forum (APRSAF) in 1993, which coordinates space programmes and enhances cooperation among the space agencies in this region. However, APRSAF only focuses on the technical and educational aspects of the space cooperation programme among different space agencies. It was also a technical "forum" organised by JAXA and MEXT without any treaty or MoU. There was no involvement of the Ministry of Foreign Affairs of Japan until recently. In this regard, Japanese leadership through APRSAF can be characterised as cooperative leadership.

However, the emergence of Chinese influence in the region has made the cooperative leadership model into something much closer to hegemonic leadership. In 2005 at the APRSAF meeting in Fukuoka, JAXA launched an initiative on the Sentinel-Asia programme. This was inspired by the EU-ESA-sponsored GMES (Global Monitoring for Environment and Security) programme to provide regional imagery data for environment and disaster management. It uses the Japanese Earth observation satellite "Daichi (ALOS)" and NASA's MODIS as well as software and the Digital Asia network that was developed by Keio University of Japan. Furthermore, JAXA initiated the SAFE (Space Application For Environment) programme to analyse climate change by monitoring water resources, sea levels, forest degradation, and agricultural data.

However, these programmes were not enough to satisfy other Asian partners because they demanded further technology transfer. Thus, in 2009, JAXA initiated the STAR (Satellite Technology for the Asia-Pacific Region) programme, which mimicked APSCO's Small Multi-Mission Satellites (SMMS). The STAR programme includes Malaysia, Thailand, India, South Korea, Indonesia and Vietnam for developing small satellites (Micro-STAR and EO-STAR) together with JAXA. This was a big step ahead for APRSAF to become a more technology-oriented forum, but the commitment of JAXA to this programme seems to be ambivalent. Because of security concerns, JAXA is not free to transfer all necessary technologies for developing small satellites under various legal frameworks on export control and the "Exclusively Peaceful Purpose" clause in the Law to Establish JAXA. Although the Basic Space Law of 2008 allows JAXA to engage in intensive international cooperation including security related affairs, export controls over space technology to other Asian states remains rigid and stringent. Thus, in 2012, JAXA had to reorganise the STAR programme into the UNIFORM programme, which puts much less emphasis on technology transfer but more on capacity building.[11]

Furthermore, APRSAF launched a new initiative called the Regional Readiness Review for Key Climate Missions (Climate R^3) in 2012.[12] This is a joint initiative launched by Australia and Japan (JAXA) to build space data capabilities and regional readiness for climate change. The Climate R^3 initiative was a sign that Japan is coming back to cooperative leadership, together with a partner state.

6.4 Impact of Space Governance in Asia

The increasing competition between Japan and China for leadership in this region has several implications. First, it impacts on the global commercial space market. The hegemonic leadership exercised by China enables developing states to build their own capabilities to construct satellites and launch them on Chinese launchers. This means that one of the fastest growing markets in the world is dominated by China even under strict export control regimes such as ITAR. The launch of STAR by Japan was a counter-measure to compete against the extension of Chinese influence, but it has failed to capitalise. However, the demand for satellites and satellite-based services, which cannot be fulfilled by domestically developed satellites, remains high. The shift of Japan from hegemonic leadership to a cooperative one, as seen in the shift from STAR to the UNIFORM programme, can be considered to aim at penetration into the satellite-based services market.

However, the impact on space governance in Asia also includes more geostrategic issues. There are many territorial disputes in the region. Monitoring of the movements of vessels and troops is critically important for avoiding escalation of the tension in these disputes. Satellite-based Earth observation services will ultimately contribute to the stability of the region. Both Japan and China are developing not only their capabilities but also the capabilities of other states to interpret and utilise Earth observation imageries and remote sensing data through APRSAF and APSCO.

[11] "Transfer from STAR project to UNIFORM (STAR keikaku no UNIFORM he no ikou ni tsuite: Available only in Japanese)", JAXA press release, 7 March 2012.
http://www.jaxa.jp/press/2012/03/20120307_sac_star_j.html.

[12] Information on Climate R^3 can be found at: http://www.aprsaf.org/initiatives/climate/.

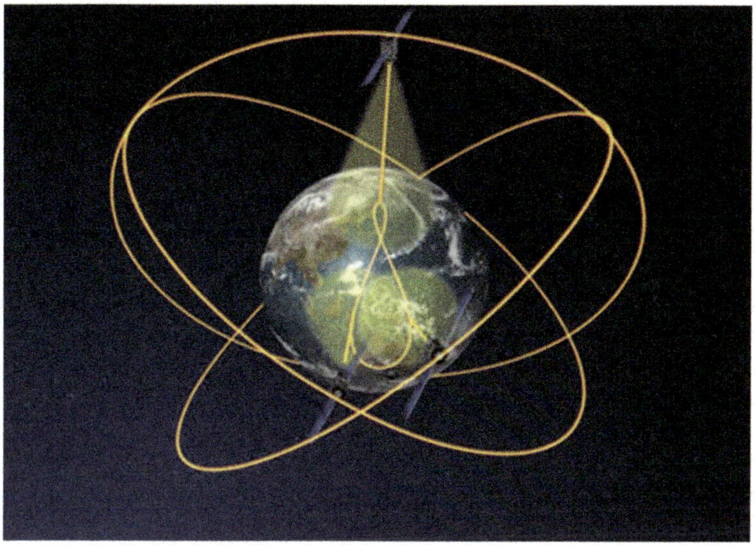

Fig. 6.1 Orbital image of QZSS (Source: JAXA)

Particular emphasis is on oceanography and maritime surveillance capabilities since most of the territorial disputes are focused on the ownership of islands.[13] This emphasis on maritime surveillance also requires navigation and communication capabilities derived from space. Determination of geo-spatial location is extremely important for avoiding unnecessary confrontation, which can only be done through satellite-based information. Maritime patrolling and real time communication is also vitally important to maintaining command and control over vessels on front-line. Since these locations are far from terrestrial communication networks, only satellite-based telecom services can fulfil these geostrategic needs.

In this regard, competition in the development and application of navigation satellite systems in Asia is particularly interesting. Japan is developing the Quasi-Zenith Satellite System (QZSS) while China is developing the Beidou (Compass) system. QZSS uses a unique orbit that allows a satellite to hover around the Zenith of Japanese cities, so that satellites can be seen in very high elevation (Fig. 6.1). It requires at least three satellites to provide continuous navigation signal from quasi-zenith positions. In other words, QZSS itself cannot provide sufficient signal to

[13] Warren Ferster, "Japan Considers Regional Maritime Surveillance Satellite Constellation", Space News, 11 April 2013.
http://spacenews.com/34795japan-considers-regional-maritime-surveillance-satellite-constellation/#sthash.kVyjMrKQ.dpuf;
"China to launch new marine surveillance satellites in 2019", Xinhua News Agency, 10 October 2014.
http://www.spacedaily.com/reports/China_to_launch_new_marine_surveillance_satellites_in_2019_999.html.

autonomously identify the location. Thus, QZSS has to be complemented by GPS signals and compatible with the US system. Although the Japanese government decided to launch the seven-satellite QZSS system in 2015, three on QZSS orbits and four on geostationary orbit for autonomous services, QZSS requires a GPS signal for greater accuracy. Furthermore, unlike GPS, QZSS is a regional system centred on Japan. The navigation signal can only be received in the East, South East Asia and Oceania regions. Thus, QZSS-based services are limited to regional services, which means that it is not applicable for the navigation of aircraft or maritime vessels.

Meanwhile, the Chinese Beidou system is being developed to give China an independent capability for satellite navigation. It is designed to be a constellation of 35 satellites for global and autonomous navigation services. Although it has similar functions to GPS, its orbital design is a bit different. It consists of 5 geostationary satellites and 30 medium Earth orbit (MEO) satellites. In the initial phase, China launched geostationary satellites and some MEO satellites to provide regional services in Asia. This was an attempt to directly compete with Japanese QZSS services. The initial deployment of these satellites was completed in 2012 and entered into service. Although the application and devices for Beidou services are not yet fully functional, China entered into service much ahead of the Japanese QZSS (Fig. 6.2).

As seen in the case of navigation satellite systems, Japan and China are in direct competition for gaining a share in the applications and services market, and also

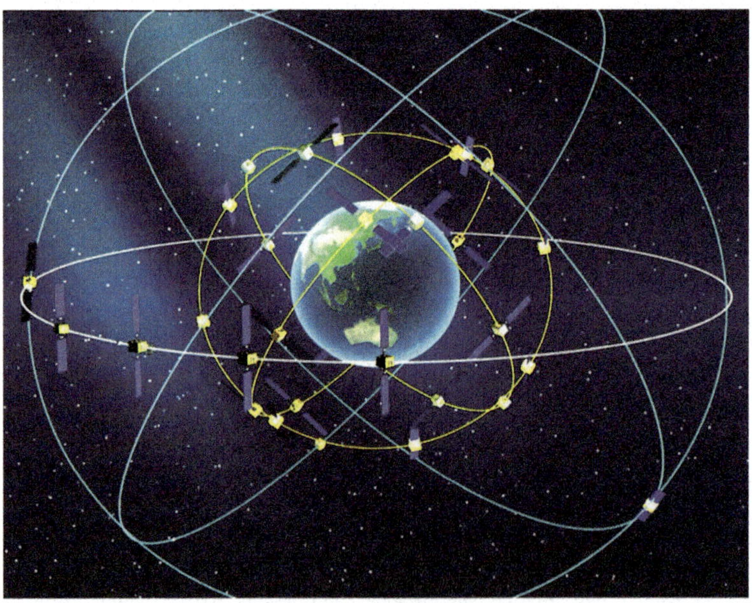

Fig. 6.2 Orbital image of Beidou (Compass) (Source: GPS World)

regional influence. At this moment, China seems to be ahead of Japan in launching its services, but Japan also has potential to regain its advantage.

The potential comes from the Japanese relationship with the United States. Based on an alliance with the United States, Japan enjoys a good relationship with the GPS operator, the Department of Defense, and developed QZSS as a compatible system with GPS. If successful, QZSS will provide added value to current GPS services, higher accuracy and additional messaging services, while Beidou has to develop a completely new system and new devices that overlap with the services already provided by GPS. Although Beidou has services additional to GPS, such as messaging, it will be difficult to expand its market by switching from current GPS users to Beidou.

Since QZSS is complementary to GPS, the United States is also expecting QZSS to play a role as a regional back up system in case of GPS failure. After consecutive Chinese ASAT tests, U.S. authorities considered multiple scenarios in which one or several GPS satellites become disabled. GPS-derived services are not only deeply embedded in socio-economic activities all over the world, but they are critical to U.S. military operation and strategic capabilities. If some GPS satellites were disabled, it would have significant implications for the civilian and military users of these services. Thus, QZSS as back up to GPS would provide a valuable addition to the American strategy.

6.5 Conclusions

Conventional wisdom tells us that it is geopolitics that defines space governance. During the Cold War, the governance of space was developed under the geopolitical structure of competition between the two superpowers. Such geopolitical competition was clearly reflected in the "space race". The race to land on the Moon, the race to longer space flights, and the race to explore Mars and other planets drove space technology much further at a very fast pace. Also numerous international collaborations took place under the Cold War framework, such as the International Space Station, Intelsat and Intersputnik.

International governance in Asia is also contested. Several territorial disputes and confrontations between national interests and nationalistic sentiments are raising tension among the major states in the region. Such confrontation is also reflected in space governance in Asia. Japan and China are now competing for influence and leadership in the region. However, there are differences in the nature of the leadership—Japan tends to exercise cooperative leadership whereas China prefers to exercise hegemonic leadership. And this creates fluid dynamics in the region. On the one hand, Chinese leadership provides more opportunities for the "have not" states to access advanced space technologies and learn from China. On the other hand, Japanese leadership provides more opportunities for "have not" states to access space-based data and applications for the benefit of their own social welfare.

The rivalry between Japan and China will is creating more opportunities and benefits for states that have not had the opportunity to develop autonomous capabilities for space. Geopolitical confrontation and competition for leadership could thus result in the greater good for the entire region. In this regard, the "Asian space race" or space race of any kind is not something to worry about, but rather it is a blessing for the "have nots". Space governance, based on the absence of territorial jurisdiction and the nature of the Global Commons, may provide an opportunity to rethink the geopolitical structure of international governance.

Chapter 7
The Optimum Role of Governments in Space

Gérard Brachet

7.1 An Historical Perspective

Historically, governments have played the major role in initiating and funding space exploration and space applications, including defense systems. The sector of satellite-based telecommunication is the only one where the commercial market started playing a strong role as early as the 1980s. Ever since, this sector developed independently of government funding but Governments continued playing a strong regulatory role, for example by disbanding former national telecommunication monopolies in Europe and by encouraging increased competition. This led in the 1990s to the transformation of Intelsat, Inmarsat and Eutelsat from international government organisations to private sector commercial operators.

All other areas of satellite-based applications areas continued to develop until recently thanks to significant government funding, be it satellite meteorology, remote sensing, navigation and positioning, search and rescue, etc.

Launch services developed mostly within a tight control by governments and it is only recently, with the California-based Space-X company and to a lesser extent with Orbital Sciences Corporation, that industry initiatives have demonstrated their ability to provide a reliable service with relative autonomy with respect to Government technical control.

In satellite-based applications, the role of industry, commercial operators and new entrepreneurs has developed tremendously over the last 15 years:

G. Brachet (✉)
Space Policy Consultant, 37 rue Tournefort, 75005 Paris, France
e-mail: Sicitur@wanadoo.fr

C. Al-Ekabi et al. (eds.), *Yearbook on Space Policy 2014*, Yearbook on Space Policy, DOI 10.1007/978-3-7091-1899-3_7

- Introduction of new concepts, particularly in Low Earth Orbit, in the area of telecommunication and broadcasting (Iridium, Globalstar, Orbcomm, O3B, etc.)
- Industry-initiated ventures in very high resolution remote sensing, encouraged in the United States by major purchase commitments from the US Government (GeoEye, Digital Globe)
- Development of new or emerging markets (zero G experiments, short manned flights to the boundary of outer space, etc.)
- Technology initiatives
- Public-Private Partnerships (PPP) and Private Funding Initiatives (PFI)

7.2 Role of Governments

This very welcome development of private sector initiatives does not mean that the role of Governments will disappear, but instead that it will evolve from a funding and space missions-oriented role toward the traditional role of governments in all matters, i.e. Policy formulation, Security, Regulatory, International relations, etc.

Indeed, the prime role of Governments in the area of space exploration and space exploitation can be summarised as follows:

- Definition of a nation' space policy;
- Setting up the regulatory framework, including licensing of commercial operators and establishing appropriate export control rules;
- Negotiating international conventions/treaties and bi- or multilateral cooperation agreements;
- Expressing the requirements for government-funded space missions, such as in defense and security, in "public good" applications such as meteorology, and in scientific endeavours and exploration;
- ...And of course in allocating an appropriate funding level for upstream research and for the development and operation of governmental space missions.

7.3 Where Are We Today?

In the field of satellite telecommunication and broadcasting, the evolution from government-funded programmes to commercial systems took place a long time ago, mostly in the 1990s and early 2000s. Today, it is a sector which is 100 % market driven, with little input from governments apart from some technology research funding. Dedicated telecommunication systems for the defense sector however remain largely in governments' hands, with one exception, the successful implementation of a "Private Funding initiative" (PFI) in the United Kingdom for the Skynet 5 military satellite telecommunication system via the Paradigm Secure Communication Company, a British registered company established as a 100 %

subsidiary by Airbus Defense and Space. No doubt that similar PFIs will be set up in Europe for military telecommunication satellite systems, provided that the military community in France, in Germany, in Italy, make up their mind between the "buy and own and the "lease" approaches. Obviously, the "lease" approach is easier to plan in budgetary terms and is widely used for other defense equipment such as refuelling aircraft tankers. It remains to be shown that it can be less costly than the "buy and own" approach.

In other areas of space activities, some governments have well understood the new opportunities that are offered by the commercial sector, as shown by the examples in the United States such as the wide use of commercial satellite telecommunication services by the US Department of Defense for non-sensitive communication or the very extensive purchase of satellite imagery from US commercial operators (Digital Globe) by the National Geospatial-Intelligence Agency.

In the field of launch services, as noted above, the successful upstart of the Space-X company, while strongly supported financially by its contract with NASA to service the International Space Station, demonstrates that an industry-led launcher development is not only feasible, but can be very competitive with government backed launch systems. This is a lesson for the European actors in this field; they may have relied too long on the relative comfort of the Ariane 4 and Ariane 5 successes in the 1980s, 1990s and 2000s. The reliability of the Ariane launchers allowed Arianespace to gain a good 50 % of the world open market for commercial telecommunication satellites but this undeniable success may have reduced its shareholders' ability to sense the winds of changes...

This led to some years of soul searching at government and at industry level in order to agree on what to do next. The decision taken by the European governments at the ESA ministerial Council meeting in Luxembourg on Dec 2, 2014, to go ahead with the funding of the development of a new Ariane 6 launcher, together with a decision to switch to an industry-led development and production, will be an excellent test of the European Governments' willingness to relent the tight control that they have had so far on launch systems development and production.

In other areas of space activities, the European space industry has shown its dynamism and competitiveness by gaining a large share of the export market of telecommunication satellites, including constellations (60 %) and a very large share of the export market of observation satellites (75 %). However, with the exception of the Skynet 5 example quoted above, the industry-led initiatives to develop new space programmes have been relatively few and generally found it difficult to convince European governments of the soundness of their proposals. For example, a major initiative was taken by Astrium—now the space component of Airbus Defense & Space—in the late 1990s when they purchased from CNES, the French Space Agency, its 39 % share of Spot Image S.A., the commercial operator of the SPOT system, thereby taking full control of the company, and decided to invest in the funding of the SPOT-6 and SPOT-7 satellites. This decision was taken without any purchase commitment of satellite images from either the French government nor from the European Commission. Indeed, attempts to convince the French government to commit to a minimum volume of purchase were made by the management of Astrium but without success. The risks taken by EADS (now Airbus), the mother company of Astrium, were therefore quite high, in fact possibly unreasonable. This is a rather sad story since a more modest initiative in this area of remote sensing from space had been successfully implemented in the late 1990 when Astrium, CNES and Spot image had put together a public-private partnership to co-fund (55 % Astrium, 45 % CNES) the development of the new High Resolution Stereoscopic (HRS) instrument which was carried on board the SPOT-5 satellite, launched in 2002. This industrial initiative was backed by a significant purchase agreement from the French Ministry of Defense of Digital Terrain Models (DTM) derived from the stereoscopic images collected by HRS. It was a very successful operation, contributing in a significant manner to the large increase in the Spot Image's turnover following the launch of SPOT-5.

7.4 Historical Space Agencies Need to Adapt

As more industrial initiatives are taking place, sometimes in areas where a commercial approach had not even been considered by the policy making circles in governments, it is clear that established space agencies need to adjust their mission and mode of operation accordingly.

In the United States, this applies to NASA, in particular for the servicing of the International Space Station, but will soon affect other areas of the NASA historical responsibility, including research in microgravity and space exploration. It will affect also NOAA in the way they build and operate their dedicated weather spacecraft, with several private ventures proposing to establish the space-based observation systems and selling the service to the federal government in a similar fashion to the Skynet 5 model in telecommunications.

In Europe, the established space agencies such as ESA, ASI, CNES, DLR, etc. are already facing this new paradigm, for example in the field of satellite telecommunications and launch systems and need to adapt to a situation where their degree of control is less than before, with industry taking more risks. The various partnerships that ESA has developed with industry such as the Alphasat partnership with Inmarsat or the ERDS data relay satellite partnership with Airbus Defense & Space, which involves the European Commission via its Copernicus programme, are good example of such new approaches.

The space agencies will also be called upon by the national administrations in charge of regulating the private sector commercial space activities to provide their expertise in the licensing process and in monitoring the private operators' compliance with the regulations. This will require some re-allocation of resources within space agencies and possibly hiring of more personnel having both a technical and a legal background.

7.5 The Political Level

The political level needs to understand the new situation created by industry's initiatives in space activities and be made aware of their Country's responsibility as per the Outer State Treaty of 1967 and the space-related international conventions that followed. States are ultimately responsible for all space activities conducted by their nationals, be they government entities or private operators under their jurisdiction. This calls for more attention to be paid to the legal and regulatory framework required for the private sector's activities to develop. The United States clearly leads the way in this area but European countries are also actively setting the regulatory framework, as shown for example by the national space laws adopted by several European states in the last few years. In France, for example, the national space law "Loi relatives aux opérations spatiales" adopted by the French Parliament in June 2008[1] is a good illustration of the necessary steps that need to be taken to create the stable legal framework for commercial operators to develop their activities.

The need to define a clear national regulatory framework in each country is not immediately obvious for policy makers that have generally a very superficial knowledge and understanding of the space environment and of space activities. This is where space agencies have an important role to play: educating the decision makers, administrators, members of parliament, ministers, etc. As noted above, their role also can be very significant in the implementation of the regulatory framework because of the highly technical nature of the work that needs to be done: technical analysis of filing by operators, monitoring of compliance, providing technical expertise to inquiry boards set up by governments when mishaps happen, etc.

7.6 Conclusion

In conclusion, the optimum role of governments in space will evolve, or more precisely is evolving already, in a significant manner: it will no longer be primarily as promoter of space technology and space missions - apart from some government-dedicated space missions—but rather as a regulator of an economic activity where industry is taking the lead and will no doubt come up with very innovative and unproven concepts, thereby increasing the risks of mishaps, some of which will involve the state's responsibility.

[1] Loi n° 2008-518 du 3 juin 2008 relative aux opérations spatiales.

Chapter 8
What Place for the People? The Role of the Public and NGOs in Space Innovation and Governance

Amy Kaminski

8.1 Introduction

A small set of actors has dominated civil space policy and programme development in the United States since the inception of the National Aeronautics and Space Administration (NASA). NASA officials, major aerospace firms, space scientists in academia, U.S. presidential administrations, members of Congress, and leaders of other U.S. government offices—often working in concert with or in reaction to the policies of foreign national space agencies—have guided decisions concerning what space projects the nation has undertaken. Walter McDougall's *...The Heavens and the Earth*, W. D. Kay's *Can Democracies Fly in Space?*, and many other works describe the powerful state, industry, and university actors who contributed to NASA's rise in the late 1950s and helped to shape the agency's sense of identity, organisational culture, programme choices, and external relationships, which have been focused so centrally on human space flight, over its decades of existence.[1]

The views expressed herein are those of the author. They do not necessarily reflect the views of NASA or the U.S. government

[1] McDougall, WA (1985) ...The heavens and the earth: a political history of the space age. Johns Hopkins, Baltimore; Kay, WD (1995) Can democracies fly in space? The challenge of revitalising the U.S. space programme. Praeger, Westport. See also, for example, Bromberg, JL (1999) NASA and the space industry. Johns Hopkins, Baltimore; Launius, RD, McCurdy, HE (eds) (1997) Spaceflight and the myth of presidential leadership. University of Illinois, Chicago; Logsdon, JM (2010) John F. Kennedy and the race to the moon. Palgrave Macmillan, New York; Logsdon, JM (2015) After Apollo? Richard Nixon and the American space programme. Palgrave Macmillan, New York; Smith, RW (1989) The space telescope: a study of NASA, science, technology, and politics. Cambridge University, Cambridge.

A. Kaminski (✉)
Office of the Chief Scientist at NASA, Washington, DC, USA
e-mail: amy.p.kaminski@nasa.gov

C. Al-Ekabi et al. (eds.), *Yearbook on Space Policy 2014*, Yearbook on Space Policy, DOI 10.1007/978-3-7091-1899-3_8

In contrast, historical accounts depict NASA's interactions with American citizens beyond government agencies and aerospace firms since the 1950s and 1960s as efforts to "sell" its human space flight initiatives and to position external publics as would-be observers and supporters of such activities. These portrayals of NASA's engagement of American citizens during the Mercury, Gemini, and Apollo human space flight programmes comprise reasoned observations and meticulous research conducted by well-regarded scholars. Only a few works, including Howard McCurdy's *Space and the American Imagination* and Michael A.G. Michaud's *Reaching for the High Frontier*, have suggested that citizens outside of the identified sphere of space programme developers have played more active roles in the American space programme.[2]

A close examination of NASA's history to the present day, however, indicates that the U.S. space agency has grappled throughout its decades of existence with how best to engage and involve citizens with the governance and accomplishment of its space activities. Increasing pressures from a variety of quarters since the end of the Apollo programme have spurred NASA to consider, and to accommodate, more active roles for new entities, operating either individually or as part of non-governmental organisations (NGOs), in the nation's space programme. This paper analyses this participatory trend's emergence. I examine the reasons for this trend's development and show how NASA has facilitated it. In addition, I flag the limits and considerable challenges that NASA has faced and continues to encounter in opening opportunities for external groups to contribute directly to shaping the space programme. I argue that NASA, and quite possibly other national space agencies, must pay close attention to the issue of participatory public engagement and embrace the involvement of their publics in their work in order to preserve the legitimacy of their space activities among the citizenries that sustain them.

8.2 From "Consumer-Product" to More Substantive Public Engagement

At the beginning of the Space Age, the role of most American citizens in the governance of the U.S. space programme was relatively simple and straightforward. During the Mercury, Gemini, and Apollo eras of space flight, government, industry, and university stakeholders shaped space policy. NASA officials and their counterparts in developing space policy tended to regard "ordinary" Americans as

[2] McCurdy, HE (1997) Space and the American imagination. Smithsonian, Washington; Michaud, MAG (1986) Reaching for the high frontier: The American pro-space movement, 1972–1984. Praeger, New York.

witness to and consumers of the space programme, focusing on how to demonstrate accountability and to garner public appreciation of their achievements.[3]

This relationship between NASA and American citizens is deeply rooted in the agency's origins in the mid-twentieth century. During that time, many U.S. government leaders, techno-scientific experts, and citizens alike subscribed to the belief advanced after World War II by President Franklin D. Roosevelt's science advisor, Vannevar Bush, that the results of government-funded scientific research and development activities would ultimately serve the nation's people.[4] The U.S. government created several new agencies focused on techno-scientific advances, and the technocrats and experts who managed them and conducted research under their auspices aspired to attain public appreciation of their research and development work via displays of their achievements and informing their citizenries about the value of their efforts.[5]

NASA was founded and moulded according to these principles. President Eisenhower responded to the perceived technological and ideological threats of the Soviet Union's deployment into Earth orbit of *Sputnik* in 1957 by establishing NASA and authorising the agency to embark on a programme of space research. Eisenhower and the Congressional authors of NASA's originating legislation, the National Aeronautics and Space Act of 1958, mandated that NASA, a publicly funded organisation, accordingly work "for the benefit of all mankind." Concomitantly, the drafters of the Space Act recognised that attaining legitimacy of their vision of America made stronger via a national space programme depended on ensuring both a global and national public awareness of the agency's activities. Thus, the legislation mandated that NASA "provide for the widest practicable and appropriate dissemination of information concerning its activities and the results thereof." This seminal phrase served from the outset to guide and define NASA's relationship with those outside the government and the aerospace industry. NASA

[3] Kauffman, JL (1994) Selling outer space: Kennedy, the media, and funding for project Apollo, 1961–1963. University of Alabama, Tuscaloosa, AL; Byrnes, ME (1994) Politics and space: image making by NASA. Praeger, Westport; Starr, KA (2008) NASA's hidden power: NACA/NASA public relations and the cold war, 1945–1967. Dissertation, Auburn University; Lewenstein, BV (1993) NASA and the public understanding of space science. Journal of the British Interplanetary Society 46: 251–254; Billings, L (2010) Fifty years of NASA and the public: what NASA? What Publics? In: Dick, SJ (ed) NASA's first 50 years: historical perspectives. U.S. Government Printing Office, Washington, pp 151–181; Scott DM, Jurek, R (2014) Marketing the moon: the selling of the Apollo lunar programme. MIT, Cambridge, MA; Kaminski, AP (2015) Sharing the shuttle with America: NASA and public engagement after Apollo. Dissertation, Virginia Polytechnic Institute and State University.

[4] Bush, V (1945) Science: the endless frontier: a report to the president. U.S. Government Printing Office, Washington.

[5] See, for example, Ezrahi, Y (1990) The descent of Icarus: science and the transformation of contemporary democracy. Harvard, Cambridge, MA; Macdonald, S (1998) The politics of display: museums, science, culture. Routledge, New York; Schmid, SD (2006) Celebrating tomorrow today: the peaceful atom on display in the Soviet Union. Social Studies of Science 36 (3):331–365; Sarewitz, D (1996) Frontiers of illusion: science, technology, and the politics of progress Temple, Philadelphia.

personnel committed to using all available communications modes to share openly with American and global citizens one successful advance in human space flight after another throughout the 1960s in the pursuit of an eventual Moon shot—the ultimate display of the nation's prowess. The agency developed positive images of space flight, forged strong relations with the news media, welcomed public visitors to NASA facilities and launches, and conducted extensive public outreach and education campaigns.

Many Americans were quite satisfied with NASA defining the public experience with space, thrilling at and swelling with patriotism on the occasion of space launches. By the late 1960s, however, problems with this model of public engagement became evident. For the millions of Americans who got caught up in the excitement of the space race and wrote "fan" letters to NASA's astronauts, there were just as many who frowned on the nation's commitment to human space exploration. By that time, the use of chemical weapons, nuclear reactors, pesticides, and other contentious creations of research and development activities revealed that science and technology did not de facto benefit all segments of society or the environment. Social activists and scholars consequently began to question the propriety of according scientific experts and technocrats unchecked authority, with many demanding for citizens no longer to be viewed as passive consumers of these entities' actions but as participants in deciding the rightful place of science and technology within society.[6] Where NASA was concerned, the spacemen's socially homogeneous origins as white, male, military test pilots seemed unrepresentative of the diversity of the United States and out of touch with contemporary public pursuits of civil rights and equal opportunity.[7] Moreover, with social and economic turmoil at home and an unpopular war in Vietnam taking a toll on American lives and finances, many citizens vocally opposed human space flight activity's quest to demonstrate technological and ideological superiority over the Soviet Union with no clear long-term benefits for individual American citizens.

These public sentiments reached the agency through opinion polls, letters, and media critiques. Regarding citizen endorsement of its activities as crucial to the agency's legitimacy and continued Congressional funding, NASA officials responded by striving to make space activities more directly relevant to society. Long emphasising the agency's human space flight achievements, they realised this shift in part by focusing new attention on initiatives like Earth observations satellites and explaining the "spinoff" benefits of space activities in promotional materials and exhibitions. Further, NASA reframed the function of the human space programme from primarily serving geopolitical and national security aims to making space accessible for all sorts of uses and users. Indeed, officials branded

[6] See, for example, Price, DK (1967) The scientific estate. Harvard, Cambridge, MA; Nelkin, D (1975) The political impact of technical expertise. Social studies of science 5(1):35–54; Irwin, A and Wynne, B (eds) (1996) Misunderstanding science? The public reconstruction of science and technology. Cambridge, Cambridge.

[7] For more on the evolution of the astronaut profession, see Hersch, MH (2012) Inventing the American astronaut. Palgrave Macmillan, New York.

the new Space Shuttle as a "space truck" and cast astronauts who would ride aboard the vehicle as workers who would routinely deliver the benefits of space to society.

NASA put this notion of a more utilitarian and democratised space programme into practice by expanding the ranks of space programme participants. Beginning in the Shuttle's development period, NASA established rules and regulations to promote use of the Shuttle and in turn generated diversity in who could put payloads into space and who could ride into space. The Shuttle served NASA and other U.S. government agencies' satellite delivery needs but offered affordable flight opportunities and thus also accommodated satellites and research payloads owned by corporations, foreign space agencies, school students, and individuals. In an effort to comport to social expectations for demographic diversity in the workplace, NASA also expanded the diversity of its astronaut corps by actively recruiting women and racial and ethnic minorities with science, engineering, medical, or military credentials. The agency invited "guest" flyers from other nations and corporations and also began a programme to fly "ordinary citizens" in space in response to extensive public enthusiasm and interest in personal space travel. Further, NASA partnered with various NGOs to offer American and worldwide citizens opportunities to engage closely with the Shuttle programme. The agency, for example, worked with the American Radio Relay League and the Amateur Radio Satellite Corporation to enable schoolchildren to dialogue with Shuttle astronauts during their missions while teaming with a grassroots National Space Institute to allow anyone to dial a phone number to hear communications between the astronauts in flight and ground controllers.

Opening space flight to new groups of users and flyers was not easy for NASA. Proponents of these initiatives encountered resistance from others within the agency and among traditional space policy developers who were concerned about these initiatives' ramifications for the priorities of professional scientists, astronaut safety, and NASA's image of competence and seriousness of purpose. The explosion of the Space Shuttle *Challenger* in 1986 in particular dampened NASA's ability and willingness to expand such opportunities greatly. But a model of "democratisation"—seeing citizens as not just passive supporters but active participants—began to emerge alongside the consumer-product model as one means of public engagement as top officials desired to make NASA relevant.

8.3 An Increasing Participatory Ethos

Attention within NASA and other parts of the U.S. government on broadening engagement with space activity continued to grow in the 1990s and 2000s, extending to programmes and topical areas beyond the Space Shuttle. One reason why is that citizen space enthusiasts have continued to make known their desires to realize specific visions for space exploration and exploitation and to fly in space themselves. Such enthusiasts have banded into a variety of grassroots space advocacy groups which have for decades endeavoured to influence government space programmes with varying degrees of success. Many new NGOs and companies have emerged and strived to make their own marks in space. The X Prize Foundation, for example, announced in 1996 its intent to offer a prize of ten million dollars to stimulate private teams to build and fly a reusable vehicle capable of carrying humans to space and returning them safely to Earth and then sending the craft into space again within two weeks. More recently, a citizen-driven team undertook a crowd-funded effort to reboot the International Sun Earth Explorer 3 (ISEE-3) spacecraft, launched by NASA in the 1970s and since dormant due to budget availability. These privately initiated activities have pressed NASA to consider new partnerships and capabilities in achieving goals in space. The agency has, for instance, entered into agreements with companies such as Nanoracks, which has developed and leases hardware to support educational and commercial research aboard the International Space Station. The emergence of private space activity has also engaged other U.S. government institutions such as the Federal Aviation Administration, which has jurisdiction over commercial space flights in order to ensure their compliance with safety regulations as well as international laws and treaties.

Approaches to space activities are also being shaped by evolving ideas about the nominal public role in science and technology activities. Over the past few decades, the idea that citizens should be regarded as not only supporters or consumers of science and technology but also active participants in shaping these enterprises has gained traction among governments, techno-scientific experts, and citizens. Their beliefs have shifted regarding the locus of knowledge and propriety of expert-dominated governance of issues involving science and technology, particularly where choices have profound social implications. Indeed, citizens lacking scientific or technical credentials have proven on multiple occasions that they nonetheless often possess important knowledge, skills, and values that professionals may lack but bear on the ability to address effectively issues involving science and technology. For example, AIDS activists have teamed with physicians and researchers to develop more effective clinical drug trial protocols through the combination of medical expertise and patients' knowledge of their disease.[8] Similarly, local awareness and concern led citizens in Woburn, Massachusetts, to identify an unusually high incidence of cancer in their region and, working in partnership with credentialed experts, to link them to the presence of nearby areas of industrial chemical contamination.[9]

The proliferation of the Internet and social media has catalysed these trends. These tools have provided a platform and empowered people to voice their views and concerns and to band with others to champion their causes. They have also proven the "power of the crowd," the ability of a broad range of people all

[8] Epstein, S (1996) Impure science: AIDS, activism, and the politics of knowledge. University of California, Berkeley.

[9] Fischer, F (2001) Citizens, experts, and the environment: the politics of local knowledge. Duke, Durham.

considering an issue, to generate innovative solutions for tough sociotechnical problems. Increasingly, scholars, government policy makers, scientific professionals, and concerned citizens around the world have begun to promote the idea that citizens should be not just informed of policy decisions but also involved in setting agendas and contributing to solving challenges faced by institutional stewards of scientific and technological activities.

Consequently, governments and other institutions have undertaken initiatives to include citizens in decision-making and problem-solving processes involving techno-scientific matters. Europe has been a leader in using "citizen juries" to evaluate technologies such as genetically modified foods and nanotechnologies to decide whether and how they should best be welcomed and used in society. In the United States, the Obama administration introduced the Open Government Initiative, which directs U.S. government agencies to act transparently and to collaborate wherever appropriate with citizens and NGOs. As part of this initiative, the www.challenge.gov web portal has offered more than 300 prize competitions offering more than $64 million in awards to invite citizens to bring their knowledge, skills, and creativity to bear in helping U.S. government agencies solve problems.

Given these many circumstances plus a continued desire to remain relevant and legitimate in the eyes of American and global citizens, NASA has continued to open itself to various types of public involvement in its work. The agency has included declarations of this commitment in planning documents. NASA's Open Government Plan, a document required by all U.S. government agencies under the Obama administration, acknowledges that the public will "directly contribute to expanding human presence into the solar system."[10] Meanwhile, the 2014 NASA strategic plan states that the agency regards individual citizens as partners in its work, much as it has long depended on industry, academia, and international space agencies to advance goals in space.[11] Accordingly, NASA has been recognised by the White House Office of Science and Technology Policy as a leader within the U.S. government in embracing new forms of public involvement.[12]

On one level, NASA's efforts to engage with citizens have remained a form of promoting public awareness and understanding through greater immersion with NASA content and people thanks to the availability of new information technologies. The agency is using social media extensively to share its messages. Many of the astronauts use Twitter accounts to communicate with hundreds of thousands of followers during missions. NASA also hosts regular "NASA Socials," in which social media users apply to be selected to visit a NASA facility or attend a launch to meet some NASA staff and learn more about the agency and its work. The aim of

---

[10] NASA (2010) NASA open government plan. http://www.nasa.gov/open/plan/index_prt.htm. Accessed 3 May 2015.

[11] NASA (2014) NASA strategic plan. https://www.nasa.gov/sites/default/files/files/FY2014_NASA_SP_508c.pdf. Accessed 3 May 2015.

[12] Office of Science and Technology Policy (2013) Implementation of federal prize authority: fiscal year 2012 progress report. https://www.whitehouse.gov/sites/default/files/microsites/ostp/competes_prizesreport_dec-2013.pdf. Accessed 3 May 2015.

such events is to raise participants' consciousness of NASA as well as that of their social media contacts. In addition, the agency conducts contests and offers opportunities for citizens to name spacecraft, send their own names into space aboard certain missions, or otherwise participate in some simple fashion in its activities.

On another level, however, NASA, like other techno-scientific institutions and observers, has recognised the value of engaging a wide range of American and global citizens in its work. Some officials have begun to champion the idea that individual citizens and NGOs can contribute to the advancement of scientific knowledge and the development of data applications as well as hardware, software, and algorithms to support future space missions. Rather than relying only on traditional procurement methods for scientific and technical work including grants and contracts, NASA is also using prize competitions and crowdsourcing to bring ideas and technological solutions into the agency and further space capabilities. In doing so, the agency is tapping into the talents of all sorts of people, from computer coders to garage tinkerers to students and others. NASA also sponsors several opportunities for citizens with little technical background or skill to assist NASA in collecting or analysing scientific data.

Rather than only announcing opportunities through channels familiar only to universities and corporations, NASA is looking for prospective participants where they are online, using social media, challenge.gov, and other broadly accessible and widely visited platforms to share opportunities. As a result, the reach of these initiatives and their outcomes have been staggering. For example, the third annual International Space Apps Challenge, pioneered by NASA, attracted more than 8000 volunteers in 95 cities across 46 countries to participate in creating solutions to 40 data applications challenges supporting NASA's and other space agencies' goals. In the programme's 3-year span, more than 2000 crowd-sourced solutions have resulted in methods to monitor air, water, and urban pollution, track environmental mishaps, alert citizens of weather or health-related disasters, and track the stars. NASA's Disk Detective project invites participants to follow a short online tutorial and then begin categorising images of stars taken by space- and ground-based telescopes with the goal of finding candidate stars with debris disks, indicating potential locations of planet formation. In less than a year, Disk Detective volunteers classified one million images and the most avid participants were invited by the project's organizers to work alongside the science team on planning for follow-up observations of the candidate stars.

While these approaches are not appropriate for every research or technological challenge NASA wishes to solve, results to date show that they offer numerous benefits when applied to suitable problems. Prize competitions linked to particular challenges or problems can allow NASA to evaluate a wide range of potential solutions, bring out-of-discipline perspectives to bear, avoid needing to predict in advance which team or approach is most likely to succeed, and make payments only after participants demonstrate results. Inviting citizens to aid in data collection or analysis enables NASA researchers to harness the cognitive power of the crowd and gather more data in more locations and analyse them in shorter periods of time and at far lower costs than possible with small teams of experts.

To be sure, obstacles remain to NASA's widespread acceptance of and reliance on public involvement in research and innovation. Legal and procurement procedures hinder the agency's ability to incorporate readily into the agency's sphere of use the technologies citizens have developed through prize competitions. Meanwhile, some scientists are not fully convinced that citizens lacking technical degrees can generate or analyse data of suitable quality to produce valid results. A re-examination of policies and perceptions will be necessary to take full advantage of these approaches.

8.4 The Future of Public Involvement in Space Governance

One might ask, however, how far NASA or other space agencies will allow public involvement in their work to extend. Should citizens and NGOs be invited, for example, to aid in setting goals for space activity?

Governments and space agencies have, to a limited degree, been supportive of this kind of public engagement. Occasionally since the mid-1980s, NASA and various commissions established by U.S. presidents and congresses to help recommend directions for the space programme have reached out to solicit citizen views.[13] In 2006 the European Space Agency conducted a citizen jury to gauge public interest in its own space activities.[14] In November 2014 NASA partnered with an NGO, the Expert and Citizen Assessment of Science and Technology network, to convene forums of citizens in two American cities to hear their views and values regarding NASA's plans to search for potentially hazardous asteroids and to relocate an asteroid or a piece of an asteroid to lunar orbit for humans to explore.

Making space policy decisions more democratic is not simple in practice. Soliciting public views requires staffing and budget, which are both limited resources. It requires a space agency to consider logistics such as how to provide mechanisms for citizen input that are socially and culturally appropriate for the population. Citizens will be empowered to express their views only if they understand and recognize the significance of the issues in question and are provided with accessible, unintimidating platforms for registering their opinions. But beyond issues of accessibility, agency and government decision-makers must determine whose preferences prevail. What weight should they accord to citizen inputs versus the perspectives of longstanding stakeholders and experts, whose livelihoods depend on space programmes? Further, what happens if a sizable proportion of

---

[13] See, for example, National Commission on Space (1986) Pioneering the space frontier: the report of the National Commission on Space. Bantam, New York; NASA (1993) Toward a shared vision: 1992 town meetings. U.S. Government Printing Office, Washington.

[14] Safwat, B, Stilgoe, J, Gillinson S (2006) Open space: a citizen's jury on space exploration. http://esamultimedia.esa.int/docs/exploration/Public/DEMOS_Space_Jury_final_report_v5.pdf. Accessed 3 May 2015.

citizens disagrees with a direction that elected officials regard as important for the national good? Or if they prefer a direction that policy officials consider detrimental?

For all of these reasons, NASA and U.S. decision-makers have rarely involved citizens beyond inviting them to become aware of or advance already-set goals. However, there are numerous questions having budgetary, social, and ethical implications that would benefit from being openly and broadly deliberated in society. For example, what should be priority aims and objectives for government-sponsored space exploration? What roles should be left to the private sector? Should taxpayer funds be directed to nurturing new space industries? What should be the relative emphasis on using space applications to protect and sustain life on Earth versus exploring space and attaining scientific knowledge? What investment should governments make in human space flight? How important is it to be mindful of the environmental impacts and cultural sensitivities concerning activities in space, including the exploitation of other planets? How important is it to accept risks and press boundaries versus trying to assure reliability and success with each and every project? What sorts of timeframes and funding commitments should be established for federal space exploration efforts?

Citizen involvement in space governance will almost certainly complicate policy decisions, but multi-way engagement is the future of our hyper-networked societies. Government institutions now solicit public views on a variety of policy issues in the United States as well as in other countries. Polls in the United States have consistently shown that Americans are divided on the question of whether the returns of NASA's space programmes justify the billions of dollars the nation invests annually in them.[15] Not asking for or ignoring input means government-run space programmes risk becoming perceived as more esoteric and irrelevant than many people already perceive them to be. Inviting broad views could make space programmes more meaningful to citizens.

8.5 Conclusion

This paper has considered how and why NASA has engaged with the citizens of the United States over its 56 years of existence. Since the Apollo era, NASA has endured increasing pressures to involve citizens more actively in its work. Given this persistent trend, one is likely to see the U.S. space agency continue to strive to make headway against present obstacles to accommodate and include more individuals and NGOs as participants in the quest to achieve its goals. Indeed, these approaches are bound to prove vital to NASA not only as it aims to act in response

[15] Launius, RD (2003) Public opinion polls and perceptions of U.S. human space flight. Space Policy 19:163–175.

to these pressures but also as the agency strives to achieve more with limited budgets.

How government-run space programmes compare to the NASA case and to one another in their involvement of their citizenries and NGOs and in the benefits they do or would accrue from such engagement is also a topic worthy of examination. The insights gained should prove instructive to space policy development for national space programmes around the globe. Specifically, the findings should help space programme officials examine their assumptions about their publics and assess how best to conceptualize them, space programme choices, and public engagement alternatives in order to serve them most effectively in the present day. As with the United States, taking such measures may well be crucial to preserving the relevance of other national space programmes in the face of evolving external attitudes about the place of governments in space activities.

Illustrations and Links

NASA conceptualised its Space Shuttle as a "space truck" that would open space to new user groups and uses:

Source: http://www.nasa.gov/centers/goddard/images/content/542464main_SSPP bridge.jpg

NASA's Open Government Plan lays out the agency's plans and mechanisms for involving members of the public in space activities: http://www.nasa.gov/open/plan/index_prt.htm

NASA's Centennial Challenges programme invites individuals and teams to propose solutions to tough problems in space technology development, such as robots that can locate and return samples from a terrain without human intervention, as shown here: https://www.flickr.com/photos/nasahqphoto/14383756106/

A consortium of academic, informal education, and policy analysis institutions partnered with NASA to conduct two forums in November 2014 bringing together citizens of many backgrounds to solicit their views on NASA's plans to find potentially hazardous asteroids and move an asteroid or piece of an asteroid to lunar orbit for human exploration.

http://ecastnetwork.org/2015/02/18/consider-it-the-final-phase-of-online-forum-to-inform-nasas-asteroid/

Source: https://ecastnetwork.files.wordpress.com/2015/02/p1060131.jpg?w=624

Chapter 9
The International Code of Conduct and Space Sustainability

Rajeswari Pillai Rajagopalan

Outer space security is generally construed as a security challenge of the future. However, in reality, it poses itself as a clear and present danger already. Space security and sustainability of outer space have thus begun to assume a new currency in the international discourse on global security, at least within a smaller community. Yet there are no effective outer space regimes in place. There is an effort within this backdrop to write new rules of the road for outer space activities. This has been driven primarily by the fact that the number and types of players in this domain are changing dramatically and that outer space has become crowded, congested and contested.

Meanwhile, China's anti-satellite (ASAT) test in January 2007 has had a compelling impact on countries to review their perspective on the "peaceful use vs. militarisation of outer space" debate. There are certain treaties and legal instruments in this domain, however, these measures have not been able to deal with some of the challenges of the day. Efforts must be taken either to tweak the current instruments or frame new ones in recognition of the emerging landscape as it pertains to outer space. Any new instrument that might take shape must factor in the following: define and communicate clear boundaries of responsible behaviour, and develop norms of responsible behaviour, while working towards more binding mechanisms.

9.1 Why Emphasise Responsible Behaviour?

Challenges facing outer space domain have grown immensely in the last decade, with changes in the nature of space programmes and capabilities being the most significant. Countries are increasingly pursuing advanced military space

R.P. Rajagopalan (✉)
Observer Research Foundation, New Delhi, Delhi, India
e-mail: rajeswarirajagopalan@gmail.com

C. Al-Ekabi et al. (eds.), *Yearbook on Space Policy 2014*, Yearbook on Space Policy, DOI 10.1007/978-3-7091-1899-3_9

programmes, including ASAT capabilities, which are inherently destabilising in nature. Potential destructive use of ASAT weapons "for interfering with enemy satellites" under conditions of hostility is well-known.[1] Past literature on the subject examined such use in the US-Soviet context. However, a study done in the current context will feature China as having such destructive capabilities. China's ASAT test in January 2007 became a wake-up call for India on the kind of destabilising threats in its neighbourhood. Emergence of new actors and technologies is also beginning to challenge the sustainability of outer space. The number of actors has grown from a small group of 2–3 countries to more than 60 players, including non-state actors. This has created its own dynamics. Along with the growth in the number of players, there is also the phenomenon of proliferation of small satellites—mini, micro, nano satellites, making the outer space not only crowded but also making the process of tracking and detection difficult. Also, new technologies in the form of counter-space technologies are entering the realm, producing new uncertainties. Cyber-enabled space threats are also becoming significant, challenging the sustained use of outer space. With these new threats, the theatre of conflict, means and impact are across the domains.

A related development is the debate between weapons in outer space and ground-based weapons that have the same destructive capabilities as those of those placed in outer space orbits. A certain group of countries argue that weapons placed in outer space pose as the biggest threat today, but in reality the potency and dangers of weapons based on the ground that are used for targeting assets in outer space, such as ASAT weapons, are far greater. A third major challenge in outer space is the issue of spectrum allocation and radio frequency interference. Radio frequency spectrum like most other resources is a finite commodity and more number of space actors means overcrowding and congestion, which could itself lead to radio frequency interference, even if unintentional in nature. Space weather and other natural events could also result in radio frequency interference. But increasingly, states are developing capabilities for intentional jamming and frequency interference as a means of "denying or degrading an adversary's ability to use space capabilities."[2]

While spectrum allocation has not become a serious issue yet, the growth in the number of actors mean that it has the propensity to become one in the coming years. Space debris constitutes the fourth major challenge for the sustainable and uninterrupted use of outer space. Rising levels of space debris are already risking space assets. Ecuador lost its sole satellite, a cubesat named NEE-01 Pegasus, hit by a Russian piece of space debris, SCC-15890.[3] Today, there are more than 21,000

[1] U.S. Congress, office of Technology Assessment, Anti-Satellite Weapons, Counter-measures, and Arms Control. OTA-1S (Washington, Government Printing office, September 1985), http://www.princeton.edu/~ota/disk2/1985/8502/8502.PDF.

[2] "Radio Frequency Spectrum, Interference and Satellites Fact Sheet," Secure World Foundation, Updated June 25, 2013, http://swfound.org/media/108538/swf_rfi_fact_sheet_2013.pdf.

[3] Ronnie Nader, "It Happened to Us," in Rajeswari Pillai Rajagopalan and Daniel A Porras, *Awaiting the Launch: Perspectives on the Draft Code of Conduct for Outer Space Activities*

pieces of debris that are larger than 10 cm, approximately 500,000 of size between 1 and 10 cm and smaller than 1 cm exceeds 100 million (see chart). The growth of population is best illustrated in the following pictorial illustration.

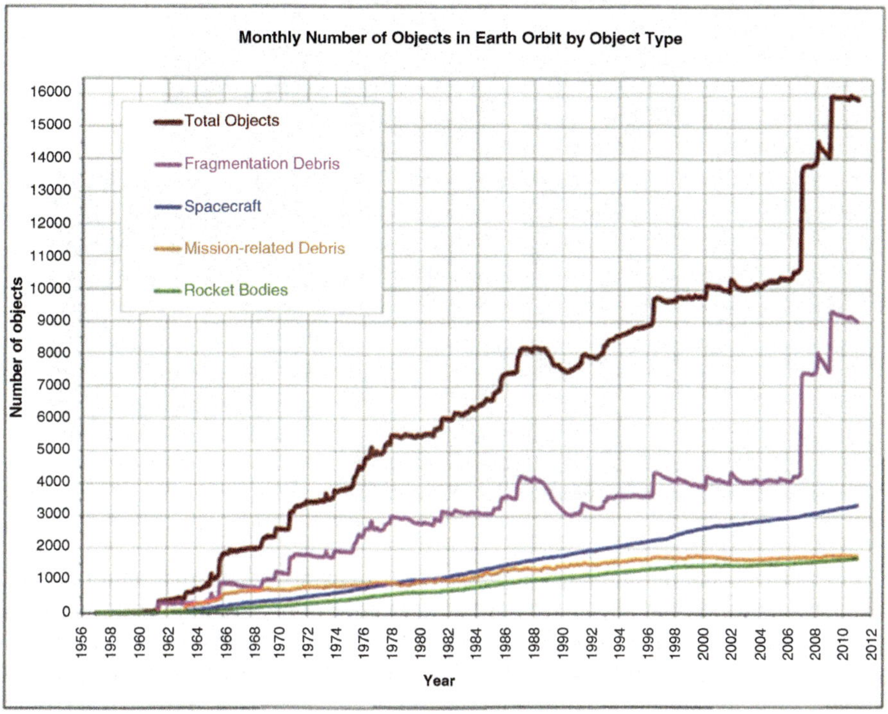

Monthly Number of Cataloged Objects in Earth Orbit by Object Type: This chart displays a summary of all objects in Earth orbit officially cataloged by the U.S Space Surveillance Network "Fragmentation debris" includes satellite breakup debris and anomalous event debris, while "mission-related debris" includes all objects dispensed, separated, or released as part of the planned mission.

(Observer Research Foundation, New Delhi, 2010), http://orfonline.org/cms/export/orfonline/modules/report/attachments/AwaitingLaunch_1397728623369.pdf.

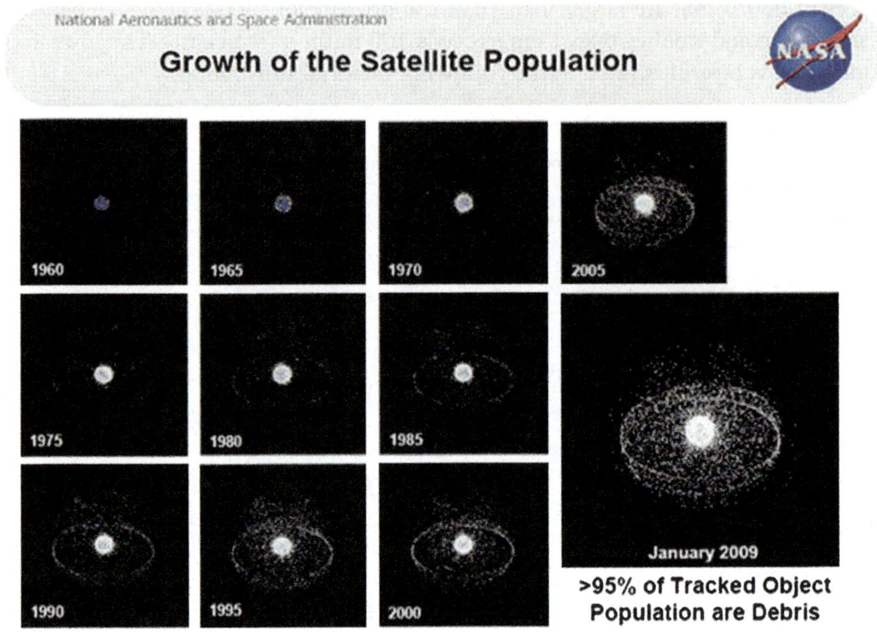

Greater reliance on outer space means that regional and international cooperation are becoming inevitable. However, cooperation that is unregulated in nature could further insecurities at multiple levels. Asia, which is already going through a phase of uncertainty and competition, is likely to see greater insecurity in the absence of broad set of rules for cooperation. This calls for a review of the outer space regime as it exists today. In fact, the state of the regime has become the biggest challenge in the outer space domain. Lack of consensus among major powers and political difficulties are hampering the process of tweaking the current outer space instruments or establishing new ones.

9.2 What Should an Outer Space Regime Do?

The objectives of an outer space regime must be three: ensuring security of outer space, ensuring order and stability, and ensuring sustainability. In ensuring security of space, states and other players must safeguard and prevent any danger or threat to outer space utility. Ensuring order and stability would mean that freedom of action should not create adverse effects that would restrict other players' options in using outer space. Ensuring sustainability would finally mean that outer space should be used for peaceful purposes alone and that space is kept clean, unpolluted for sustainable use into the future. These objectives are accepted by most states, at least in rhetoric. However, the state of the regime, as it exists today, is far from satisfactory. Lack of consensus among major powers to deal with the challenges is contributing to slow development of outer space norms. There are several difficulties that explain the current situation, although not unique to outer space security. Changing balance of power equations with the "rise of the rest" is a phenomenon that is affecting today's rule-making process. The relative decline of the US (or at least the perception of it) has cast a big shadow on the US efforts to play a dominant role in this exercise. On most global security developments, there are two opposing camps among the major powers—the US-led western group and a Russia-China bloc—that have continued to counter each other, making the regime creation process challenging. These developments are also taking place in the backdrop of a new vibrant Asia that lays an increasing emphasis on hard power and competition for the same including in the space domain is spurring new insecurities. Also, proliferation of technologies, including outer space capabilities, across a large number of players has added to the difficulties in establishing a regime. The US, UK and the USSR that had the most advanced technologies in the 1960s had an inherent interest in controlling these technologies. But the fact that technology is no more controlled by a handful of these states makes their case weak. However, given the insurmountable nature of the challenges facing the outer space domain, countries need to work on what is feasible in the meantime.

9.3 State of the Outer Space Regime

There are a few good measures managing space security, however, they have become insufficient. While COPUOS (UN-Committee on the Peaceful Uses of Outer Space) is a good measure, but one big lacuna is that military space activities are not part of the COPUOS mandate. The Outer Space Treaty (OST) is yet another excellent measure, formulated in 1967 after detailed negotiations among the then three major space powers—the United States, USSR and the UK. But there are loopholes in OST that could be exploited by states to pursue their own narrow interests. The treaty and some of the definitions therein have become far too expansive in their understanding and application. Secondly, a treaty formulated in the 1960s is far too distant from the range of new and emerging threats. The lack of definitional clarity of many space security-related terms has provided ample scope for mischief. Terms and concepts such as 'peaceful activity', 'space weapon', 'defensive actions in space', and 'space weaponisation', among others, need to be clarified. Lastly, negotiating the PAROS (Prevention of An Arms Race in Outer Space) has not been easy. A resolution on PAROS was passed in the UN General assembly in 1981, but in the three decades since then the CD has yet to have a productive session on PAROS. While there is near global consensus on the fact that outer space should be used solely for peaceful purposes (and hence the need for PAROS), the political difficulties among the major powers have come in the way of making PAROS a reality. PAROS has also come into criticism for the fact that the focus is entirely on arms race and with no mention of the debris issue. However,

PAROS was debated in a different context of US-USSR competition and whether it requires the same urgency needs to be debated.

Meanwhile, even as countries support different mechanisms for different reasons, the challenges facing outer space are significant. And therefore, one has to look at what is the least controversial and minimally acceptable set of measures one could start with. One needs to start with the least common denominator and these will include norms of responsible behaviour, TCBMs and GGE. These steps could potentially get states to move toward more binding, legal and verifiable instruments. However, one must acknowledge the political difficulties that have become serious hindrances to building effective space security regime. Therefore, while countries try and work out some of their political difficulties in building consensus, promoting TCBMs is key for establishing greater trust between nations that might help in mitigating the political difficulties among states. This could also potentially lead the way for more binding and verifiable mechanisms.

9.4 What and Why TCBMs?

TCBMs are simply means to strengthen dialogue and interactions while encouraging openness, greater transparency and information-sharing. These are voluntary measures that countries join and adopt and are usually not legally binding. They are, like I said, least controversial measures that can be agreed upon to start addressing challenges with information sharing as a key component that would result in better understanding among countries, potentially reducing wariness, competition and rivalry.

These steps are gaining certain amount of criticality given that the number of players in outer space has gone up in huge numbers. There are not just state actors but non-state players in the form of private sector firms, universities and such other institutions, all making outer space crowded and congested. In such a scenario, multiple levels of dialogue and interaction are of importance because states would have a broader understanding of what other countries are doing, what are the different orientations of their space programmes and so on, reducing misperception, wariness and miscalculations between states.

In addition, TCBMs are important measures that can address the growing number of challenges in outer space such as space orbital debris, collision avoidance, while actively promoting space situational awareness (SSA), planetary defence. SSA is the ability to monitor, understand, and predict the changing physical environment of outer-space. This includes the location and movement of natural and manmade objects in an effort to prevent collisions and accidents.

According to NASA, there are over 500,000 pieces of debris in Earth's orbit, some of which are large enough to cause serious damage to satellites and spacecraft. Space weather, caused by charged solar particles and the Earth's magnetic field, also create disturbing conditions for space objects.

Improving space situational awareness can help in alleviating some of these concerns. Simply put, SSA aims at creating a constant understanding of the space environment by keeping a close watch on developments, including tracking of space objects, debris, and space weather. This includes predicting collisions in orbit, detecting launches of new space objects, predicting re-entry of space objects into the atmosphere, and detecting threats and attacks on spacecraft. This can be done by radar, optical telescopes, electronic signals sensors, infrared sensors, or other spacecraft.

The United States already has the largest SSA mechanism, which is called the Space Surveillance Network (SSN), followed by Russia and European Union. Even the US network is not comprehensive, given that its coverage of the southern hemisphere or South America, Africa or Asia is not extensive.

Similarly, most recently, the February 2013 events of meteor hit in Russia and Asteroid 2012 DA 14's close miss, have shown that planetary defence is no more a fictional theme but they present the reality of threat to human life from asteroids and meteors.

However, no one state is in a position to address these challenges effectively. There are both the technological availability and financial wherewithal issues and, therefore, cooperation and collaboration among states is of great importance.

TCBMs also provide as important forum to develop familiarity and undertake activities that would build trust and transparency among space actors. These are measures that can be taken at bilateral, regional and global levels. And successful TCBMs could promote significantly higher level of openness and transparency while beefing up confidence levels of states, alleviating tension and ultimately strengthening regional and global peace and stability. Once trust is established through such TCBMs, they could eventually lead to more binding legal commitments.

Why do countries support TCBMs? TCBMs are a good intermediate measure between a functional need and a binding instrument. The need for binding and non-binding instruments has been debated for quite some time without much success. However, the recognition among states on the challenges is an important enough step to think of more practical ways of addressing them and TCBMs does serve that purpose of a bridge idea of a binding instrument and the fructification of one, by identifying practical, near term steps to improve transparency and openness. TCBMs let the countries open a forum for talking to each other, building up the confidence in each other that could potentially result in more binding measure.

GGE is another good political initiative under the UN that would promote mutual trust and encourage cooperation and openness, while reducing tensions and misperceptions as a means to prevent intended or unintended conflicts. This group, again, explores voluntary and pragmatic TCBMs in space to ensure space security and sustainability. However, we need to wait and see the GGE outcomes in the second half of this year.

So, why should states pursue TCBMs? What are the advantages? There are some legal advantages to TCBMs. First, these are voluntary commitments and therefore easier to reach. Presumably, there is sufficient interest in both sides about the outlines of an agreement. Second, they are less complicated. It is far easier to reach an agreement on TCBMs because of this. There are no complications on what needs to be enforced and verified and there are no extended discussions of highly technical nature because these are more political agreements than legal ones. A formal treaty, on the other hand, would require extensive verification, as plenty of examples would indicate.

So the question is if a measure is just a declaratory statement, what are the advantages and disadvantages?

One of the legal disadvantages is that because it is non-verifiable, it could be broken without penalties. However, it is easier to sell such agreements domestically when it is not a verifiable commitment.

A second disadvantage may be that there is no way to know that countries are not undermining it. If countries violate, there is no way of knowing it. The only way to prevent from cheating it is deterrent. That is if you do it, others will do it. For instance, on the various TCBM measures between India and China, both sides know that if they cheat, the other side could do the same. Therefore, both sides may have an incentive to abide by their agreement, even if it is not verifiable.

Lastly, the biggest advantage is that TCBMs are excellent measures targeting the political aspects of addressing outer space challenges. Any instrument that comes about goes through three stages—technical, legal and political. While the technical side of solutions is relatively easy, the political aspects of finding a solution remain the trickiest ones. TCBMs are excellent platforms as it opens up dialogue and periodic interactions between states at the political level, potentially reducing the wariness and misperception.

9.5 Why a Code?

A code is seen as less controversial because it is not a legal instrument. However for the same reason, it has also its opponents. Those who have argued against a code made two key arguments: why should states adopt and institutionalise a code with no enforcement element and whether a code beefs up state security in significant terms. A second argument says that the principles contained in a code exist in various forms already, such as national space policies, statements in Parliaments or in the form of CBMs (Confidence Building Measures). These countries articulate that a code does not bring a particular advantage. Nevertheless, there is utility to a code and this must be explained as well. Codifying these principles in one single argument will bring greater cohesiveness and endorsement of such an instrument by a large number of states is significant.

There are a few other considerations because of which a code must be pursued. Firstly, a code becomes an intermediate step between a functional need and a binding instrument, which could possibly be an early step to get all sides talking and build up the necessary confidence in each other. Code, in addition, can be a useful instrument that institutionalise best practices and can eventually lead to more legal measures. Code, however, does not guarantee responsible behaviour. On the other hand, code could bring about international peer pressure to abide by commitments. Lastly, code is certainly a risk-reduction measure, and could form the basis for greater understanding and dialogue among states, a pre-requisite for addressing political challenges.

9.6 Reactions to the EU ICoC

The EU-initiated International Code of Conduct (ICoC) is potentially a good instrument but has received mixed reactions at this stage. The EU developed an initial code in 2008 and a revised version was released in October 2010. As for the content of the code, most countries saw it as an innocuous measure because it listed a few desirable steps to be taken up by states that would ensure the sustainable use of outer space and avoid possible accidental and intentional mishaps. Nevertheless, a large majority of states outside Europe were sceptical of the ICoC, especially during the 2011–2012 period. For most states, what the code contained was not so much an issue as much as the process, because many of the even established spacefaring powers were not part of the process that developed the code. This factor seriously impeded progress on the ICoC with many viewing the EU effort as presumptuous. Due to this oversight, many countries in Asia and Africa objected to "an otherwise acceptable document."[4] Had the EU made an effort to reach out to all the established spacefaring states, the impact of the Code could have been different. The EU in that regard lost an opportunity to connect with the non-European space powers because having the non-European bloc support for such an initiative could have been significant. Asia, Africa and Latin America are important in this regard because newer space powers are going to be coming from these regions, and not from Europe. Hence, there is a need to have these countries on board, without which one may end with an instrument that may have a significant number of countries but the critical players that will make a difference stay outside. Europe and the West in general need to acknowledge that it is in these regions that new challenges are going to be coming from. Without an inclusive process, the EU ICoC does face the possibility of meeting the similar fate of the Hague Code of Conduct against Ballistic Missile Proliferation (HCOC). While the HCOC can take credit in enlisting the membership of around 140 countries, it lacks some of the critical players in this regard. Countries such as Iran, North Korea, Pakistan and China remain outside this mechanism, thus questioning the validity of the HCOC because challenges in terms of missile proliferation come from these countries.

---

[4] For the initial responses to the ICoC, see Rajeswari Pillai Rajagopalan, "Debate on Space Code of Conduct: An Indian Perspective," *ORF Occasional Paper* #26, Observer Research Foundation, October 2011, http://www.observerindia.com/cms/export/orfonline/modules/occasionalpaper/attachments/ocp26_1319777951241.pdf.

There have also been substantial differences in addition to the process issue. Starting from what the code should contain and whether it should deal with only peaceful or military activities have been widely contested. There have also been divisions between major spacefaring powers on whether it is space arms race or space debris that constitute a bigger threat. For the Russia-China combine, space arms race has remained the most serious threat whereas most other spacefaring powers see space debris as the more severe issue that needs immediate attention. Ecuador losing the only satellite hit by satellite debris reflects the urgency around this issue. Whereas arms race in space and weaponisation of outer space are issues that will become serious in the coming years. It is true that there is the PAROS (Prevention of Arms Race in Outer Space) being debated from the late 1980s but there was a different context (one of the US-USSR competition) that provided for such a measure. Therefore, it is difficult to see the urgency as yet on the space arms race issue.

Emphasising on the arms race aspect, Russia and China have submitted a draft Treaty on Prevention of the Placement of Weapons in Outer Space and of the Threat or Use of Force against Outer Space Objects (PPWT), first submitted in 2008 and thereafter a revised version submitted to the Conference on Disarmament (CD) in Geneva in June 2014. While it rightly emphasises on the arms race as it has the potential to become a major threat in the future, the biggest drawback of the PPWT is that it does not even mention space debris as a problem. Also the over-emphasis on the placement of weapons in outer space is a stretch because one can be reasonably certain that no country is going to place in the orbit. But the more significant challenge is the ability of ground-based weapons such as the ASATs to target assets in outer space, which are again ignored in the draft PPWT. PPWT is yet to measure up to many states' expectations in terms of addressing all the challenges.[5] Meanwhile, the West has generally pursued Transparency and Confidence Building Measures (TCBMs) as a means of instilling confidence and gradually building on the political consensus among major powers that has been an issue in the recent years, for establishing a more binding mechanism. The manner in which the Outer Space Treaty was formulated is a reflection of this thinking.

There are also divisions between established space powers and new entrants on the question of the right to self/collective defence, enshrined in article 51 of the UN Charter. The obvious reference to the right in the code has been resented by a number of newcomers, particularly the Latin American ones. A few African countries too have criticised the categorical reference of this right in the code. These countries argue that this "reiterates the sovereign right to national self-defense."[6] They also argue that this obvious reference will strengthen the process

[5] For a review on the 2014 draft of the PPWT, see Michael Listner and Rajeswari Pillai Rajagopalan, "The 2014 PPWT: A New Draft But with the Same and Different Problems," *The Space Review*, August 11, 2014, http://www.thespacereview.com/article/2575/1.

[6] Rajeswari Pillai Rajagopalan and Daniel A Porras, "EU Courts Support for Space Code of Conduct," Commentary, *Space News*, July 14, 2014, http://spacenews.com/41254eu-courts-support-for-space-code-of-conduct/#sthash.8rySTUVi.dpuf.

of space weaponisation and is seen as a legitimate way through which the established space powers would weaponise their capabilities. Since these countries are relatively new players in the domain, they do not yet possess effective counterspace technologies required to defend themselves should a situation arise. The US, India and much of Asia has not made a big issue to this reference, since it is in line with the UN Charter commitments. However, in real, an obvious reference of this right or absence of it in the code will make little difference since this is part of international law, recognised in the UN Charter. Nevertheless, states on either side have been inflexible on this aspect. For much of the West, they find it difficult to sell the code to their domestic audience once the reference is taken out.

Another major issue relates to the perception that code is restrictive in nature. Many developing countries, which are not spacefaring at the moment but intend to do so in the coming years, look at the code as something that will slow down their development or something that might impose additional costs in order to ensure clean and responsible space activities. Many see close similarities between the debates on climate change, on the regulation of carbon emissions and outer space sustainability. This perception has gained greater currency, especially since the code has been proposed by a set of developed countries. However, the code has made categorical references to international cooperation that could aid in transfer of technology, among other benefits. The EU can be benevolent and make statements to the effect that states that endorse the code will stand to gain, particularly in relation to technology transfer. The code could see endorsement in greater numbers if the EU were to come up with innovative solutions and incentivise their approach.

This leads to another concern on the code that it does not spell out any rules of engagement for regional and international cooperation. The fact that the EU Code specifies no regulations has been seen problematic in certain regions such as Asia, where there is competition for acquiring hard power capabilities. While cooperation in general is seen as positive, as mentioned earlier, unregulated cooperation has the potential to fuel serious security concerns. The EU could consider outlining certain broad political arrangements that would guide international space cooperation.

Also the manner in which the EU has made a distinction between countries that sign the code and those that do not is problematic. For instance, Clause 5.1 states, subscribing states "resolve to notify, in a timely manner, to the greatest extent practicable, all potentially affected Subscribing States of any event related to the outer space activities they are conducting which are relevant for the purposes of this Code, including: scheduled manoeuvres that could pose a risk to the safety of flight of the space objects of other Subscribing States." Similarly, Clause 5.2 states that "The Subscribing States resolve to provide the notifications on any event related to the outer space activities described above to all potentially affected Subscribing States." Also Section 6 on information sharing makes a distinction. In fact, the new text actually omits the reference that was contained in the previous text to share information with non-subscribing states. The EU might use this as a bargaining tactic in order to get more states to sign, which is understandable but if the EU were to leave this in the final text, it goes against principles enshrined in several UN agreements on international cooperation. Lastly, space is a global common and a debris collision will result in further debris creation, which is detrimental in more ways than one.

Meanwhile, the EU has undertaken a larger outreach effort to strengthen its case for the code. One of the first multilateral efforts in this regard was the open meeting in Vienna in June 2012. Further, the UN Institute for Disarmament Research (UNIDIR) was brought in to start a series of regional meetings to understand the regional perspectives on the code. The UNIDIR held regional seminars in Malaysia, Ethiopia, Mexico and Kazakhstan. This indeed became the first step in understanding the regional variations in the understanding around the ICoC, providing the EU with substantial inputs to review and revise the code. In addition, the EU organised three open-ended consultations, in an effort to make the process as inclusive and international as possible. These meetings were held in Kiev (May 2013), Bangkok (November 2013) and Luxembourg (May 2014). Nevertheless, these were truly consultative in nature, with states stating their views on the draft code and not negotiations/ drafting process, which many states were hoping to do. To be fair, the EU has done an excellent job in incorporating most views that were offered in these consultations. The process also had become quite comprehensive with the involvement of 63 countries and 111 participants. There were more than 80 countries that participated in these open-ended consultations. The EU should be releasing a new draft of the code in the coming months.

Are the countries satisfied with the process and the content? The EU needs to be applauded for the open-ended consultations and today there are many more countries that are engaged constructively in the process of establishing the code. However, it is a long way before the EU can bring some of the critical players like Russia and China on board. Several countries including Russia, China, Ukraine, Mexico, Thailand, Brazil and Ethiopia have stated that the ICoC should be limited to peaceful uses of outer space and that the title must be suitably modified. The EU, however, articulated the need to institute a measure that is more comprehensive bringing together the peaceful and the military aspects of outer space utilities. India has also articulated along these lines because the line that makes a distinction between peaceful and military uses of outer is increasingly blurred and that it is time to have one instrument that would combine the two profiles. Countries like the US, Japan, the UK, Italy, Germany and Canada too have made a strong pitch for a comprehensive instrument because of the dual-use nature of the space technology. Several countries have called for some sort of UN endorsement, in order to make it truly multilateral. While there is merit to such an articulation, the logistics around it is not clear. One can rule out the UN-COPUOS since it has a limited mandate (only peaceful uses). Should it be taken to the UN General Assembly?

9.7 Way Forward

Space is truly a global common as one state's actions can affect a large number of other states. Space debris, for instance, does not make a differentiation between different states' space objects. Space is also a limited commodity and thus, there has to be greater appreciation for the finite space that states are dealing with. Accordingly, states must make earnest efforts in defining and devising what responsible behaviour in outer space is, draw clear red lines that must not be broken in order to keep the outer space sustainable and accessible to all.

The treaty making was a relatively easier process in the 1960s and 1970s but today technology diffusion and political difficulties among major spacefaring powers make that process extremely challenging. The challenges are far too many and if we have to ensure that outer space remains secure and sustainable. There is a need to regulate activities and capabilities, while encouraging dialogue as a way of reducing misperception and wariness. Instituting TCBMs and norms of responsible behaviour may be a good stepping stone before realising more binding legal instruments. In the absence of consensus among major powers in identifying challenges to ideating solutions, states should start with the least common denominator in the form of TCBMs and other mechanisms that would gradually ease the mind-set of countries and become accommodative. TCBMs are good starting points in this regard, helping establish certain parameters of responsible behaviour, encourage initiatives to strengthen SSA and so on.

Instituting an inclusive process will ensure greater acceptability and compliance, thus contributing to the effectiveness and longevity of the code. Ensuring the participation of new and upcoming states is critical because the challenges are going to be coming from these new regions that have so far remained outside the space age. An inclusive process that brings together all the different stakeholders is equally important in order to reach greater consensus and compliance.

Chapter 10
Creating ESA

Thomas Hoerber

10.1 Introduction

In the aftermath of the Second World War, space technologies became an important area of development for the leading world powers. The Sputnik shock in 1958—when the Soviet Union proved that they could put a satellite into orbit—and the moon landing achieved by the United States were important technological milestones. They were also major events in the Cold War context, serious evidence that both sides had access to, and knew how to use, cutting-edge technology, strengthening the leadership status of both sides vis-à-vis their allies. Such endeavours required substantial resources which were not available to medium powers such as France and Britain. They also understood clearly that the formidable costs of a fully fledged space programme could not be borne by any one of them alone. However, just as for nuclear capabilities, both Britain and France realised that such key technologies were indispensable for them, for reasons which will be analysed in detail in the main text. Against this background, this paper will trace out the progress of the early bilateral cooperation which finally led to the wider project of the European Space Agency.

The methodology of this contribution follows essentially a historical analysis of primary sources taken from parliamentary debates in France and Britain from 1970 to the signing of the European Space Agency (ESA) convention in 1975, with some insights into German space policies from the outside perspective of these sources. The main points which these primary sources produced are budgetary constraints, independence and technological advance in space. The contribution will thus be structured according to these main issues. These focal points will be used to find out what developments led to the foundation of the ESA?; what were the necessities nation states could not meet from their own resources?; and what was that the rationale which impelled them to seek a European solution?

T. Hoerber (✉)
École Supérieure des Sciences Commerciales d'Angers, Angers, France
e-mail: Thomas.HOERBER@essca.fr

C. Al-Ekabi et al. (eds.), *Yearbook on Space Policy 2014*, Yearbook on Space Policy, DOI 10.1007/978-3-7091-1899-3_10

10.2 National Politics

Each of the three nations had different priorities in national politics which led them go into the direction of advocating the development of a European space policy under ESA. Having said that, the ways and means how to get there were points of major contestation between national governments. In the British case, the main concern which comes out in the sources are budgetary considerations and constraints. The gist of the financial argument is well summarised by a British minister.

> There is no economic case for an independent European launcher. Launchers are the most expensive and least profitable items of space technology. Satellites, on the other hand, are relatively inexpensive and their applications in commercial ventures offer the best prospects of commercial return. The resources available in Europe are, by common agreement, small by comparison with those of the United States, and the Americans have an impregnable lead in launcher technology, while there is good prospect that Europe can compete successfully in satellites.[1]

Spending issues were consequently the major concern in the British discourse, particularly how much Britain was supposed to contribute to the future ESA budget and how the ESA budget would be constituted.

> Mr. Michael McNair-Wilson asked the Secretary of State for Industry on what the European Space Agency's budget of £185 million is being spent; on what basis Great Britain's contribution has been decided; which countries are paying as much or more; and what work is being carried out in the British aerospace industry as a result of our membership of the ESA.
>
> Mr. Kaufmann: About 85 per cent of the budget of the European Space Agency is devoted to applications programmes covering development and proving of communication satellite system technology, and meteorological satellite, a spacecraft launching vehicle, and a space laboratory to be carried in the United States space shuttle. The remainder is being spent on scientific research in space; the United Kingdom contribution to this is the responsibility of the Science Research Council. Contributions to the majority of the ESA programmes [are] on a gross national product basis, although the three most recently adopted applications programmes are being funded according to the national interests of the countries participating in each specific programme. Thus, the United Kingdom does not participate directly in the .French-led launcher project and makes only a 6.3 per cent, contribution to the German-led Spacelab programme, but bears just over half of the cost of the maritime communications satellite development. France and Germany are each contributing about twice as much as the United Kingdom to the total ESA budget for 1975.[2]

For Britain, independence in the development of European space capabilities comes right after budgetary considerations; and competition with the major Western space power, the US, was an important element of the British discourse.[3] Such competition, if it was meant to be serious, necessitated a comprehensive space programme on the European side, from space ports to launchers, satellite

[1] Onslow, Weekly Hansard, Under-Secretary of State for Trade and Industry, 1973: 842.

[2] Weekly Hansard, Under-Secretary of State for Trade and Industry, 1975: 563–564.

[3] McNair-Wilson, Weekly Hansard, 1973: 828–829.

technology and even the capability to maintain humans in space. Naturally, Britain wanted to see its technologies used in any such comprehensive space programme. For that purpose the British space industry would receive corresponding funding out of the European space budget.[4,5] However, Britain also realised that an overarching European programme was the key to the successful implementation of such a programme, as can be seen in the following quotation.

> (...) I find it difficult to see how Europe as an entity can operate a total space programme without having a total space capability (...) This means that Europe will not be able to compete on a comparable basis with the United States industry. I do not find it surprising, therefore, that the French and the Germans are now considering a launcher programme. What I do find surprising and disappointing, however, is that our own first stage rocket launcher, Blue Streak, is about to come to the end of its active life after only 11 firings. (...) If Europe is determined to have a launcher programme, I wonder whether there is not some way by which the Blue Streak at least might find its way back into that programme and be used again. I believe that the Minister for Aerospace is right in his attempts to create a European space agency. Perhaps because space is such a new industry, it is the first that can be set up on a European basis.[6]

Space capabilities were seen as a driver of technological advance, in the British case particularly in satellites, but in the context of a European space programme Britain saw room for the development of other technologies, such as launchers.

In the French case, the economic and the ideological argument for independent access to space for Europe was the first priority in debates in the Assemblée Nationale.[7,8] France was able to proudly present its launcher programme as the most sophisticated in Europe and thus Ariane was seen as the leading candidate for a European launcher programme. This element of a European response to the ensuing space age was in fact a response to American urging, but in the wake of the creation of ESA the launcher programme was not supported by all European partners.

> We felt that no consensus could be reached on a coherent European [space] programme, because of the hesitation of certain of our partners, who seem to prefer to abandon the project of developing a large European launcher, as called for by the Americans, in order to join in the Post-Apollo programme.[9]

Nevertheless, European autonomy in space affairs—as can be seen in the following quotation—became, important. And that resembled to some extent a "European response" in space affairs, as demanded by the Americans, and acting as European 'federator', again, who was not prepared to deal with individual European space programmes.[10]

---

[4] McNair-Wilson, Weekly Hansard, 1973: 830.

[5] Onslow, Weekly Hansard, Under-Secretary of State for Trade and Industry, 1972: 177.

[6] McNair-Wilson, Weekly Hansard, 1973: 831–832.

[7] Boulloche, Journal Officiel, Assemblée Nationale, Paris, 1974: 6376II.

[8] d'Ornano, Minister for Industry and Research, Assemblée Nationale, Paris, 1974: 6400II.

[9] Charbonnel, Journal Officiel, Minister for Development of Industry and Research, 1972: 4467II.

[10] Dalyell, Weekly Hansard, 1973: 825.

> The Government, following my proposal and that of the Prime Minister, has decided that the essential objective of our space effort must be for Europe to enjoy autonomy in the field of space applications. In telecommunications, air and sea traffic control, in meteorology or for defence purposes, we cannot rely solely on the US or the USSR. An exclusively national effort in this field does not make sense. Our choice, therefore, must be a European one. (. . .) the European Space Agency, which will be the framework for cooperation will be an instrument of the highest standards in terms of organisation and structures.[11,12]

In addition to the French push for space independence, the track record of previous launcher also showed that rocket science might not be exactly new, but it certainly needed development. The L III S launcher, the French abbreviation for the third-generation substitution launcher later named Ariane, came in the aftermath of a disastrous serious of failures of the Europa II and III launchers.[13,14] This was one of the reasons why European partners such as Britain, but also Germany, foresaw serious overspending on launcher programmes and eventually dropped out, very much to the frustration of France. After the early demise of British Blue Streak rockets, France, with its Diamant rockets, was the only European country in a position to deploy launcher capability.[15] Against this background, France went into the negotiation on the European space agency to found a comprehensive European space programme, including a large launcher.[16] Eventually, Ariane became the European launcher which would ensure independent access to space for European countries.[17,18]

> Europe has the resources, technologies and know-how needed for the development of its own launchers. We are also aware of the influence France possesses in this field, because of the vision and the unwavering determination which we have shown in our national space programme. To abandon our claim to build a launcher for Europe would not have been a mere misapprehension of the repercussions of that technology for other applications, but utter technological and political abdication.[19]

Very much in contrast to Britain, in the French official discourse, budgetary considerations came after independence and technological advance as policy objective for a French and respectively for a European space programme. Proportional

[11] d'Ornano, Journal Officiel, Quatrième législature, Assemblée Nationale, Paris, 1974: 6364I.

[12] For further French arguments on independence see Charbonnel, Journal Officiel, Minister for Development of Industry and Research, 1972: 4927II.

[13] Cousté, Journal Officiel, Quatrième législature, 1972: 4468I.

[14] Charbonnel, Journal Officiel, Minister for Development of Industry and Research, 1973: 1299I-II, 1300I, 5371II-5372I.

[15] McNair-Wilson, Weekly Hansard, 1973: 831–832.

[16] Charbonnel, Journal Officiel, Minister for Development of Industry and Research, 1972: 4467II.

[17] Charbonnel, Journal Officiel, Minister for Development of Industry and Research, 1973: 1299I-II, 1300I, 5371II-5372I.

[18] d'Ornano, Journal Officiel, Minister for Industry and Research, 1974: 6364I.

[19] Charbonnel, Journal Officiel, Minister for Development of Industry and Research, 1973: 5371II-5372I.

spending in the French budget shows, nevertheless, that for France, the development of Ariane was key.

- 120 million FF for Ariane
- 9 million FF for a space laboratory
- 7 million FF for the British maritime navigation satellite Marots.[20]

Ariane was supposed to deliver independent access to space for Europe, and, of course, France, despite the fact that the innovative character of the development of a heavy launcher was doubted.[21,22] France could not finance Ariane alone. Therefore, European cooperation on the project was vital.[23]

In the German case, technological advance was the first priority. This played out in different areas, such as the development of the telecommunication satellite Symphony, or the Space Laboratory which was another of the early programmes, led by Germany.[24] It was developed within the American post-Apollo programme.[25] It would finally find its application in the Space Shuttles. Budgetary considerations, naturally come after the objective of technological advance, but this tied in with the British main concern of who would pay, and how much.

> On Spacelab we took the view that, provided there is general agreement on the ESA framework for Europe's future space co-operation, we should be prepared to take a share of up to 10 per cent, in the project. Because of the ceiling on our space expenditure, we can take a share only in so far as we are relieved of the cost of our maritime communications satellite by contributions from other member States. On the maritime satellite we have proposed to our European partners that the geostationary satellite which is currently being studied in the United Kingdom industry should be adopted as an experimental maritime communications satellite to meet the recently defined European requirement.[26]

Germany concurred with Britain on the importance of satellite programmes, also because considerations of independence were clearly less important for Germany, than for France, for example. Against the background of the particular German situation in the post-war period, the Atlanticists in the German political establishment would also strongly argue for a rather stronger engagement with the United States, for reasons of the need for security against the Soviet threat, but also because of the important role the United States had played in the rebuilding of West Germany.[27]

The European Space Research Organisation (ESRO) had developed a number of satellites, e.g. Meteosat for meteorology, a European telecommunications satellite

[20] Charbonnel, Journal Officiel, Minister for Development of Industry and Research, 1973: 5372I.

[21] d'Ornano, Journal Officiel, Minister for Industry and Research 1974: 7101I.

[22] Boulloche, Journal Officiel, 1974: 6376II.

[23] d'Ornano, Journal Officiel, Minister for Industry and Research, 1974: 6364I.

[24] Kaufmann, Weekly Hansard, 1975: 563–564.

[25] Heseltine, Weekly Hansard, 1973: 693–694.

[26] Onslow, Weekly Hansard, Under-Secretary of State for Trade and Industry, 1973: 839.

[27] Volker Hentschel, *Ludwig Erhard Ein Politikerleben*, Ullstein, Berlin, 1998: 608, 663, 709.

and a navigation satellite for air traffic.[28] The problem in the period before ESA was that most of these European satellites, notably the Franco-German telecommunications satellite Symphony, would be sent into orbit either by American or Soviet launchers. This might not seem such a big problem for civil application, but once power considerations come into play, antagonistic rationales vis-à-vis the other major powers were invoked, which ultimately led to the French claim for independent access to space for Europe.[29,30]

10.3 Common Denominators

However, the creation of a European space policy was not an easy process. There were a number of predecessor organisations to ESA which acted as steps towards a European solution, but fell by the wayside in the process. It is nevertheless important to understand the rationale for their creation in order to get a better understanding of the process which led to the creation of ESA.

10.3.1 European Space Research Organisation (ESRO)

Established in 1964, ESRO was one of the first attempts to bring the diverse and often overlapping European space programmes under a shared umbrella. Such national space programmes were finally grouped under three main branches of satellite development: telecommunications, air traffic control and meteorology. France was the lead nation in this organisation, with a clear agenda of Europeanisation of space assets across Europe.[31] Britain and its space industry followed this lead. The hope was that, in ESA, duplication of space research could be avoided across Europe without overbearing bureaucracy.[32] Technological development seemed to warrant hope for growth in the space industry, because, particularly in the satellite programmes under ESRO, there was maturation in the early 1970s in a shift from using satellites for scientific observation to using them for commercial services in communications and other applications. This change offered strong prospects that the industrial significance of space activities would increase further.[33]

[28] Charbonnel, Journal Officiel, Minister for Development of Industry and Research, 1973: 5372I.

[29] Boulloche, Journal Officiel, 1974: 6376II.

[30] d'Ornano, Journal Officiel, Minister for Industry and Research, 1974: 7101I.

[31] Charbonnel, Journal Officiel, Minister for Development of Industry and Research, 1972: 4467II.

[32] Onslow, Weekly Hansard, Under-Secretary of State for Trade and Industry, 1973: 840.

[33] Onslow, Weekly Hansard, Under-Secretary of State for Trade and Industry, 1973: 836.

10.3.2 European Launcher Development Organisation (ELDO)

ELDO, founded in 1962 and charged with the development of a European launcher, was also an ESA forerunner. Its history was marred by failures of the Europa rockets, based on the "Cecles-Eldo engines".[34] Overspending and successive failures of these launchers led to the demise of ELDO in 1973.[35,36] For the short period until the advent of ESA in 1975, ESRO took over its tasks, administration and facilities. ELDO is a good example that during this early period of space endeavours, the nationalist impulse can still be felt quite strongly. A European spirit of collaboration in space technologies still had to develop.

The forum where such ideas found the right environment was the European space conference, held on 20 December 1972, followed by a further such meeting on 31 July 1973, clear evidence of the urgency, perhaps the real importance of space affairs at the time.[37] ELDO was by that time already close to burn-out, with Italy withdrawing as early as 1969 and the main contributors, France, Germany and Britain squabbling about its future.[38] The European space conference brought the presentation of a new launcher programme, under several names, i.e. L III S, substitute launcher for the European programme, or finally the Ariane launcher. And the conference yielded agreement to establish a European space agency which would encompass the work of both ELDO and ESRO.[39]

Thus, the main aspirations inscribed to a European space policy in the wake of the creation of ESA, were, first, to avoid duplication between existing national space programmes in Europe; secondly, the development of a European space industry, based on existing national space capabilities; and finally this led to the development of a European space programme embodied in the creation of ESA.

10.4 What Kind of ESA?

In order to answer the question, what kind of ESA, we can return to the priorities of the founding nations in the space sector, as analysed in the previous section on National Politics, i.e. the budget, technological advance and independence.

---

[34] Charbonnel, Journal Officiel, Minister for Development of Industry and Research, 1972: 4467II.

[35] McNair-Wilson, Weekly Hansard, 1973: 831–832.

[36] Cousté, Journal Officiel, 1972: 4468I.

[37] Heath, Weekly Hansard, Prime Minister 1972: 1399.

[38] Charbonnel, Journal Officiel, Minister for Development of Industry and Research, 1973: 1299I-II, 1300I.

[39] Charbonnel, Journal Officiel, Minister for Development of Industry and Research, 1973: 5371II-5372I.

10.4.1 Budget

ESA was meant to become an effective executive agency of the European space programmes, avoiding duplication between them.[40,41,42] The staffing was to be kept very small in order to avoid overspending as in ESRO and particularly ELDO, but also to prevent the development of an authority with mind of its own, independent of the member states.[43] A good insight into British thinking at the time is given by the following quotation, which begins by making the idealistic point that there should be a European space programme, but then turns to real interests in the British space industry and the British funding that would have to be found for ESA.

> What I think my hon. Friend the Minister [Member for Tavistock (Mr. Michael Heseltine)] is rightly trying to do is to make Western European countries think in European rather than in national terms. As we are such a new member of the EEC, it is heartening to see the younger Ministers pushing forward with a concept of European agencies rather than struggling on with national concepts that do not, and cannot measure up to those of our competitors, in particular our North American competitors. (...) It is not unreasonable to suppose that (...) we could leapfrog from a national space agency and go straight to a European Space Agency. (...) there is some way to go before a European Space Agency gets off the ground (...) should [we] not consider again the Select Committee's recommendation that a national space agency bears serious consideration. (...) In other words, Britain may have a number of separate projects in hand, but she has no overall space programme as such. As we are spending more than £30 million a year on space, one may wonder whether we are getting value for money.[44]

The key organisational question in this respect was whether the budget should be made up of fixed contributions based on the GDP (previously, still GNP), or whether the countries should pay only for the programmes they were interested in—often called an *à-la-carte* approach.[45] This question was settled in the more general question on the ESA budget and the British contribution to it.[46] Budgetary flexibility was achieved in the ESA Convention through the distinction of compulsory and optional programmes. However, flexibility here is also meant in the sense of *juste retour*, i.e. funding of British space projects roughly equal to what Britain was to be paying in, and therefore a financial contribution entailing little in the way of budgetary strain.

[40] Charbonnel, Journal Officiel, Minister for Development of Industry and Research, 1973: 5372I.

[41] d'Ornano, Journal Officiel, Minister for Industry and Research, 1974: 6364I.

[42] Heseltine, Weekly Hansard, 1973: 693–694.

[43] McNair-Wilson, Weekly Hansard, 1973: 829.

[44] McNair-Wilson, Weekly Hansard, 1973: 828–829.

[45] Onslow, Weekly Hansard, Under-Secretary of State for Trade and Industry, 1973: 840.

[46] Weekly Hansard, 1975: 563–564.

10.4.2 Independence

The main French concern was independence, mainly in political terms, but also in terms of developing a space industry which would be competitive in cutting-edge technologies.

> France has thus deliberately chosen—I can confirm this to the House (...) the road to European independence in this field [space]. We fully realise that this is a narrow and difficult road, but it is the only one which has a future and corresponds to the dignity of our continent.[47]

The development of a European launcher programme, first in the Europa rockets, later supplemented by Ariane, fitted well with the argument of European independence in that such launchers would give Europe independent access to space, vital for its political freedom to pursue its space aspirations unhindered.[48,49] The problem was that 'European' (in the sense of the European Communities) was still a very risky term to use in space affairs, because the Treaties of Rome provided no specific powers for the European Institutions in this field, and thus it remained up to the will—or whim—of the member states to cooperate, or not.[50] ESA was set up on an intergovernmental basis, and this relatively unstable arrangement ultimately turned out to be the organisation's Achilles heel. One may see this as a shortcoming resulting from the political will, not least of France, at the time. Independence in the space sector as France's main objective has, however, been achieved.[51]

10.4.3 Technological Advance

The common denominator between the founding Member States was the development of a European space industry, still based on national interest, because the companies are inevitably located in the member states, but these companies became internationally competitive, not least through the European space policy under ESA, as was the original objective.[52] Since the foundation of ESA, this spirit continued in the foundation of European companies, such as 'Arianespace', founded in 1980. The French aspiration to independence was thus just as much an assertion of political independence as a claim to technological independence, which would be embodied in a European space industry producing satellites,

[47] Charbonnel, Journal Officiel, Minister for Development of Industry and Research, 1972: 4927II, see also, 4467II.

[48] Cousté, Journal Officiel, 1972: 4468I.

[49] Charbonnel, Journal Officiel, Minister for Development of Industry and Research, 1973: 1299I-II, 1300I; see also 5371II-5372I.

[50] Cousté, Journal Officiel, 1973: 1300I.

[51] d'Ornano, Journal Officiel, Minister for Industry and Research, 1974: 6364I, 7101I.

[52] d'Ornano, Journal Officiel, Minister for Industry and Research, 1974: 6400II.

launchers and technology for research. British commitment to involvement in this endeavour was also forthcoming.[53]

> We are in the EEC. If we have to accept this kind of marriage, it is not good enough for the United Kingdom not to take a lead in these important matters. (. . .) Many people will not be able to see the relevance of putting men on the moon. But in this matter we are not concerned with such ambitious enterprises. However, there is an enormous spin-off of high technology from post-Apollo projects and the Government and the industry must spell out the benefits of space participation in terms of medicine, biology, counter-pollution measures, meteorology, weather forecasting, telecommunications, crop control and many other aspects which will help not only the developed, but the under-developed, countries. This is a job for the Government and industry.[54]

And the progress European space efforts had made since the inception of the early space organisation ESRO and ELDO was substantial. The shift from theory to applications made the commercialisation of space services possible and held out the prospect of a growing space industry in Europe.[55]

Both France and Britain wanted their fair share of this growing sector. In the principle of *juste retour* one can discern the safeguard that France and Britain wrote into the ESA Convention that their taxpayers' money would be returned in due course through the development of their home-grown space industries.

10.5 Conclusions

This contribution has shown that France, Britain and Germany agreed on the foundation of ESA. It has also shown, however, that national interests were still very much present. The enthusiasm for the development of a European space policy was not shared everywhere. For Britain the main concern remained the question of costs; for France the main concern remained the possession of independent facilities, notably independent of the USA. However, both find a common denominator in the development of a viable European space industry. ESA has developed this technical expertise to manage a European space programme. It has used and fostered national space companies against that backdrop and it has provided and environment for the Europeanisation of the European aerospace industry.

In political terms, it has been shown that the European integration process of the early 1970s was no longer the idealistic brave-new-world of the 1950s, when Jean Monnet cried out for the creation of a United States of Europe.[56] In the structure of ESA is reflected the changed spirit bought about in France by the return of a passionately nationalist de Gaulle, the Fifth Republic and the new leader's idea of

[53] Heseltine, Weekly Hansard, 1972: 29–30.

[54] Bishop, Weekly Hansard, 1973: 834.

[55] Onslow, Weekly Hansard, Under-Secretary of State for Trade and Industry, 1973: 836.

[56] Jean Monnet, *Mémoires*, Fayard, Paris, 1976.

a 'Europe des Patries'. ESA was designed as an intergovernmental institution, bereft of political aspirations of any kind and with strong national interests at play within the organisation. Hopes that ESA might develop into another supranational institution were soon dashed and with them the claim to direct political legitimacy, as opposed to indirect legitimacy as in an ESA intergovernmental Council of Ministers where appointees from the member states meet. This has not been a problem in the history of ESA as a technical executive agency. But it is becoming more and more of a problem with the increasing importance which European space policy has experienced over the years.

This paper has sketched out the potential of the space sector as a component of the European integration movement. This has already led to the foundation of the European Space Agency (ESA), because the nation states could not afford viable national space policies of their own, thus being impelled towards a European solution, albeit not a political one, much less a supranational one. ESA was deliberately left apolitical and intergovernmental, though this might have to change, in the future, if space policy assumes an even more prominent role beyond its technical aspects. This will require direct political legitimacy which only supranational institutions can provide. This rationale can be seen in the current political debate on the relationship between ESA and the EU.[57] Rather than considering whether ESA should become part of the EU—which would be one solution for the legitimacy problem which ESA may face with rising budgets, such as under Galileo—this contribution argues that ESA should try to reform itself and get over one of its founding mistakes, i.e. to leave it apolitical and intergovernmental. The adoption of supranational elements in ESA could provide a way forward, such as the introduction of more frequent and more coherent rules for majority voting in the Council of Ministers. A connection to the European Court of Justice for the ESA the arbitration procedure might also be considered. And even stronger links to technical supranational institutions, such as the European central bank or the European investment Bank could become interesting on the way towards more supranationalism in ESA. However, there is no easy way to achieve this, not least because the only true supranational institutions in existence which can provide the need direct political legitimacy are the European Commission and the

[57] Carl Bildt and Jean Peyrelevade, *Towards a Space Agency for the European Union*, ESA Publications Division, Paris, 2000; Peter Creola, 'Some comments on the ESA/EU space strategy', in: *Space Policy*, 17 (2001); Frans von der Dunk, 'Towards one captain on the European spaceship—why the EU should join ESA', in: *Space Policy*, 19 (2003); Stephan Hobe, 'Prospects for a European space administration', on: *Space Policy*, 20 (2004); James Rogers, 'From 'Civilian Power' to 'Global Power': Explicating the European Union's 'Grand Strategy' Through the Articulation of Discourse Theory', *JCMS*, 2009, vol. 47, No. 4, pp. 831–862 Günter Verheugen, 'Europe's space plans and opportunities for cooperation', *Space Policy*, 21 (2005); Alain Gaubert, André Lebeau, 'Reforming European space governance', in: *Space Policy*, 25 (2009); Nicolas Peter, Katharina Stoffl, 'Global space exploration 2025: Europe's perspectives for partnerships', *Space Policy*, 25 (2009); Thomas Hoerber, 'The European Space Agency (ESA) and the European Union (EU)—The next step on the road to the stars', in: *Journal of Contemporary European Research (JCER)*, Vol. 5, No 3 (2009).

European Parliament. A duplication of these well established institutions is not desirable. Therefore the only way for more supranationalism in ESA and thus more political legitimacy, beyond the compromise solution of introducing more supranational elements in the Council of Ministers, is to find a working relationship between ESA, the European Commission and the European Parliament.[58]

[58] More detail about the exact location of primary sources can be provided by the author, e.g. exact dates of quotations or volume numbers of the Journal Officiel, Hansard's or Weekly Hansard's.

Chapter 11
Legitimising the European Space Policy

Emmanuel Sigalas

11.1 Introduction

The European Space Policy (ESP) may seem as a self-evident necessity to stakeholders, experts or enthusiasts of European cooperation, but to most Europeans it is an obscure policy area. Few people know what the ESP does, and even fewer how it works.[1] This is problematic. A policy area that is or appears to be unknown, irrelevant or unpopular will find it difficult to flourish in the short run and consequently it will have a hard time surviving in the long run.

The peculiarities of space raise the stakes for the ESP further. Outer space is dark, expensive and dangerous. It is difficult to predict investment returns, it is very costly and it has a security dimension to it. As a result, it is a potential political minefield. No political leader is keen on jeopardising their political capital in an investment with a high marginal cost and low marginal benefit. To put it differently, policies that are easier to 'sell' to the citizens stand a better chance for attracting political attention and subsequently financial capital.

Despite these peculiarities, or rather because of them, space has been gradually Europeanised. The competences are now shared between the national, international and supranational level. Since the Lisbon Treaty the European Union (EU) acquired formal powers on space upgrading the role of the EU's supranational institutions in this respect. Assuming that the role of the EU in space affairs will continue to grow, and this would be my personal prediction, the influence of the EU institutions over the development of the ESP will increase.

[1] In 2012 only 38 % of Europeans had heard of GMES (recently relabeled Copernicus) and 57 % of Galileo (Flash Eurobarometer no. 355). Up to date data are unfortunately not available.

E. Sigalas (✉)
Carleton University, Ottawa, ON, Canada

Institute of International Relations, Prague, Czech Republic
e-mail: emmanuelsigalas@gmail.com

C. Al-Ekabi et al. (eds.), *Yearbook on Space Policy 2014*, Yearbook on Space Policy, DOI 10.1007/978-3-7091-1899-3_11

The EP has veto power in the EU's ordinary legislative procedure, which is more than sufficient reason for weighing its intentions regarding the ESP. But there is another reason equally important. The EP is the only directly elected body of the EU. Its members (MEPs) represent all the European citizens, even if they are divided along partisan and national lines. Hence, the EP is in a position to offer democratic legitimacy to the decisions of the EU, including of course to the decisions related to the ESP. In other words, the EP is instrumental in 'selling' the ESP to policy-makers, stakeholders and the public alike.

The best way of exploring the EP's views is by studying its resolutions. Unlike EU documents such as legislative texts or communications, which bear primarily the European Commission's stamp, the EP resolutions are more reflective of its own positions, notwithstanding some Commission influence on them too.

The present chapter looks at all the non-technical EP resolutions from 1979 since 2013, to outline how the EP views the ESP, and how it justifies its existence and its progress. Given the space limitations, only a brief presentation is possible here, but it is enough to demonstrate that the EP is a staunch ally and supporter of Europe's common endeavors in outer space.

11.2 European Space Policy and Political Support

Public policy decisions have to be publicly justifiable, at least in democracies. The bolder, riskier or dearer a policy innovation is, the greater the need of the government to show that it acts in line with the public's preferences. High costs and potential mistakes are easier to forgive when public support abounds, and the opposition has a harder time to criticise the government. Thus, if it can be shown that the citizens, i.e. the electorate, support (or are likely to support) a particular policy choice, then it becomes much easier to convince the government to act accordingly.

It follows, that important policy initiatives or changes have to be publicly justifiable. In other words, they have to be convincing and acceptable to as wide an audience as possible. Ideally, the same set of arguments should appeal to policy-makers, stakeholders and the electorate. If it does, policy innovation and change are likely to go through fairly smoothly. In contrast, if resistance by policy-makers and stakeholders, or reluctance by the citizens is anticipated, then the argumentation has to be flexible yet potent enough, to win the impressions and help overcome any reservations or skepticism.

In the case of the ESP it is obvious that public justifiability is absolutely necessary. As I already mentioned, space is by default a sensitive policy area. On top of that, national governments are accountable to their parliaments and to their voters for transferring part of national sovereignty on space affairs to the supranational level.

Advances in European integration have long ceased to be the exclusive jurisdiction of political or bureaucratic elites.[2] The outcome of EU treaty referenda in the 1990s (Denmark, France) and in the new millennium (France, Ireland, Netherlands) sent political shockwaves throughout Europe with a clear message: public attitudes toward European integration can no longer be ignored.[3] As a result, national governments and EU institutions have to tread carefully when new policy areas, such as outer space, are added to the list of EU competences.

Whilst there is still no clear answer as to why and how space found its way in the Lisbon Treaty, national governments are no longer the only ones who have a say on space-related matters. The contribution of EU's supranational institutions to ESP precedes the Lisbon Treaty, but since 2009 they no longer have to hide behind kin EU policy areas, such as transport or research, to influence its development.[4] Now they can do it openly. The Commission has the exclusive right of legislation initiation, the EP is a veto player, and as far as EU law is concerned, the Court of the EU can overrule national courts.

In the ordinary legislative procedure, according to which EU legislation on ESP is adopted, the only real veto player is the EP. The Commission cannot refuse to draft legislation (but it cannot be dictated its content either), and the Court intervenes, if at all, after the legislation is in place. The EP, however, can always block a legislative proposal, even if on average the Council is the more influential co-legislator.[5] The EU's supranational institutions are well known for maintaining a pro-integrationist stance, but there is no guarantee that the EP will, always or unconditionally, support any legislative proposal on the ESP.[6] This is a crucial reason for studying what the EP thinks of the ESP, but there is another one. I address it below.

11.3 The European Parliament and the Legitimisation of the European Space Policy

I mentioned that the European integration process is much more politicised today than it was in the past. Consequently, securing the conditions for the growth of the new and politically sensitive ESP is a challenging task to say the least. According to

---

[2] Laffan, Brigid (1996) "The Politics of Identity and Political Order in Europe", *Journal of Common Market Studies* 34 (1): 81–102.

[3] Startin, Nick and André Kruwel (2013) "Euroscepticism Re-galvanized: The Consequences of the 2005 French and Dutch Rejections of the EU Constitution", *Journal of Common Market Studies* 51 (1): 65–84.

[4] Sigalas, Emmanuel (2012) "The Role of the European Parliament in the Development of EU Space Policy", *Space Policy* 28 (2): 110–117.

[5] Napel, Steffan and Mika Widgrén (2006) "The Inter-institutional Distribution of Power in EU Codecision", *Social Choice Welfare* 27 (1): 129–154.

[6] Sandholtz, Wayne and Alec Stone Sweet (1998) *European Integration and Supranational Governance*, Oxford: Oxford University Press.

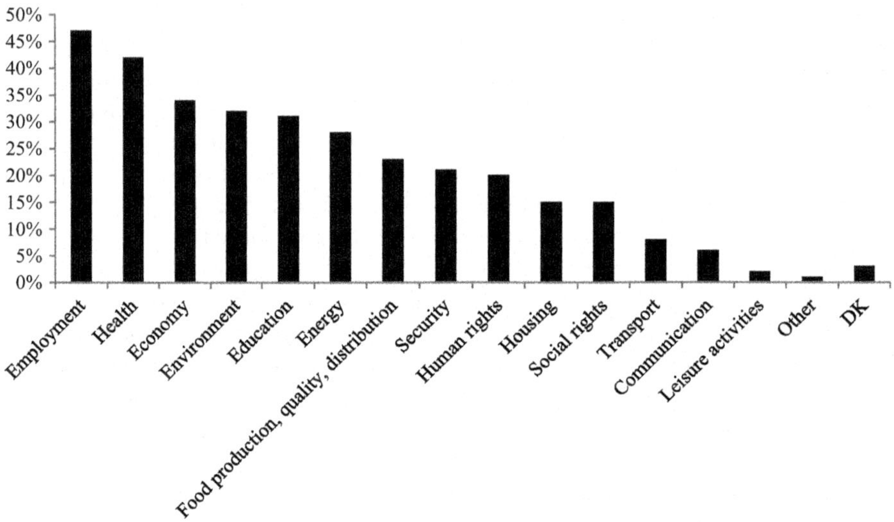

Fig. 11.1 "Which of the following, in 20 years' time, do you think should be the main priorities of the EU?" (EU-27 average). Source: Special Eurobarometer 403 (Jan. 2014)

the latest Eurobarometer data (January 2014), Europeans do not necessarily wish to see the EU playing a greater role on outer space. Instead, as Fig. 11.1 illustrates, in the future they would like the EU to prioritise more down to earth issues, such as employment, health or the economy. Given the repercussions of the global and European economic crisis, this is not surprising.

The existing Eurobarometer surveys do not ask directly if Europeans are in favour of the ESP, or if they would like to see it developing in the future, but if we take the attitudes towards space exploration as a proxy, the data are not very encouraging. Only 47 % of the respondents think that the EU should invest in space exploration, and even fewer (35 %) think that Europe should be participating in an international mission to explore Mars (Special Eurobarometer 403). In short, public support for developing the ESP is relatively weak.

To pursue the growth of the ESP, especially if the EU has a substantial role to play in it, decision-makers have to build up a stronger case. First of all, they need to legitimise their intentions or decisions to grant more powers to the EU and to increase the scope of the ESP. The EP has an important role to play in this respect. Since it is the only directly elected institution of the EU, and since it enjoys more public trust than the Council or the Commission, the EP can legitimately claim that it represents the views of the citizens and that it fights for their interests.[7] The direct democratic link is what gives the EP its political weight. It is the main reason why the proponents of the ESP need the EP on their side, in addition of course to its

[7] Whilst the EP's latest (Standard Eurobarometer 82) trust rate is not impressively high (42 %), it remains higher than public trust toward the Commission (38 %) or the Council of the EU (33 %).

involvement in the formal decision-making process. In short, the EP is not only a veto player, but also a player that can invest the EU's decisions with democratic legitimacy. Therefore, it is very useful to know what it thinks of the ESP.

11.4 The Reflection of the European Space Policy in the EP Resolutions

An efficient and innovative method to determine the EP's position on space is to study its resolutions. The resolutions are non-legislative, i.e. legally non-binding, documents that reflect the EP's opinions as an institution on a given topic. The details of the drafting and adoption process need not concern us here.[8] What is useful to know is that there are no restrictions in what goes in these documents, other than those prescribed in the EP's rules of procedure. If the EP feels that it needs to put its official position on record, so that others know where it stands, then voting upon and subsequently publishing a resolution serves this purpose.

The analysis of all available EP resolutions on space allows me to make two general observations. Firstly, over the years the space resolutions have been growing in length and sophistication. The first resolution was a very short document containing only nine paragraphs. Today it is not uncommon to read a resolution that contains more than 60 lengthy paragraphs where not only general principles, but also very specific details are discussed.[9] Secondly, notwithstanding some variation over time, there is a general pattern in the structure of these documents. The first part contains references to EU legislation or other official EU documents; the second part contains the context (or the rationale); and the last part, which may also be called the main, includes the EP's standpoint and calls for action.

Until 2013 the EP had issued as many as 27 space resolutions, of which 5 are technical in content.[10] One may be surprised to read that the first resolution dates back to 1979. At the time the EU had nothing to do with outer space, which is precisely the reason why the EP prepared a report that eventually resulted in a resolution titled the "Community's participation in space research".

Obviously, it is impossible here to present all, or in any detail, the EP resolutions on space.[11] What is possible, though, is to present their substance in a compact yet revealing way. The two sub-sections below do just that. The first gives a snapshot of why the EP believes that action on space-related matters should be taken. In other words, it offers a synopsis of the EP's public justification arguments for a European

---

[8] The interested reader may wish to consult Corbett, Richard, Francis Jacobs and Michael Shackleton (2011) *The European Parliament*, London: John Harper.

[9] Sigalas, Emmanuel (2015) "Europe in Space: The European Parliament's Justification Arsenal", in T. Hoerber and P. Stephenson (eds.) *The European Space Policy*, London: Routledge.

[10] Ibid.

[11] A more detailed analysis of the EP resolutions and of the methodology I used in my analysis can be found in Sigalas (2015).

space policy. The second looks at the development of the justification strategy over time. To analyse the EP's argumentation, I resorted to a hybrid (qualitative and quantitative) manual content analysis of all the non-technical space resolutions. Naturally, the content analysis concentrated only on the second (context/rationale) and third part (standpoint/calls for action) of each document.

The results reveal a European Parliament that has always been supportive of European cooperation in space affairs, many years before the term 'European Space Policy' was coined.[12] Here I show that, in order to promote European cooperation on space matters and to legitimise the EU's involvement, the EP adopted a varied and multi-dimensional public justification strategy.

11.4.1 The Justification Arguments for a European Space Policy

They say that a picture is worth a thousand words, but Fig. 11.2 is worth more. The depicted word-cloud is a reconstruction of the EP space resolutions condensed according to the frequency of the justification arguments. To put it simply, keywords that have been showing up more frequently in the text corpus are represented

Fig. 11.2 Why should action on space matters be taken? Source: Author's own data

[12] Ibid.

with bigger letters.[13] Thus, thanks to Fig. 11.2 one can easily see that the EP has been using very often the adjective 'important' in its official texts.

To argue that space, space policy or something related to either, are important sounds like a truism, but it is not. In politics and in diplomacy words are chosen very carefully, especially in written communication. The term 'important' implies a sense of urgency that related concepts not appearing in the text corpus, such as 'useful' or 'valuable', do not. Thus, it is very likely that the EP has chosen this particular word, in order to emphasise how important space is for Europe, and consequently how important the creation and development of a European space policy is.

The second most frequent term is 'public' followed by other keywords, such as 'applications', 'access', 'infrastructure', 'industry' or 'independent'. These words resemble dots on a white canvass. If we find the pattern that connects them, they start making some sense, even if the final picture is only a sketch. Taking into account that the keywords are justification arguments for legitimating a European space policy, it is not too difficult to discover the pattern. To cut a long story short, the EP's understanding of ESP is, that it is a public policy that will bring benefits to the public, but it is also in need of public funding and public support.

The other five keywords are more interesting. Their relatively frequent use in the space resolutions suggests, that the ESP is promoted and legitimised primarily in utilitarian terms. According to the EP, building the appropriate space infrastructure is important, because it will enable Europe to become an independent space power. It will no longer have to rely on the USA for space-related data and services, and it will have little to fear from rising space powers such as India or China. Furthermore, Europe's industry will benefit from building the high-tech space infrastructure, and businesses and citizens will benefit from the applications that will follow.

11.4.2 The Justification Strategy for a European Space Policy

Given what I just said, one may be misled to believe that the EP constantly emphasises the economic dimension of space; that a common space policy is a good idea, because of the economic benefits it will deliver. This is not quite right, although economic considerations figure prominently in the space resolutions. The EU is primarily an economic union and if the EP arguments are to have any appeal they have to reflect this. However, the resolutions also reflect that a European space policy is necessary for non-economic reasons too.

[13] The text corpus contains all the quasi-sentences (part or whole sentences containing a single argument) of the 22 resolutions addressing the imaginary question 'why should action on space matters be taken'.

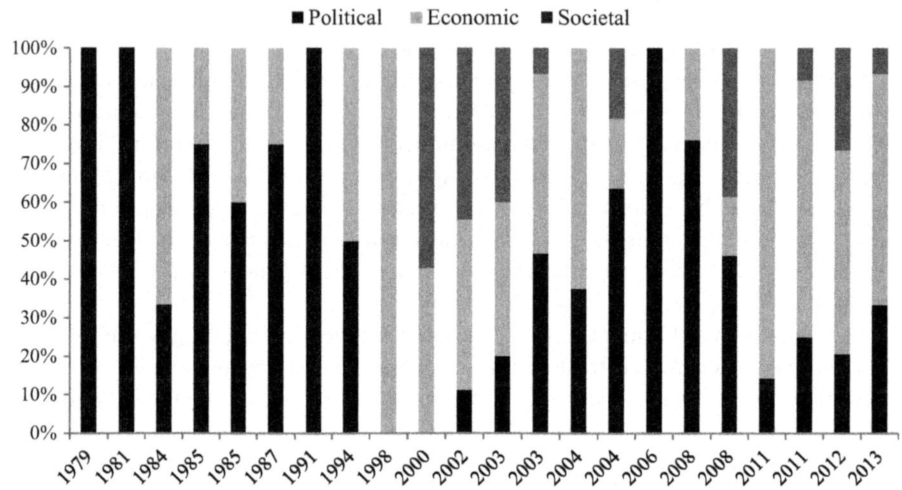

Fig. 11.3 The Dimensions of the Reasons for a European Space Policy. Source: Author's own data

In Fig. 11.3 I have divided the EP public justification arguments for a European space policy along three broad categories: economic, political and societal. The economic dimension refers to arguments that legitimise common endeavours on space because of the economic benefits this will deliver, or because of the economic problems it will solve. The second dimension (political) includes arguments that make no direct references to economic considerations and are broadly speaking political. I say broadly speaking, because they contain mainly, albeit not exclusively, security-related arguments, such as the need for independent access to or presence in space, the threat or pressure of international competition, or the potential for contributing to Europe's foreign, security and defense policies. The third category (societal) is also very broad. It contains arguments that have mostly to do with non-economic and non-political social benefits, such as space's potential for the environment, climate change or scientific research. There is also a residual category ('other'), but it is omitted from the analysis here.

Figure 11.3 suggests that the EP is resourceful in its legitimation strategy of the ESP. As the figure shows, the EP provides not only economic, but also political and societal reasons for developing space policy. Secondly, there is variation over time, which implies that the EP adapts its argumentation to respond to the changing circumstances. This in turn should help maintaining the momentum for a European space policy, despite changes in the priorities of the EU and of its member states as the national and international environment evolves.

Once again, space limitations do not allow me to go into depth, but there are a few interesting observations to make regarding Fig. 11.3. First of all, the economic dimension is not as dominant as one might expect. With 42 % of all justification arguments having primarily an economic dimension they rank first, but they still form less than half of the total. The political justification of the ESP follows closely with 38 % of all arguments, and the societal dimension of the ESP legitimation argumentation with a meagre 13 %.

As already noted, the ESP justification arguments are not evenly distributed across time. The resolutions up to 1991 rely predominantly on political reasons to justify the need for EU initiatives on outer space. The Cold War provides both the historical context and the explanation for this choice, since comparisons to the two superpowers abound. For example, the 1981 resolution emphasises, that compared to the USA and the Soviet Union, European financial commitment on space is limited, and the 1985 resolution deplores that the European Community lacks a coordinated space policy. During the same period, many of the EP resolutions stress the need for European independence on space affairs, making it the oldest argument in favour of an ESP that is still in use.[14]

From the mid-1990s and for the next ten years economic argumentation overshadows political in the EP space resolutions. The palette of the economic arguments is very extensive, but market failure and the need for public investment on space stand out. According to the EP, Europe's space industry and infrastructure cannot grow without support from the public purse. That is so, because the space sector is not profitable enough. However, given the lessons from the Kosovo war, it is highly important to shake off dependence from the USA and at the same time reap the many benefits space has to offer.[15] Consequently, public action and funding are needed, to build and consolidate a European space policy.

Most of the societal arguments are concentrated in the 2000–2003 resolutions, but they reappear in 2012. Investing in space is promoted in terms of benefits on the citizens' daily life, scientific progress, disaster management and environmental protection. Between 2003 and 2008 political considerations return in the EP resolutions. This time there is a heavy presence of security concerns and how space can help in this respect, in particular in terms of realising the goals of the EU's Common Foreign and Security Policy and European Security and Defense Policy.

Economic reasons for supporting a European space policy start dominating again the EP's justification strategy after 2008. This was to be expected. With the global economic crisis hitting the shores of Europe the EU's priorities were inevitably recalibrated. Nonetheless, the EP did not abandon the goal of ESP. Instead, it adjusted its justification rhetoric, to persuade policy-makers, stakeholders and the citizens how much Europe's competitiveness would improve, and how cost-effective and cost saving a single European space policy would be, compared to the expensive and overlapping national policies.

[14] Sigalas, Emmanuel (2015) "Europe in Space: The European Parliament's Justification Arsenal", in T. Hoerber and P. Stephenson (eds.) *The European Space Policy*, London: Routledge.

[15] During the Kosovo War (1998–1999) the US cut the GPS signal making European policy-makers painfully aware of the risks associated with depending on a foreign government for the availability of a crucial technology such as the GPS [Jones, Seth (2007) *The Rise of European Security Cooperation*, Cambridge: Cambridge University Press. p. 164].

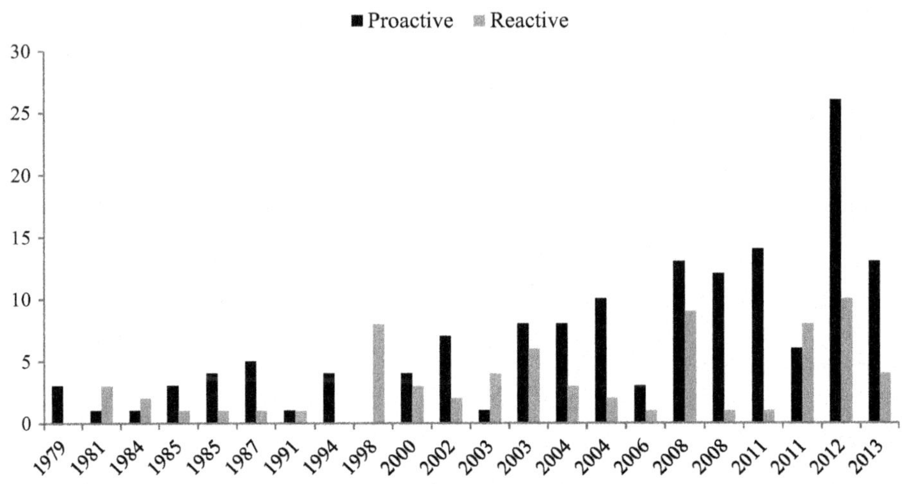

Fig. 11.4 The Outlook of the Reasons for a European Space Policy. Source: Author's own data

A way of confirming how determined the EP is to promote a supranational space policy is to study the outlook of the public justification arguments. My analysis shows that the EP deploys both proactive and reactive arguments, to maximise the appeal of ESP. By reactive I mean that the EP often presents space (or space policy) as an answer to something, usually to a problem. By proactive I mean that the EP portrays space (or space policy) as an end in itself, as a means to an end or, interestingly enough, as a feasible goal. To use an analogy, space as a solution to a problem would make a reactive justification argument, whereas space as a solution in search of a problem would make a proactive.

Figure 11.4 illustrates that the EP has been relying on both proactive and reactive arguments, to justify why a European space policy is a sensible idea. However, the use of proactive arguments is comparatively more common. This is interesting, because it suggests that the EP does not wait for problems or challenges to emerge to promote the cause of ESP, although it does not hesitate to use them for its purposes when they do. Instead, the EP is keen to emphasise not only that a European space policy would be beneficial for many and diverse reasons, but also that Europe, and the EU in particular, can succeed in having its own and independent space policy. The message is clear: Europe can and should take a European space policy off the ground.

11.5 Conclusion

The EU's official involvement in outer space is now a reality. Galileo, EGNOS and Copernicus are all EU programs and any changes to them require the prior approval of the EP. Any addition of new EU space programs also requires the approval of the

EP. Any change in the relations between ESA and the EU again requires the approval of the EP. Any EU treaty changes related to space require the approval of the EP. In short, the EP is now a veto player and no changes to the EU's space policy can materialise without its support.

My analysis, however, has shown that the EP is not an opponent of the ESP, but one of its keenest supporters. That is good so, because the still young common space policy needs all the support it can get. Financial and political investment in the ESP cannot be taken for granted. Actually, given the scarcity of resources, the security implications, and political sensitivities about national sovereignty, it is impressive that the EU can count space among its official policies. Therefore, if the ESP is to develop and mature, it needs to become even more attractive to decision-makers. As I have shown in this chapter, the EP and in all likelihood the European Commission need little convincing. It is the EU's national governments that need to be persuaded to invest in the ESP politically and economically. In order to do that, sufficient reasons have to be provided; reasons that the national governments can use, in order to convince their own domestic audiences, including of course their electorate.

In this respect the role of the EP has been instrumental. In all its space resolutions since 1979 the EP has been producing reasons that justify the need for a European policy on space. These range from concrete advantages to idealistic principles. Furthermore, the EP has been adjusting its legitimisation rhetoric, to ensure that the goal of ESP never loses relevance and therefore momentum. What is equally striking and possibly offers some insight about what the future holds for the ESP is the changing focus of the space resolutions. In the earlier years the EP was concentrating on getting the member states interested in space. As time progresses the attention shifts toward helping them to overcome any hesitations against EU involvement. It is only natural, therefore, to wonder if all this is part of a wider evolutionary process. A process whose next phase involves the EP promoting a European space policy that takes shape first and foremost at the EU level.

Part III
Facts and Figures

Chapter 12
Chronology: 2014

Cenan Al-Ekabi and Blandina Baranes

12.1 Access to Space[1]

All launch dates are calculated using Greenwich Mean Time (GMT), hence the date at the launch site may differ from the date listed here by 1 day.

Europe	Other countries
Launch log	
January 2014	
	05 GSLV Mk 2—Gsat 14 (C)
	06 Falcon 9 v1.1—Thaicom 6 (C)*
	09 Antares 120—Cygnus CRS-1 (ISS)*, and 33 cubesat secondary payloads: Flock-1 1-to-28 (D)*, ArduSat 2 (D), LituanicaSAT 1 (D), LitSat1 (D), SkyCube (D), UAPSat 1 (D)
	23 Atlas V 401—TDRS L (C)
February 2014	
06 Ariane 5 ECA—Athena-Fidus (C)*, and ABS 2 (C)	05 Soyuz U—Progress 54P (ISS), and 1 cubesat secondary payload: Chasqui 1 (D)
	14 Proton M—Turksat 4A (C)*
	20 Delta 4 Medium +(4,2)—Navstar GPS 2F-5 (N)

(continued)

[1] Federal Aviation Administration. The Annual Compendium of Commercial Space Transportation: 2014. Washington DC: FAA, Feb. 2015: 190+.

C. Al-Ekabi (✉) • B. Baranes
ESPI, Schwarzenbergplatz 6, 1030 Vienna, Austria
e-mail: cenan.al-ekabi@espi.or.at; blandina.baranes@espi.or.at

C. Al-Ekabi et al. (eds.), *Yearbook on Space Policy 2014*, Yearbook on Space Policy, DOI 10.1007/978-3-7091-1899-3_12

Europe	Other countries
Launch log	
	27 H-IIA 202—GPM Core Observatory (R), and 7 cubesat secondary payloads: STARS 2 (D), ShindaiSat (D), TeikjoSat 3 (S), KSAT 2 (S), OPUSAT (D), INVADER (D), IFT 1 (D)
March 2014	
22 Ariane 5 ECA—Astra 5B (C)*, and Amazonas 4A (C)*	15 Proton M—Express AT1 (C), and Express AT2 (C)
	23 Soyuz 2.1b—Glonass M (Cosmos 2194) (N)
	25 Soyuz FG—Soyuz TMA 12M (MF/ISS)
	31 Long March 2C—Shijian 11-06 (D)
April 2014	
03 Soyuz 2.1a—Sentinel 1A (S) 30 Vega—KazEOSat-1 (R)	03 Atlas V 401—USA 249 (DMSP-5D3-F19) (M)
	04 PSLV XL—IRNSS 1B (N)
	09 Soyuz U—Progress 23M (ISS)
	09 Shavit—Ofeq 10 (I)
	10 Atlas V 541—NROL-67 (I)
	16 Soyuz U—EgyptSat 2 (R)
	18 Falcon 9—SpaceX CRS 3 (ISS)*, and 5 cubesat secondary payloads: ALL-STAR (D), KickSat 1 (C), SporeSat 1 (S), TestSat-Lite (D), PhoneSat 2.5 (D)
	28 Proton M—Luch 5V (C), and Kazsat 3 (C)
May 2014	
	15 Proton M—Express AM4R (C) (launch failure: unspecified failure of the launcher's third stage engine)
	17 Delta 4 Medium +(4,2)—GPS 2F-6 (N)
	22 Atlas V 401—NROL-33 (I)
	23 Rockot—Cosmos 2496 (C), Cosmos 2497 (C), Cosmos 2498 (C), and 1 cubesat secondary payload: RS 47 (D)
	24 H-IIA 202—ALOS 2 (R), and 4 cubesat secondary payloads: RISING 2 (R), UNIFORM 1 (R), SOCRATES (R), SPROUT (D)
	26 Zenit 3SL—Eutelsat 3B (C)*
	28 Soyuz FG—Soyuz TMA 13M (MF/ISS)
June 2014	
	14 Soyuz 2.1b—Glonass M (Cosmos 2500) (N) 19 Dnepr—Deimos 2 (R)*, KazEOSat 2 (R), SaudiSat 4 (D), and 35 cubesat secondary payloads: TabletSat Aurora (D)*, BugSat 1 (R), Hodoyoshi 3 and 4 (R), AprizeSat 9 and 10 (C), UniSat 6 (D), BRITE-CA 1 and 2 (S), Perseus M1 and M2 (R), Flock-1c 1-to-11 (R)*, POPSAT HIP 1 (D)*, Tigrisat (R), Lemur

(continued)

Europe	Other countries
Launch log	
	1 (D)*, QB50 P1 and P2 (S), PACE (D), ANTELSAT (D), DTUSat 2 (D), Duchifat 1 (S), NanoSatC-Br 1 (S), PolyITAN 1 (D), and AeroCube 6A and 6B (D) 30 PSLV CA—Spot 7 (R)*, and 5 cubesat secondary payloads: CanX-4 (D), CanX-5 (D), AISat 1 (D), VELOX 1-NSAT (D), VELOX 1-PSAT (D)
July 2014	
10 Soyuz 2.1b—O3b FM3 (C)*, O3b FM6 (C)*, O3b FM7 (C)*, O3b FM8 (C)* 29 Ariane 5 ES—ATV 5 (ISS)	02 Delta 2 7320-10—OCO2 (R) 03 Rockot—Gonets M8 (C), Gonets M9 (C), and Gonets M10 (C) 08 Soyuz 2.1b—Meteor M2 (M), SkySat 2 (R)*, TechDemoSat 1 (D)*, MKA PN2 (Relek) (D), and 3 cubesat secondary payloads: AISSat 2 (I), DX 1 (D)*, Ukube 1 (D) 13 Antares 120—Orb 2 (ISS)*, and 32 cubesat secondary payloads: Flock-1b 1-to-28 (R)*, MicroMAS (D), TechEdSat 4 (D), Lambdasat (D), GEARRSAT (D) 14 Falcon 9—Orbcomm 2 F3 (C)*, Orbcomm 2 F4 (C)*, Orbcomm 2 F6 (C)*, Orbcomm 2 F7 (C)*, Orbcomm 2 F9 (C)*, Orbcomm 2 F11 (C)* 18 Soyuz 2.1a—Foton M4 (S) 23 Soyuz U—Progress M24 (ISS) 28 Delta 4 Medium +(4,2)—GSSAP 1 (I), GSSAP 2 (I), and 1 cubesat secondary payload: ANGELS (D)
August 2014	
22 Soyuz 2.1b—Galileo FOC 1 (N), Galileo FOC-2 (N) (launch failure: malfunction of the launcher's third stage engine—recovery of satellite orbits in process)	02 Atlas V 401—Navstar GPS 2F-7 (N) 05 Falcon 9—AsiaSat 8 (C)* 09 Long March 4C—Yaogan 20A (I), Yaogan 20B (I), Yaogan 20C (I) 13 Atlas V 401—WorldView 3 (R)* 19 Long March 4B—Gaofen 2 (R), and 1 cubesat secondary payload: BRITE-PL-2 (S)
September 2014	
11 Ariane 5 ECA—MEASAT 3B (C)*, and Optus 10 (C)*	04 Long March 2D—Lingqiao (C), and 1 cubesat secondary payload: CX 1-04 (C) 07 Falcon 9—AsiaSat 6 (C)* 08 Long March 4B—Yaogan 21 (I), and 1 cubesat secondary payload: Tiantuo 2 (D) 17 Atlas V 401—CLIO (USA 257) (I) 21 Falcon 9—SpaceX CRS 4 (ISS)*, and 1 cubesat secondary payload: SpinSat (S) 25 Soyuz FG—Soyuz TMA 14M (MF/ISS) 27 Proton M—Olimp K (C) 28 Long March 2C—Shijian 11-07 (D)

(continued)

Europe	Other countries
Launch log	
October 2014	
16 Ariane 5 ECA—Intelsat 30 (C)*, and ARSAT 1 (C)*	07 H-IIA 202—Himawari 8 (M) 15 PSLV XL—IRNSS 1C (N) 20 Long March 4C—Yaogan 22 (I) 21 Proton M—Express AM6 (C) 23 Long March 3C—Chang'e 5 (D), and 1 cubesat secondary payload: 4M (D) 27 Long March 2C—Shijian 11-08 (D) 28 Antares 130—Cygnus CRS 3 (ISS)*, and 29 cubesat secondary payloads: Flock-1d 1-to-26 (R)*, Arkyd-3 (D), RACE (D), GOMX 2 (D) (launch failure: anomaly encountered during the launcher's first stage engine) 29 Soyuz 2.1a—Progress 25M (ISS) 29 Atlas V 401—Navstar GPS 2F-8 (N) 30 Soyuz 2.1a—Maridian 7 (C)
November 2014	
	06 Dnepr—ASNARO 1 (D)*, and 4 cubesat secondary payloads: ChubuSat 1 (D), Hodoyoshi 1 (R), QSAT-EOS (D), and Tsubame (S) 14 Long March 2C—Yaogan 23 (I) 20 Long March 2D—Yaogan 24 (I) 21 Kuaizhou—Kuaishou 2 (D) 30 Soyuz 2.1b—Glonass K1 (N)
December 2014	
06 Ariane 5 ECA—DirecTV 14 (C)*, and Gsat 16 (C)* 18 Soyuz 2.1b—O3b FM9 (C)*, O3b FM10 (C)*, O3b FM11 (C)*, O3b FM12 (C)*	03 H-IIA 202—Hayabusa 2 (S), and 3 cubesat secondary payloads: PROCYON (S), DES-PATCH (D), Shin'en 2 (S) 05 Delta 4 Heavy—Orion EFT-1 (D) 07 Long March 4B—CBERS 4 (R) 10 Long March 4C—Yaogan 25A (I), Yaogan 25B (I), and Yaogan 25C (I) 13 Atlas V 541—NROL-35 (I) 15 Proton M—Yamal 401 (C) 19 Strela—Kondor E (R) 23 Angara A5—IPM (mass simulator) (D) 25 Soyuz 2.1b—Lotos S (I) 26 Soyuz 2.1b—Resurs P2 (R) 27 Long March 4B—Yaogan 26 (I) 27 Proton M—Astra 2G (C)* 31 Long March 3A—Fengyun 2G (M)

C Communications, *D* Development, *I* Intelligence, *ISS* International Space Station, *M* Meteorological, *MF* Manned Flight, *N* Navigation, *R* Remote sensing, *S* Scientific, *EW* Early warning system

12.2 Space Science and Exploration

Europe	Other countries
Earth sciences	
3 April—Launch of Sentinel 1A (ESA, Europe) radar imaging satellite contributing to the EU Copernicus Earth Observation flagship programme.[a] *19 June*—Launch of QB50 P1 and P2 (von Karman Institute, Belgium) scientific cubesats.[a] *19 August*—Launch of BRITE-PL-2 (BRITE-PL consortium, Poland/Canada) scientific cubesat.[a]	*27 February*—Launch of TeikyoSat 3 (Teikyou University, Japan), and KSAT 2 (Kagoshima University, Japan) scientific cubesats.[a] *3 April*—Launch of USA 249 (USAF, USA) meteorological satellite.[a] *19 June*—Launch of BRITE-CA1 and BRITE-CA2 (UTIAS, Canada), Duchfat 1 (Space laboratory of the Herzliya Science Center, Israel), and NanoSatC-BR 1 (INPE, Brazil) scientific cubesats.[a] *8 July*—Launch of Meteor M2 (Hydrometeorological Center of Russia, Russia) weather satellite; and MKA PN2 scientific cubesat.[a] *18 July*—Launch of Foton M4 (Roscosmos, Russia) microgravity research satellite.[a] *7 October*—Launch of Himawari 8 (Japan Meteorological Agency, Japan) meteorological satellite.[a] *6 November*—Launch of Tsubame (Tokyo Institute of Technology, Japan) scientific cubesat.[a] *31 December*—Launch of Fengyun 2G (China Meteorological Administration, China) meteorological satellite.[a]
Astronomy	
17 June—COnvection, ROtation and planetary Transits (COROT) space telescope (CNES, France) ended its mission to probe the inner structure of stars using stellar seismology and to detect extrasolar planets.[b] *25 July*—Gaia (ESA), following a series of tests and calibrations, began its 5-year scientific mission to measure the positions and motions of a billion stars of the roughly 100 billion stars in our galaxy to create the most accurate map yet of the Milky Way.[d]	*1 January*—new mission of the Wide-field Infrared Survey Explorer (NEOWISE) (NASA, USA) to discover and characterize near-Earth objects with infrared light begins at the start of 2014.[c] *16 May*—Kepler 2 mission (NASA, USA) approved which repurposed Kepler to point at target sky fields along the ecliptic plane of Earth's orbit, after losing two of its four reaction wheels resulting in the spacecraft losing its ability to precisely point at the original field of view.[e]
Exploration	
20 January—Rosetta spacecraft (ESA) exited 31 month deep space hibernation.[f] *4 April*—final telecommand sent to PICARD mission (CNES, France) which aimed to improve knowledge of how the Sun functions and the influence of solar activity on the Earth's climate.[h] *06 August*—Rosetta arrives at comet 67P/	*15 September*—Lunar Reconnaissance Orbiter (LRO) (NASA, USA) beginning its second 2 year extended science mission slated for completion in September 2016.[g] *21 September*—Mars Atmosphere and Volatile EvolutioN (MAVEN) (NASA, USA) successfully entered Mars' orbit to explore the planet's upper atmosphere, ionosphere and

(continued)

Europe	Other countries
Churyumov-Gerasimenko and begins global mapping.[j] *12 November*—Rosetta lander Philae successfully lands on comet 67P/Churyumov-Gerasimenko, utilizing its suite of sampling instruments over 56 h prior to hibernation.[l] 28 November—Last contact made with Venus Express (ESA) which studied the dynamics and chemistry of Venus' atmosphere (including surface interactions and interactions with solar wind) before the 8 year mission came to an end in December 2014.[n]	interactions with the Sun and solar wind.[i] *24 September*—Mars Orbiter Mission (MOM) (ISRO, India) successfully entered Mars' orbit to observe Mars' surface, atmosphere and exosphere for a detailed understanding of the planet's evolution.[k] *23 October*—launch of Chang'e 5-TI (CAST, China) which looped around the Moon as a test-run for an eventual lunar sample return mission (the mission also accommodated a the cubesat 4M).[m] *3 December*—Launch of Hayabusa 2 (JAXA, Japan), PROCYON (JAXA, Japan), and Shin'en 2 (Kagoshima University) to fly by an asteroid, land and return samples.[o] *15 December*—new frontiers mission Juno (NASA, USA) to study and reveal the origin and evolution of Jupiter approximately 565 million km from Earth, and with another 430 million km still to travel before its expected arrival at Jupiter's orbit in July 2016.[p]
Manned Spaceflight and Cargo Transfers	
28 May–10 November—Alexander Gerst (Germany) participated in ISS Expeditions 40 & 41 as a flight engineer for ESA's Blue Dot mission.[q] *29 July*—Launch of ATV-5, George Lemaître (European Space Agency), ISS resupply vehicle.[a] *24 November*—Samantha Cristoforetti (Italy) began a five-month mission on ISS Expedition 42 & 43 as a flight engineer for ESA's Futura mission.[r]	*9 January*—launch of Cygnus Orb-1 (Orbital Sciences, USA) commercial resupply services mission to ISS.[a] *5 February, 9 April, 24 July, 29 October*—launches of 4 Progress (Roscosmos, Russia) cargo transfer vehicles to the ISS.[a] *26 March*—launch of Soyuz TMA-12M (Roscosmos, Russia) with three ISS Expedition 39/40 crew. *18 April*—launch of Dragon CRS-3 (SpaceX, USA) commercial resupply services mission to ISS.[a] *28 May*—launch of Soyuz TMA-13M (Roscosmos, Russia) with three ISS Expedition 40/41 crew. *13 July*—launch of Cygnus Orb-2 (Orbital Sciences, USA) commercial resupply services mission to ISS.[a] *21 September*—launch of Dragon CRS-4 (SpaceX, USA) commercial resupply services mission to ISS.[a] *26 September*—launch of Soyuz TMA-14M (Roscosmos, Russia) with three ISS Expedition 41/42 crew. *28 October*—launch of Cygnus Orb-3* (Orbital Sciences, USA) commercial resupply services mission to ISS.[a] * (*launch of Antares-130 failed due to anomaly during the first stage of its launch.*)[s]

(continued)

Europe	Other countries
	23 November—launch of Soyuz TMA-15M (Roscosmos, Russia) with three ISS Expedition 42/43 crew.

[a]Federal Aviation Administration. The Annual Compendium of Commercial Space Transportation: 2014. Washington DC: FAA, Feb. 2015: 190+

[b]"CoRoT Events." Cnes 2 Mar. 2015. http://smsc.cnes.fr/COROT/GP_actualite.htm

[c]"NEOWISE Celebrates First Month of Operations After Reactivation." 23 Jan. 2014. NASA 9 Mar. 2015. http://www.jpl.nasa.gov/wise/newsfeatures.cfm?release=2014-023

[d]"Gaia overview." 23 Apr. 2014. ESA 9 Mar. 2015. http://www.esa.int/Our_Activities/Space_Science/Gaia/Gaia_overview

[e]"Kepler Begins K2 Mission Field 1 Observing." 30 May 2014. NASA 9 Mar. 2015. http://kepler.nasa.gov/news/nasakeplernews/index.cfm?FuseAction=ShowNews&NewsID=341

[f]"ESA's 'sleeping beauty' wakes up from deep space hibernation." 20 Jan. 2014. ESA 19 July 2015. http://www.esa.int/Our_Activities/Space_Science/Rosetta/ESA_s_sleeping_beauty_wakes_up_from_deep_space_hibernation

[g]Keller, J. "The Lunar Reconnaissance Orbiter and the New Moon: Mission Highlights and Two More Years of Science From Lunar Orbit!" 22 Nov. 2014. LRO—Lunar Planetary Institute 16 Mar. 2015. http://www.lpi.usra.edu/meetings/leag2014/presentations/keller.pdf

[h]"PICARD News." Cnes 1 Mar. 2015. http://smsc.cnes.fr/PICARD/GP_actualites.htm

[i]"MAVEN spacecraft enters orbit around Mars." 22 Sept. 2014. University of Colorado Boulder 1 Mar. 2015. http://lasp.colorado.edu/home/maven/2014/09/22/maven-spacecraft-enters-orbit-around-mars/

[j]"Rosetta's arrives at comet destination." 6 Aug. 2014. ESA 19 July 2015. http://www.esa.int/Our_Activities/Space_Science/Rosetta/Rosetta_arrives_at_comet_destination

[k]"Mars Mission: India creates history as Mangalyaan successfully enters Mars orbit in first attempt." 24 Sept. 2014. The Economic Times 22 Mar. 2015. http://economictimes.indiatimes.com/news/science/mars-mission-india-creates-history-as-mangalyaan-successfully-enters-mars-orbit-in-first-attempt/articleshow/43299562.cms

[l]De Selding, Peter B. "Philae Managers Say Recovery Possible as Comet Approaches Sun." 18 Nov. 2014. SpaceNews 19 July 2015. http://spacenews.com/42586philae-managers-say-recovery-possible-as-comet-approaches-sun/

[m]Spudis, Paul D. "China is Now Positioned to Dominate the Moon." 6 Nov. 2014. Air & Space Smithsonian 28 Feb. 2015. http://www.airspacemag.com/daily-planet/china-now-positioned-dominate-moon-180953267/?no-ist

[n]"Venus Express goes gently into the night." 16 Dec. 2014. ESA 1 Mar. 2015. http://www.esa.int/Our_Activities/Space_Science/Venus_Express/Venus_Express_goes_gently_into_the_night

[o]"Hayabusa 2." 4 Dec. 2014. Gunter's Space Page 24 July 2015. http://space.skyrocket.de/doc_sdat/hayabusa-2.htm

[p]"MISSION STATUS." 8 Jan. 2015. NASA 1 Mar. 2015. http://missionjuno.swri.edu/news/juno_status_12152014

[q]"ESA astronaut Alexander Gerst returns to Earth." 10 Nov. 2014. ESA 26 May 2015. http://www.esa.int/Our_Activities/Human_Spaceflight/Blue_dot/ESA_astronaut_Alexander_Gerst_returns_to_Earth

[r]"ESA astronaut Samantha Cristoforetti arrives at Space Station." 24 Nov. 2014. ESA 26 May 2015. http://www.esa.int/Our_Activities/Human_Spaceflight/Futura/ESA_astronaut_Samantha_Cristoforetti_arrives_at_Space_Station

[s]De Selding, Peter B. "Orbital Sciences Entitled To Partial NASA Payment for Antares Failure." 25 Nov. 2014. SpaceNews 13 July 2015. http://spacenews.com/42658orbital-sciences-entitled-to-partial-nasa-payment-for-antares-failure/

12.3 Applications

Europe	Other countries
Earth observation	
3 April—Launch of Sentinel 1A (ESA, Europe) radar imaging satellite contributing to the EU Copernicus Earth Observation flagship programme.[a]	*27 February*—Launch of GPM Core Observatory (NASA, USA; and JAXA, Japan).[a]
19 June—Launch of Deimos 2 (Deimos Space, Spain) small Earth observation satellite; along with the Tigrisat (University of Rome, Italy and Iraq) cubesat to detect dust storms over Iraq.[b]	*3 April*—Launch of USA 249 (USAF, USA) meteorological satellite.[a]
30 June—Launch of Spot 7 (Azercosmos, France) Earth observation satellite.[a]	*16 April*—Launch of EgyptSat 2 (NARSS, Egypt).[a]
	30 April—Launch of KazEOSat-1 (KGS, Khazakhstan).[a]
	24 May—Launch of ALOS 2 (JAXA, Japan) radar Earth observation satellite; along with RISING 2 (Tohoku University, Japan), UNIFORM 1 (Wakayama University, Japan), SOCRATES (NICT, Japan) Earth observation cubesats.[a]
	19 June—In addition to European payloads, the launch of KazEOSat-2 (KGS, Khazakhstan) Earth observation satellite; along with BugSat 1 (Satellogic S.A., Argentina), Hodoyoshi 3 and 4 (University of Tokyo, Japan), Perseus M1 and M2 (Dauria Aerospace, Russia and USA), Flock-1c 1-to-11 (Planet Labs, USA) commercial Earth observation cubesats.[a]
	2 July—Launch of OCO 2 (NASA, USA).[a]
	8 July—Launch of Meteor M2 (Hydrometeorological Center of Russia, Russia) weather satellite; along with SkySat 2 (Skybox Imaging, USA) commercial Earth observation satellite.[a]
	13 July—Launch of Flock-1b 1-to-28 (Planet Labs, USA) commercial Earth observation cubesats (*NB*: Flock-1b 3, 4, 13, 14, 19, and 20 were returned to Earth).[a]
	13 August—Launch of WorldView 3 (DigitalGlobe, USA) for Earth observation.[a]
	19 August—Launch of Gaofen 2 (CNSA, China).[a]
	7 October—Launch of Himawari 8 (Japan Meteorological Agency, Japan) meteorological satellite.[a]

(continued)

Europe	Other countries
	28 October—Failed launch of Cygnus CRS-3 carrying Flock-1d 1-to-26 (Planet Labs, USA) commercial Earth observation cubesats.[a]
	6 November—Launch of Hodoyoshi 1 (University of Tokyo, Japan) remote sensing cubesat.[a]
	7 December—Launch of CBERS (INPE/CASC, China and Brazil).[a]
	19 December—Launch of Kondor E (Roscosmos, Russia).[a]
	26 December—Launch of Resurs P2 (Roscosmos, Russia).[a]
	31 December—Launch of Fengyun 2G (China Meteorological Administration, China).[a]
Intelligence and early warning	
8 July—Launch of AISSat 2 (UTIAS, Norway) IMINT cubesat for maritime surveillance.[a]	*9 April*—Launch of Ofeq 10 (IDF, Israel) Synthetic Aperture Radar satellite.[a]
	6 May—Launch of Kobalt-M (Russian Aerospace Defence Forces, Russia) reconnaissance satellite.[a]
	22 May—Launch of NROL-33 (NRO, USA).[a]
	28 July—Launch of GSSAP 1 and 2 (USAF, USA) space surveillance satellites[c]
	9 August—Launch of Yaogan 20A, 20B, and 20C (PLA, China).[a]
	8 September—Launch of Yaogan 21 (CNSA, China).[a]
	20 October—Launch of Yaogan 22 (PLA, China).[a]
	14 November—Launch of Yaogan 23 (PLA, China).[a]
	20 November—Launch of Yaogan 24 (PLA, China).[a]
	10 December—Launch of Yaogan 25A, 25B, and 25C (PLA, China).[a]
	13 December—Launch of NROL-35 (NRO, USA).[a]
	25 December—Launch of Lotos S (Russian Aerospace Defence Forces, Russia).[a]
	27 December—Launch of Yaogan 26 (PLA, China).[a]
Navigation	
22 August—Partially failed launch of Galileo FOC 1 and FOC 2 (ESA, Europe) satellites (orbital recovery in process).	*20 February*—Launch of the Navstar GPS 2F-5 (USAF, USA) space-based radio positioning, navigation, and time distribution system.[a]

(continued)

Europe	Other countries
	23 May—Launch of Glonass M (Cosmos 2494) (Russian Aerospace Defence Forces, Russia) satellite.[a]
	4 April—Launch of IRNSS 1B (ISRO, India) satellite.[a]
	17 May—Launch of the Navstar GPS 2F-6 (USAF, USA) satellite.[a]
	14 June—Launch of Glonass M (Cosmos 2500) (Russian Aerospace Defence Forces, Russia) satellite.[a]
	2 August—Launch of the Navstar GPS 2F-7 (USAF, USA) satellite.[a]
	15 October—Launch of IRNSS 1C (ISRO, India) satellite.[a]
	29 October—Launch of the Navstar GPS 2F-8 (USAF, USA) satellite.[a]
	30 November—Launch of Glonass K1 (Cosmos 2502) (Russian Aerospace Defence Forces, Russia) satellite.[a]
Telecommunications/Broadcasting	
6 February—Launch of Athena-Fidus (ASI/CNES, Italy/France).[a]	*5 January*—Launch of Gsat 14 (ISRO, India).[a]
22 March—Launch of Astra 5B (SES, Luxembourg) and Amazonas 4A (Hispasat, Spain).[a]	*9 January*—Launch of Thaicom 6 (Thailand).[a]
26 May—Launch of Eutelsat 3B (Eutelsat, France).[a]	*23 January*—Launch of TDRS L (NASA, USA).[a]
10 July—Launch of O3b FM3, FM6, FM7, and FM8 (O3b Networks, UK).[a]	*6 February*—Launch of ABS 2 (Asia Broadcast Satellite, China).[a]
16 October—Launch of Intelsat 30 (Intelsat, Luxembourg).[a]	*14 February*—Launch of Turksat 4A (Turksat AS, Turkey).[a]
18 December—Launch of O3b FM9, FM10, FM11, and FM12 (O3b Networks, UK).[a]	*15 March*—Launch of Express AT1 and AT2 (RSCC, Russia).[a]
27 December—Launch of Astra 2G (SES-Astra, Luxembourg).[a]	*18 April*—Launch of KickSat 1 (Cornell University, USA) communications cubesat.[a]
	28 April—Launch of Luch 5V (Roscosmos, Russia) and Kazsat 3 (JSC KazSat, Khazakhstan).[a]
	15 May—Failed launch of Express AM4R (RSCC, Russia) due to a Proton M launch failure.[a]
	23 May—Launch of Cosmos 2496, Cosmos 2497, and Cosmos 2496 (Russian Aerospace Defence Forces, Russia).[a]
	19 June—Launch of AprizeSat 9 and 10 (ExactEarth, Argentina/USA) communications cubesats.[a]

(continued)

Europe	Other countries
	3 July—Launch of Gonets M8, M9, and M10 (Roscosmos, Russia).[a]
	14 July—Launch of Orbcomm 2 F3, F4, F6, F7, F9, and F11 (Orbcomm, USA).[a]
	5 August—Launch of AsiaSat 8 (AsiaSat, China).[a]
	4 September—Launch of the Lingqiao (Tsinghua University) and the CX 1-04 (Chinese Academy of Sciences, China) communications cubesat.[a]
	7 September—Launch of AsiaSat 6 (AsiaSat, China).[a]
	11 September—Launch of MEASAT 3B (MEASAT Satellite Systems, Malaysia) and Optus 10 (SingTel Optus, Australia).[a]
	27 September—Launch of Olimp K (Russian Aerospace Defence Forces, Russia).[a]
	16 October—Launch of ARSAT 1 (AR-SAT, Argentina).[a]
	21 October—Launch of Express AM6 (RSCC, Russia).[a]
	30 October—Launch of Meridian 7 (Russian Aerospace Defence Forces, Russia).[a]
	6 December—Launch of DirecTV 14 (DirecTV, USA) and Gsat 16 (ISRO, India).[a]
Technology development	
9 January—Launch of LituanicaSAT 1 (Kaunas University of Technology, Lithuania), LitSat1 (Lithuanian Space Association, Lithuania) development cubesats.[a]	9 January—Launch of Flock-1 1-to-28 (Planet Labs, USA) , ArduSat 2 (NanoSatisfi Inc., USA), SkyCube (Southern Stars Group LLC, USA), UAPSat 1 (Universidad Alas Perunas, Peru) development cubesats.[a]
19 June—UniSat 6 (University of Rome, Italy), DTUSat 2 (Danmarks Tekniske Universitet, Denmark), and PolyITAN 1 (National Technical University of Ukraine, Ukraine) development cubesats.[a]	5 February—Launch of Chasqui 1 (Universidad Nacional de Ingenieria del Peru, Peru) development cubesat.[a]
30 June—Launch of AISat 1 (DLR, Germany) development cubesat.[a]	27 February—Launch of STARS 2 (Kagawa University, Japan), ShindaiSat (Shinshu University, Japan), OPUSAT (Osaka Prefecture University, Japan), INVADER (ARTSaT project/Tama Art University, Japan), and IFT 1 (University of Tsukuba, Japan) development cubesats.[a]
8 July—Launch of TechDemoSat 1 (SSTL, UK), and Ukube 1 (UKSA, UK) development cubesats.[a]	31 March—Launch of Shijian 11-06 (PLA, China).[a]

(continued)

Europe	Other countries
13 July—Launch of Lambdasat (Lambda Team, Greece) development cubesat.[a]	18 April—Launch of ALL-STAR / THEIA (CoSGC, USA), TestSat-Lite (Taylor University, USA), and PhoneSat 2.5 (NASA Ames Research Center, USA) development cubesats.[a]
23 October—Launch of 4M (LuxSpace, Luxembourg) development cubesat accompanying the Chang'e 5-T1 moon mission.[a]	23 May—Launch of RS 47 (ISS Reshetnev, Russia) development cubesat.[a]
28 October—Failed launch of GOMX 2 (GOMSpace, Denmark) development cubesat due to destruction of launcher.[a]	24 May—Launch of SPROUT (Nihon University, Japan) development cubesat.[a]
	19 June—Launch of TabletSat Aurora (SPUTNIX, Russia), SaudiSat 4 (KACST, Saudi Arabia), POPSAT HIP 1 (Microspace, Singapore), Lemur 1 (Spire, USA), PACE (National Cheng Kung University, Taiwan), ANTELSAT (Uruguayan Facultad de Ingenieria de la Universidad de la Republica, Uruguay), and AeroCube 6A and 6B (The Aerospace Corporation, USA) development cubesats.[a]
	30 June—Launch of CanX-4 and CanX-5 (UTIAS, Canada), and VELOX 1-NSAT and VELOX 1-PSAT (Nanyang Technological University, Singapore) development cubesats.[a]
	8 July—Launch of DX 1 (Dauria Aerospace, Russia) development cubesat.[a]
	13 July—Launch of MicroMAS (MIT SS/L, USA), TechEdSat 4 (SJSU, USA), and GEARRSAT (NearSpace Launch, USA) development cubesats.[a]
	28 July—Launch of ANGELS (AFRL, USA) development cubesat.[a]
	8 September—Launch of Tiantuo 2 (National University of Defense Technology, China) development cubesat.[a]
	28 September—Launch of Shijian 11-07 (PLA, China).[a]
	23 October—Launch of Chang'e 5-T1 (CNSA, China) toward the Moon as a test-run for an eventual lunar sample return mission.[d]
	27 October—Launch of Shijian 11-08 (PLA, China).[a]
	28 October—Failed launch of Arkyd-3 (Planetary Resources, USA), and RACE (University of Texas at Austin, JPL, USA) development cubesats due to destruction of launcher.[a]

(continued)

Europe	Other countries
	6 November—Launch of ASNARO 1 - (J-spacesystems, Japan), and development cubesats ChubuSat 1 (Nagoya University, Japan) and QSAT-EOS (Kyushu University, Japan).[a]
	21 November—Launch of Kuaizhou 2 (Chinese Academy of Sciences, China).[a]
	3 December—Launch of DESPATCH (Tama Art University, Japan) development cubesat.[a]
	5 December—Launch of Orion EFT-1 (NASA, USA).[a]
	23 December—Launch of IPM (mass simulator) (Khurnichev, Russia) to test the Angara A5 launcher.[a]
Business	
1 January—Eutelsat's acquisition of Satmex finalized.[e]	*16 September—NASA awarded 'Commercial Crew Transport Capability' (CCtCap) contracts to Boeing and SpaceX ($4.2 billion and $2.6 billion, respectively) to continue to develop commercial spacecraft that are capable of transporting NASA astronauts the ISS by 2017.*[f]
Mid-February—OHB's subsidiary CGS S.p.A was awarded a sub-contract worth 44 million euros ($ million) from principal contractor Telespazio S.p.A for the Italian defence ministry's OPSAT (OPtical SATellite) 3000 programme.[g]	
7 April—Airbus Defence and Space was selected to be the prime contractor for the Metop-SG, after beating Thales Alenia paired with OHB in guaranteeing equal work shares in France and Germany to avoid a repeated struggle in project leadership, as was experienced in the Meteosat Third Generation satellite system.[h]	
20 May—Germany's BAE System Intelligence and Security, and Airbus Defence & Space announced that they will market data from German radar satellite data to the US government, including certain commercial and international customers.[i]	
August—Thales Alenia Space was selected to develop a very-high-resolution optical payload for the United Arab Emirates Air Force' Falcon Eye programme, in a consortium led by Airbus D&S; with the contract entering into force by the end of the year.[j]	

[a]Federal Aviation Administration. The Annual Compendium of Commercial Space Transportation: 2014. Washington DC: FAA, Feb. 2015: 190+

[b]"TigriSat." 2 Mar. 2015 Gunter's Space Page 24 July 2015. http://space.skyrocket.de/doc_sdat/tigrisat.htm

[c]"GSSAP (Geosynchronous Space Situational Awareness Program)." Gunter's Space Page 24 July 2015. http://space.skyrocket.de/doc_sdat/gssap-1.htm

(continued)

[d]Spudis, Paul D. "China is Now Positioned to Dominate the Moon." 6 Nov. 2014. Air & Space Smithsonian 28 Feb. 2015. http://www.airspacemag.com/daily-planet/china-now-positioned-dominate-moon-180953267/?no-ist
[e]Reference Document 2013–2014. Eutelsat Communications 9 Feb. 2015: 14. http://www.eutelsat.com/files/contributed/investors/pdf/reference-document-2013-2014.pdf
[f]Foust, Jeff. "NASA Commercial Crew Awards Leave Unanswered Questions." 19 Sept. 2014. SpaceNews 22 July 2015. http://spacenews.com/41924nasa-commercial-crew-awards-leave-unanswered-questions/
[g]Annual Report 2014. 26 Mar. 2015. OHB 31 Mar. 2015. http://www.ohb.de/tl_files/ohb/pdf/finanzberichte_hauptversammlung/2014/OHB_GB_14_E.pdf
[h] De Selding, Peter B. "Airbus Bests Thales Alenia-OHB Team for Billion-dollar ESA Weather Satellite Contract." 7 Apr. 2014. SpaceNews 8 Jan. 2015. http://spacenews.com/40109airbus-bests-thales-alenia-ohb-team-for-billion-dollar-esa-weather/
[i]De Selding, Peter B. "BAE To Market German Radar Satellite Data to U.S. Government." 21 May 2014. SpaceNews 25 June 2015. http://spacenews.com/40644bae-to-market-german-radar-satellite-data-to-us-government/
[j]"Thales Alenia Space to supply very-high-resolution optical payload for United Arab Emirates' Falcon Eye programs." 9 Dec. 2014. Thales Group 1 Apr. 2015. https://www.thalesgroup.com/en/worldwide/space/press-release/thales-alenia-space-supply-very-high-resolution-optical-payload-united

12.4 Policy and International Cooperation

Europe	Other countries
General policy	
28–29 January 2014—Sixth high-level Conference on EU Space Policy, 'What direction for Europe in space between now and 2020?', held in Brussels, Belgium.[a] 6 February—progress report by the Commission determined that transforming ESA into an EU agency would require political consensus that may be difficult to reach in the foreseeable future.[c] *31 March*—The fifth draft of the draft International Space Code of Conduct for Outer Space Activities (ICoC) was issued.[e] *Mid-April*—French government proposed its Law on Space Operations requiring French launch providers to direct the upper stages of their launchers onto courses which would cause them to drop into open waters or disintegrate immediately after launch in Earth's atmosphere to reduce orbital debris and prevent liability from upper stages re-entering the atmosphere and causing injury or property damage.[g] *30 April*—the UK government published its	*9–10 January*—the International Space Exploration Forum (ISEF) took place in Washington, D.C. (USA), wherein government representatives from 34 countries met to further advance the international dialogue on exploration and utilisation of space and to highlight the direct benefit of space exploration to humankind.[b] *7 February 2014*—the Canadian government released Canada's Space Policy Framework, which outlines broad national goals for its space programme.[d] *19–20 April*—The annual UNIDIR space security conference on the topic "*Implementation and compliance*" aimed to provide an overview of current initiatives, an update on the implementation and adherence to existing instruments.[f] *20 August*—Japan's government released its policy report "A Comprehensive Space Strategy to Implement Japan's National Strategy" as an update to Japan's five-year Basic Plan of

(continued)

Europe	Other countries
'Government Response to the UK Space Innovation and Growth Strategy 2014—2030', where it seeks to reduce the amount of insurance needed by satellite companies before government guarantees on third-party liability take effect from £80 million ($ million) to a ceiling of £60 million ($ million) by modifying the UK's Outer Space Act.[i] *30 April*—the UK government also published its 'U.K. National Space Security Policy' stressing the need for a terrestrial alternative to space-based navigation, positioning, and timing systems in case of potential loss of access to GNSS signals, either intentionally or due to space weather.[j] *5 December*—Council conclusions of the Competitiveness Council meeting on '*Underpinning the European space renaissance: orientations and future challenges*' invited both the Commission and ESA to review the European Space Policy and to identify to what extent it may require to be updated, with regard to TFEU and ESA Convention, to appropriately meet the long-term needs of European users.[k]	2012, and to function as a blueprint to realign Japan's space activity and space spending closer in line with security and economic priorities.[h]
General cooperation	
1 January—Lithuania became a full member of Eumetsat.[l] *22 January*—Iceland became a full member of Eumetsat.[m] *30 April*—Bulgaria became a full member of Eumetsat.[n] *1 October*—Germany became the 10th member of the EDA-led programme to pool and share purchases of commercial satellite bandwidth for military and government use in the EU.[o] *2 December*—ESA Ministerial Council meeting held. *17 December*—Germany's Johanne-Dietrich Woerner was elected as Director General of ESA, to take effect at the beginning of July 2015[q]	*10–21 February*—UN COPUOS Scientific and Technical Subcommittee meeting took place in Vienna, Austria *24 March-4 April*—UN COPUOS Legal Subcommittee meeting took place in Vienna, Austria *10 April*—Iran was said to have signed a protocol on cooperation with Russia's Roscosmos on wide cooperation in space exploration, ranging from training Iranian cosmonauts in Russia to potential technology transfers in developing remote sensing and telecommunication satellites.[p] *11–20 June*—UN COPUOS Plenary Meeting took place in Vienna, Austria *29 September–3 October*—International Astronautical Congress held in Toronto, Canada
By September 2014—the US government has signed nearly 50 data-sharing agreements with other governments, e.g. the UK, Japan, Australia, Italy, France, and the Republic of Korea, as well as private sector entities.[r] *30 October*—ESA and the US Strategic Command signed a space surveillance data-sharing agreement to exchange situational data as the space environment continues to become more congested.[s]	

(continued)

[a]De Selding, Peter B. "U.K. Blocks Bid To Create Common European Imagery Policy." 30 Jan. 2014. SpaceNews 24 June 2015. http://spacenews.com/39328uk-blocks-bid-to-create-common-european-imagery-policy/
[b]European Commission. International space exploration forum (Washington DC, 9–10 January 2014)—Information from the Commission. 13 Feb. 2014, Council of the European Union 6405/14 of 13 Feb. 2014. Brussels: European Union
[c]European Commission. Report from the Commission—Progress report on establishing appropriate relations between the European Union and the European Space Agency (ESA). European Commission COM(2014) 56 final of 6 February 2014. Brussels: European Union. http://register.consilium.europa.eu/doc/srv?l=EN&f=ST%205978%202014%20INIT
[d]"Canada's Space Policy Framework | Launching the Next Generation." 7 Feb. 2014. CSA 29 June 2015. http://www.asc-csa.gc.ca/pdf/eng/publications/space-policy/canadas-space-policy-framework.pdf
[e]*See further* "Code of Conduct for Outer Space Activities." European Union External Action 24 June 2015. http://eeas.europa.eu/non-proliferation-and-disarmament/outer-space-activities/index_en.htm
[f]"Space Security 2014: Implementation and Compliance." UNIDIR 8 Apr. 2015. http://www.unidir.org/programmes/emerging-security-threats/space-security-2014-implementation-and-compliance
[g]De Selding, Peter B. "French Debris-mitigation Law Could Pose Issue for Arianespace." 10 Apr. 2014. SpaceNews 25 June 2015. http://spacenews.com/40171french-debris-mitigation-law-could-pose-issue-for-arianespace/
[h]"Recommendation for a Comprehensive Space Strategy to Implement Japan's National Strategy." 26 Aug. 2014. Hiroshi-i.net 30 June 2015. http://hiroshi-i.net/10/wp-content/uploads/b785de34e6cd85a6423eb90542522d2d.pdf
[i]"Government Response to the UK Space Innovation and Growth Strategy 2014–2030." 30 Apr. 2014. UKSA 29 June 2015. https://www.gov.uk/government/uploads/system/uploads/attachment_data/file/307347/Government_response_-_space_growth_action_plan.pdf
[j]"National Space Security Policy." 30 Apr. 2014. UKSA 29 June 2015. https://www.gov.uk/government/uploads/system/uploads/attachment_data/file/307346/National_Space_Security_Policy.pdf
[k]Council of the European Union. Underpinning the European space renaissance: orientations and future challenges. Competitiveness Council meeting of 5 December 2014. Brussels: European Union. http://italia2014.eu/media/4061/council-conclusions-on-underpinning-the-european-space-renaissance-orientations-and-future-challenges.pdf
[l]See further "EUMETSAT is a user-governed operational organisation, formed in 1986, which serves the needs of its Member States." Eumetsat 24 June 2015. http://www.eumetsat.int/website/home/AboutUs/WhoWeAre/MemberStates/index.html
[m]See further "EUMETSAT is a user-governed operational organisation, formed in 1986, which serves the needs of its Member States." Eumetsat 24 June 2015. http://www.eumetsat.int/website/home/AboutUs/WhoWeAre/MemberStates/index.html
[n]See further "EUMETSAT is a user-governed operational organisation, formed in 1986, which serves the needs of its Member States." Eumetsat 24 June 2015. http://www.eumetsat.int/website/home/AboutUs/WhoWeAre/MemberStates/index.html
[o]"Germany joins the EU Satcom Market." 1 Oct. 2014. EDA 23 July 2015. https://www.eda.europa.eu/info-hub/news/article/2014/10/01/germany-joins-the-eu-satcom-market
[p]"Russia to train Iranian cosmonauts, build recon sats—report." 7 May 2014. RT.com 20 July 2015. http://www.rt.com/news/157496-russia-iran-space-satellite/
[q]De Selding, Peter B. "How ESA's Next Director-General Got the Job." 23 Dec. 2014. SpaceNews 22 June 2015. http://spacenews.com/dlrs-woerner-to-head-esa/
[r]Gruss, Mike. "UK Joins List of US Allies Agreeing To Strengthen Space Surveillance Sharing." 25 Sept. 2014. SpaceNews 25 June 2015. http://spacenews.com/41995uk-joins-list-of-us-allies-agreeing-to-strengthen-space-surveillance/

(continued)

[s]Gruss, Mike. "U.S. Strategic Command, ESA Sign Space Surveillance Data-sharing Agreement." 31 Oct. 2014. SpaceNews 23 July 2015. http://spacenews.com/42384us-strategic-command-esa-sign-space-surveillance-data-sharing-agreement/

12.5 Country Profiles

Austria

FFG

Population[2]	8.58 million
GDP[3]	329,295.6 million euros
Responsibility[4]	The Austrian Space Program is financed by the Federal Ministry for Transport, Innovation and Technology (BMVIT) and managed by the Aeronautics and Space Agency (ALR), which is integrated into the Austrian Research Promotion Agency (FFG), the central organisation for fostering science and technology activities in Austria, and serves as a docking station to the international aerospace world for Austrian business and science.
Activities (see footnote 3)	ALR of FFG implements national aerospace policy and represents Austria in numerous European and international aerospace institutions—from ESA and EUMETSAT through to EU committees and other international and national organisations responsible for aerospace policy. Austrian representation in the European Space Agency is of special strategic importance in this context. In addition to ESA programmes, the Austrian Space Applications Programme ASAP and the Space Programmes under the 7th EU Framework Programme and "Horizon 2020" are of special relevance for space research projects.
Budget (see footnote 3)	2014 ESA contribution of 50.2 million euros; and 2014 National expenditure of 15.8 million euros
Staff (see footnote 3)	ALR: 12
Direct Employment in Space Manufacturing Industry (see footnote 3)	355

[2] "Eurostat—Tables, Graphs and Maps Interface (TGM) table: Population." 31 July 2015 <http://epp.eurostat.ec.europa.eu/tgm/refreshTableAction.do?tab=table&plugin=1&pcode=tps00001&language=en>.

[3] "Eurostat—Tables, Graphs and Maps Interface (TGM) table: GDP." 31 July 2015 <http://epp.eurostat.ec.europa.eu/tgm/refreshTableAction.do?tab=table&plugin=1&pcode=tec00001&language=en>.

[4] Information provided by the agency.

Belgium

Population (see footnote 1)	11.25 million
GDP (see footnote 2)	402,072.0 million euros
Responsibility (see footnote 3)	Two-level competence system: – Federal: space in the frame of international agreements; – Regions and communities: basic and applied research. The Belgian Federal Science Policy Office (BELSPO) with its Department of Space Research and Applications manages the scientific aspects of the space policy. The Belgian High Representation for Space Policy (BHRS), acting under the auspices of the federal minister in charge of Science Policy, defends Belgian interests in the international institutions dealing with space (ESA and EU levels). Other federal ministries involved with space are the Ministries of Defence and Foreign Affairs. On the regional level, the Ministries for Economic Development, Transport and Environment have stakes in space applications and industrial development.
Activities (see footnote 3)	Strategy and operational goals: – to create and/or strengthen expertise and industrial capabilities in high-growth areas of the space sector; – to increase the return on investment of the public money invested in space at the ESA level, in the frame of the EU as well as in bilateral cooperation projects; – to support industry to be able to tackle new markets, positive impact on economy; – to give to the public authority the space means necessary to define and to implement its policies; – to represent Belgium's interests in the definition and implementation of the European space policy. Cooperation with ESA, EUMETSAT, EC and other aerospace related agencies. Within ESA, Belgium traditionally has major interests in Launchers, Telecommunications, Integrated Applications and Technology Support programmes, Earth Observation. Bilateral programmes: – France-Belgium SPOT 1 TO 5 launched VEGETATION = instrument on SPOT 4 and 5 CTIV: Processing Centre for the Végétation images in VITO (Mol) Plate-form Protéus, myriades... Pléiades MUSIS – Argentina-Belgium Co-operation framework signed in 2000. SAOCOM: Earth Observation satellite. National programme – Research in Earth observation – Programme STEREO (4 million euros/year) Basic research Products and services R&D and public users R&D and industry

(continued)

	– Programme ORFEO (1.2 million euros) – Support for the VEGETATION data Programme VEGETATION (1 million euros) Processing centre (CTIV) (2.5 million euros/year) – EO-DESK Access to images Support for the users Education Space infrastructures in Belgium: – ESA Space station Redu Telecom, Techno, Security – CTIV (Mol) Processing of Vegetation data and EO data – Test facilities in CSL (Liège) – Radiation test facility in UCL (Louvain LN) – ESA BIC in Redu and in Mol
Budget (see footnote 3)	National Contribution to ESA Budget (2014): 188,644,759.43 euros Total. – Mandatory Activities and Associated with General Budget : 23,922,806.43 euros – Optional Programmes: 164,721,953.00 euros
Staff (see footnote 3)	No Belgian technical space agency
Direct Employment in Space Manufacturing Industry (see footnote 3)	2000

Czech Republic

Population (see footnote 1)	10.53 million
GDP (see footnote 2)	154,738.7 million euros
Responsibility[5]	The Ministry of Transport of the Czech Republic is the coordinator of all space activities in the Czech Republic. For the coordination purposes it established the Coordination Council for Space Activities under its leadership, which includes all relevant ministries and institutions. The Ministry of Transport is also directly responsible for the major space activities in the Czech Republic, including the membership in ESA, EU space policy or international cooperation in space.
Activities (see footnote 4)	The Czech Republic is subscribed in ESA Optional Programmes such as: MTG, MetOp-SG, EOEP, GSC, ARTES, EGEP, GSTP, MREP, ELIPS, SSA, FLPP and PRODEX.
Budget[6]	2014 ESA contribution to ESA 13.9 million euros
Staff (see footnote 4)	14
Direct Employment in Space Manufacturing Industry	n.s.

[5] http://www.czechspace.cz/en, 30 July 2015.

[6] European Space Directory 2014. 29th Edition. Paris: EDS Partners, 2014.

Denmark

DTU Space
Institut for Rumforskning og -teknologi

Population (see footnote 1)	5.65 million
GDP (see footnote 2)	257,443.8 million euros
Responsibility[7]	NSI (National Space Institute) is the only research organization in Denmark with the basic funding and mandate to carry out space research. Although space research is also conducted at other universities and organizations, these activities primarily rely on funding from research agencies and other external funding sources. Through national and international cooperation, NSI leverages its research capabilities to create new knowledge and technology of value for Danish research infrastructure, industry, and society at large. NSI's scientific competences further give it the ability to provide qualified advice to national and international agencies and organizations, and also forms the basis for educating researchers and engineers in both basic and specialized disciplines. The common denominator in NSI's research activities is the application of satellites for studying Earth and space physics and the development of systems, methods and instruments which support these activities. The department is characterized by a close-knit interplay between engineering disciplines and natural science. The scientific challenges require development of new technologies, and the technological research opens up new scientific opportunities. The combination of the technical and scientific competencies enables the NSI to deliver unique end-to-end solutions within selected areas, i.e. to both develop the necessary instrumentation and process the collected data to scientific results.
Activities (see footnote 6)	Within the ESA programmes, the Danish areas of strength encompass Earth observation and climate monitoring, launchers, the international space station, telecommunications (ARTES) and technology development (GSTP). In addition, Denmark has decided to contribute to the development of new meteorological satellites in cooperation with the European meteorological organisation EUMETSAT and in the ESA's new programme, Space Situational Awareness that ensures research and warnings of soar flares' effect on the security of supply infrastructure.
Budget (see footnote 5)	2014 ESA contribution of 23.4 million euros; and 2014 National expenditure of 5.0 million euros
Staff[8]	DTU: 147
Direct Employment in Space Manufacturing Industry[9]	288

[7] http://www.space.dtu.dk/english/About_NSI, 30 July 2015.

[8] http://www.space.dtu.dk/english/About_NSI/Organisation/All, 30 July 2015.

[9] ASD-Eurospace. "Facts and Figures – The European Space Industry in 2014" 19th Edition. Paris: ASD Group, June 2015.

Estonia

EAS Enterprise Estonia

Population (2015) (see footnote 1)	1.31 million
GDP (2014) (see footnote 2)	19,525.3 million euros
Responsibility (see footnote 3)	Space activities in Estonia are coordinated by the Space Affairs Council (SAC), established in 2010 by Ministry of Economic Affairs and Communications. The main task of SAC is initiation and governance space related programs and activities on national and international level and coordination of the utilization of resources. SAC is supported by secretariat, containing representatives of Enterprise Estonia and Estonian Research Council. Since 2006 Estonian public business development and support agency Enterprise Estonia (www.eas.ee) is assigned as implementing body (ESA delegation) for the ESA-Estonian agreements. Enterprise Estonia also carries out the tasks of Estonian Space Office, being intermediary between Estonian companies/institutions and international space community. Main tasks concluded mapping and auditing space related competences and capabilities, organizing trainings for entrepreneurs and public sector officials, taking companies to expositions and study trips, encouraging the technology transfer etc. Enterprise Estonia is a member of EURISY since 2008 and member of IAF since 2009. Space science activities in Estonia are carried out by Tartu Observatory (www.to.ee), having long tradition in astronomy, cosmology and Earth observation. Nowadays Tartu Observatory serves as Estonian space technology development and testing center.
Activities (see footnote 3)	Estonian space focus of space activities is on earth observation, navigation, technology, especially related to ICT and science. Estonian space activities are mainly related to Horizon 2020 projects and cooperation with ESA and PECS program, where Estonia has 27 projects. In 2015 Estonia will be 21st ESA member state. Estonia is actively preparing for Galileo and Copernicus program implementation in public and private sector.
Budget (see footnote 3)	Estonian contribution to ESA as ECS is ca 1.2 million euros per year, public funding of space science research activities is ca 1 million euros per year.
Staff (see footnote 3)	Enterprise Estonia—280, space related 10 Tartu Observatory—80
Direct Employment in Space Manufacturing Industry (see footnote 3)	About 100 personnel working in different research groups belonging to university departments and space technology and space applications related companies.

Finland

Population (see footnote 1)	5.47 million
GDP (see footnote 2)	205,178.0 million euros
Responsibility (see footnote 3)	Finland's public sector space activities are funded by the Ministry of Employment and Economy (ESA membership), Tekes (ESA and national programmes), Academy of Finland (EISCAT, ESO, NOT) under the Ministry of Education and Culture, and Finnish Meteorological Institute (EUMETSAT) under the Ministry of Transport and Communications. National coordination is provided by the Finnish Space Committee under Ministry of Economy.
Activities (see footnote 3)	In addition to cooperation with ESA and EC (e.g. Space Programme under the Horizon 2020 EU Framework Programme), Tekes funds space technologies and applications nationally.
Budget (see footnote 3)	2014 ESA contribution of 17 million euros; and 2014 National expenditure of 2 million euros
Staff (see footnote 3)	6 at Tekes
Direct Employment in Space Manufacturing Industry (see footnote 8)	144

France

Population (see footnote 1)	66.35 million (2014—INSEE)
GDP (see footnote 2)	2,132,449.0 million euros
Responsibility (see footnote 3)	Space activities in France are under parent Ministries of Higher Education and Research and of Defence. Among different actors, France relies on its national space agency CNES (Centre National d'Etudes Spatiales) for space policy proposal and application.
Activities (see footnote 3)	France inscribes its space activities in the frame of European space activities and intent to play there a leading role. Activities are conducted in strong interface with EC, ESA, Eumetsat and others Members States or at international level. National orientations include 5 acting domains, namely Ariane (autonomous European access to space), Observation, Sciences, Telecommunications and Defence.
Budget (see footnote 3)	Greater than 2000 million euros (Ministry of Higher Education and Research). Including ESA contribution, Eumetsat contribution, National Programme, Future Investment Programme (PIA) and Defence budget.
Staff (see footnote 3)	CNES # 2400 FTE
Direct Employment in Space Manufacturing Industry (see footnote 3)	# 16000 FTE i.e. 40 % of European Manufacturing Industry

Germany

Population (see footnote 1)	81.17 million
GDP (see footnote 2)	2,903,790.0 million euros
Responsibility[10]	Acting on behalf of the Federal Government, the Space Administration designs and implements Germany's Space Program, which integrates all German space activities on the national and European plane. These activities include Germany's national Space Program, DLR's 'Space' research and development program and Germany's contributions to the European Space Administration (ESA) as well as the European Organization for the Exploitation of Meteorological Satellites (EUMETSAT). In addition, the Administration designs and supervises space and security research projects under the sixth and seventh EU research framework program. The Space Administration's principal client is the Federal Ministry for Economics and Technology. However, the Space Administration also works for other ministries, mainly in application-related fields such as Earth observation, navigation and satellite communication. The Federal Government's Space Program provides both business and science with a reliable political framework for independent planning and action, thus ensuring that public funds are used efficiently. The national Program is implemented by the Space Administration. Its major function is to promote and support Germany's strategic goals in the European programs of ESA and the EU by purposefully equipping Germany's industry and scientific institutions for competition within the EU as well as for their tasks within the ESA framework.
Activities (see footnote 9)	Germany has a national civil programme, which includes bi- or trilateral co-operations, a participation in the ESA programmes, and a defence programme (e.g. SARLupe, MUSIS). The major areas of interest for Germany's domestic programmes are Earth observation, navigation, telecommunications, space transportation systems, space infrastructures, microgravity research, space science, space technologies and space exploration.
Budget (see footnote 5)	2014 ESA contribution of 765.70 million euros; and 2014 National expenditure of 450 million euros
Staff (see footnote 9)	DLR: approximately 8000 employees
Direct Employment in Space Manufacturing Industry (see footnote 8)	7269

[10] http://www.dlr.de/rd/en/desktopdefault.aspx/tabid-2099/3053_read-4706/, 29 July 2015.

Greece

Population (see footnote 1)	10.81 million
GDP (see footnote 2)	179,080.6 (p) million euros
Responsibility[11]	The General Secretariat for Research and Technology (GSRT), under the responsibility of the Ministry of Education and Religious Affairs, is responsible for Greek space activities.
Activities (see footnote 5)	Space R&D is a priority area of the Greek Space Strategy, mostly pursued through ESA's technological activities' EOEP, GSTP, ARTES-1, ARTES-3-4 and ARTES-5, ELIPS, and GNSS Evolution Programme. Greece and Cyprus have created the HellasSat consortium to launch the Hellasat telecommunication satellites. Due to the economic crisis Hellasat was sold to Arabsat in 2013. Greece is a member of Eumetsat and Eutelsat.
Budget (see footnote 5)	2014 ESA contribution of 14.5 million euros.
Staff (see footnote 10)	GSRT: 34
Direct Employment in Space Manufacturing Industry	n.s.

Hungary

Population (see footnote 1)	9.84 million
GDP (see footnote 1)	103,216.5 million euros
Responsibility[12]	The Hungarian Space Office (HSO) manages, coordinates and represents the Hungarian space activities. In 2010 the Hungarian Space Office was administratively integrated into the body of the Ministry of National Development, keeping its name, image and independence in external relations. The Minister's advisory body in technical matters is the Scientific Council on Space Research (SCSR). The SCSR provides the scientific background of all Hungarian space activities.
Activities (see footnote 5)	Participation in many different fields of space activity such as microgravity, Earth observation, life and material sciences and GSTP programmes of ESA.
Budget	n.s.
Staff (see footnote 11)	250 scientists and engineers
Direct Employment in Space Manufacturing Industry	n.s.

[11] http://www.gsrt.gr/central.aspx?sId=119I428I1089I323I488743, 29 July 2015.

[12] http://www.hso.hu/page.php?page=215, 29 July 2015.

Ireland

Population (see footnote 1)	4.62 million (2011)
GDP (see footnote 2)	185,411.7 million euros
Responsibility (see footnote 3)	Enterprise Ireland (EI) is the Irish Government's business development agency. EI manages industrial and technology development programmes relating to space activity on behalf of the Department of Jobs, Enterprise & Innovation (DJEI). EI promotes ESA and EU space programmes in Ireland, primarily to the industrial sector, as well as to the academic community and end-user organisations.
Activities (see footnote 3)	ESA programmes (launchers, telecoms, sat nav, Earth observation, science); Horizon 2020 Space; EUMETSAT
Budget (see footnote 3)	ESA: 17.2 million euros (2014) EUMETSAT: 2.9 million euros (2013)
Staff (see footnote 3)	Delegation to ESA: 5
Direct Employment in Space Manufacturing Industry (see footnote 3)	506 (2015)

Italy

Population (see footnote 1)	60.79 million
GDP (see footnote 2)	1,616,253.6 million euros
Responsibility (see footnote 3)	The Italian Space Agency, Agenzia Spaziale Italiana (ASI), defines, coordinates and manages national space programs and the Italian participation to European and international space projects, under the supervision of the Ministry of Education, University and Research and in coordination with the Ministry of Foreign Affairs and International Cooperation. ASI's main headquarter is located in Rome and the agency has also three additional centers in Matera (Space Geodesy Centre); in Malindi, Kenya (Luigi Broglio Space Centre); and Rome (ASI Science Data Centre).
Activities (see footnote 3)	Italy is the third country in terms of contribution to the European Space Agency's (ESA) programs. Italy participates in the European Union (EU) programs as well such as Galileo and GMES/Copernicus. ASI has a long tradition in space scientific missions at ESA level and in cooperation with NASA. Indeed, ASI has significantly contributed to space exploration embarking many scientific payloads aboard NASA and ESA satellites for discovering the secrets of Mars, Jupiter, Saturn and Venus and for the study of cosmology. Other scientific payloads have been carried out for the astrophysical study of high energy and black matter such as AMS onboard the ISS. Italy is the first European country

(continued)

	in terms of investment in the ISS through the participation in ESA and through the MoU with NASA. ASI is deeply engaged in international bilateral cooperation with many partners in the five continents. At multilateral governmental level Italy is member of the UNCOPUOS and GEO. ASI is also member of CEOS, IADC, ICG, ESPI, EURISY and IAF.
Budget (see footnote 3)	2014 ESA contribution of 350 million euro out of a total of 584.48 million national expenditure 2014
Staff (see footnote 3)	ASI: app. 250
Direct Employment in Space Manufacturing Industry (see footnote 3)	6125

Luxembourg

Luxembourg Space Cluster

Population (see footnote 1)	562,958
GDP (see footnote 2)	49,428.3 million euros
Responsibility (see footnote 3)	The Ministry of the Economy, Directorate of ICT & Space Affairs, is in charge of space affairs in Luxembourg. Luxinnovation, the National Agency for Innovation and Research, also acts on behalf of the Ministry, representing Luxembourg interests in the ESA and EU committees.
Activities (see footnote 3)	The Ministry of the Economy defines the National Space Policy, manages relationship with ESA, the EU and space agencies worldwide. The Ministry is responsible for financial, political and programmatic aspects related to ESA activities. Luxinnovation supports the Ministry for the implementation of the National Action Plan. Moreover, Luxinnovation is the national point of contact for space matters. Luxinnovation provides companies and research organizations with information and advices on ESA and EC programmes. Luxembourg is active in telecommunications, Earth observation, navigation, security and space technologies.
Budget (see footnote 3)	Space budget: Contribution to ESA Budget (2014): 18,300,000 euros National Civil Space Budget (2014): ~1,600,000 euros EUMETSAT Budget (2013): ~600,000 euros
Staff (see footnote 3)	5 (3 Ministry of the Economy + 2 Luxinnovation)
Direct Employment in Space (see footnote 3)	650

Netherlands

SRON Netherlands Institute for Space Research — Netherlands Space Office

Population (see footnote 1)	16.90 million
GDP (see footnote 2)	662,770.0 (p) million euros
Responsibility[13]	The Netherlands Space Office acts as the Dutch agency for space affairs. The Netherlands Space Office (NSO) was established by the Dutch government in order to develop the Netherlands' space programme and to bring that programme to action. The NSO is the face of the Dutch space community for international space organisations like ESA, NASA and JAXA as well as the central point of contact for the space community within the Netherlands. The NSO also works to innovatively and openly bring the story of spaceflight science, usage and exploration to teachers, students and the general public. The Ministry of Economic Affairs, Ministry of Education, Culture and Science, Ministry of Transport, Public Works and Water Management and the Netherlands Organization for Scientific Research (NWO) signed an agreement in October 2008 for the establishment of the Netherlands Space Office.
Activities (see footnote 12)[14]	In the ESA context, the Netherlands participate in all major programmes with a special interest in earth observation and environmental measurements and in robotics. The Netherlands Institute for Space Research (SRON) develops and uses innovative technology for ground-breaking research in space, focusing on astrophysical research, Earth science and planetary research. In addition to this, SRON has a line of research into new and more sensitive sensors for X-rays and infrared radiation. Space policy in the Netherlands is primarily focused on international cooperation in European contexts within the European Space Agency, the European Organisation for the Exploitation of Meteorological Satellites (EUMETSAT), and the European Union. The Netherlands also has a national programme guided by the ministries of Economic Affairs, of Education, Culture and Science, and of Traffic, Public Works and Water Management.
Budget (see footnote 5)	2014 ESA contribution of 125.1 million euros; and 2014 National expenditure of 25.0 million euros
Staff (see footnotes 12 and 13)	NSO: 26, SRON: app. 200
Direct Employment in Space Manufacturing Industry (see footnote 8)	983

[13] http://www.spaceoffice.nl/en/The-NSO/, 29 July 2015.

[14] https://www.sron.nl/facts-aamp-figures-about-sron-1727, 29 July 2015.

Norway

Population (see footnote 1)	5.16 million
GDP (see footnote 2)	377,008.6 million euros
Responsibility (see footnote 3)	The Norwegian Space Centre (NSC), under the Ministry of Trade, Industry, and Fisheries, is responsible for organizing Norwegian space activities, particularly with respect to ESA and the EU, and for coordinating national space activities.
Activities (see footnote 3)	In addition to ESA programmes (in particular Earth observation, telecommunications and launchers), Norway has national support programmes and commercial activities (Telenor). Moreover, Norway operates the Andøya rocket range and the Svalbard and Antarctica ground stations. Norway has also a bilateral agreement with Canada on the use of Radarsat 2 data.
Budget (see footnote 3)	2014 ESA contribution of 56.4 million euros; and 2014 National expenditure of 159.4 million euros
Staff (see footnote 3)	NSC: 40
Direct Employment in Space Manufacturing Industry (see footnote 3)	Approximately 1000

Poland

Population (see footnote 1)	38.00 million
GDP (see footnote 2)	413,133.9 million euros
Responsibility[15]	The Space Research Centre is an interdisciplinary research institute of the Polish Academy of Sciences, established to conduct scientific research and activities in order to develop the space industry in Poland. Basing on satellite techniques PAS SRC carries out basic and applied research into phenomena in interplanetary and circumterrestrial space, on Earth, and into solar physics and solar system bodies. PAS SRC develops solutions and promotes the use of satellite systems in the national economy, in areas such as navigation, telecommunications and earth observation. In the field of research and development, the PAS SRC collaborates with leading research centers and private and public companies in the aviation and defense industry. An important task is to support the Government in the organization of space activities in Poland and eventually in the establishment of the Polish Space Agency.

(continued)

[15] http://www.cbk.waw.pl/en/, 30 July 2015.

Activities (see footnote 14)	The five main SCR's research areas are: physics of the Sun, study of planets and small solar system bodies, interplanetary space physics and astrophysics, plasma physics, and planetary geodesy and geodynamics. The SRC is the body cooperating with ESA. Poland has participated or is participating in several scientific missions of ESA, such as: Ulysses, ISO, Soho, XMM, Cluster, DoubleStar, Huygens, Mars Express, Herschel, Planck, XEUS, Integral, Rosetta, ExoMars and BepiColombo.
Budget (see footnote 5)	2014 space expenditure of around million euros; 2014 ESA contribution of 28.7 million euros
Staff (see footnote 14)	Space Research Centre: 198
Direct Employment in Space Manufacturing Industry	n.s.

Portugal

Population (see footnote 1)	10.37 million (2014)
GDP (see footnote 2)	173,044.3 million euros
Responsibility (see footnote 3)	Portuguese Space institutional activities are under responsibility of the Ministry of Education and Science. The main tool to develop national space capabilities is ESA. The National Innovation Agency (ANI) coordinates the national participation in ESA, with the executive support of the Portuguese Space Office, and in close collaboration with the Foundation for Science and Technology (FCT). Most of Portuguese ESA budget is funded by FCT, and in part by the telecom regulator (Anacom), mainly for the ARTES programmes. Other Ministries and entities are also involved with ESA funding and activities according to their specific needs and priorities. ANI is also responsible for promoting the Portuguese participation in H2020 Space programme, through the Office for the promotion of EU H2020 research programme (GPPQ).
Activities (see footnote 3)	Portuguese space activities are implemented mainly through ESA. The EU space programmes, Galileo and Copernicus, alongside H2020 Space, are also of relevance for the space community. Additionally Portugal is a member state of ESO and Eumetsat.
Budget (see footnote 3)	2015 ESA contribution of 16.7 million euros
Staff (see footnote 3)	4 (FTE in Space Office, including external support)
Direct Employment in Space Manufacturing Industry (see footnote 3)	250 (estimation of jobs in the space industry)

Romania

Population (see footnote 1)	19.86 million
GDP (see footnote 2)	150,018.5 (p) million euros
Responsibility[16]	The Romanian Space Agency (ROSA) is the coordinator of Romania's national and international space activities. The Romanian Space Agency (ROSA) is a public institution entirely self-funded, operating under the decisions of the Ministry of Education and Research—National Authority for Scientific Research and Innovation (A.N.C.S.I.). As a coordinator of national space research and applications programme, ROSA designs and coordinates the implementation of the National Space Programme. Following its objectives, the Agency is authorised to establish research and development centers.
Activities (see footnote 15)	Following the accession to ESA PECS, Romania has conducted the following projects: ROKEO—Romanian Centered Knowledge Earth Observation LEOWorks—Image processing/GIS software for educational purposes ECSTRA—Energy Conversion and Transfer in the Solar Wind–Magnetosphere–Ionosphere System (within ESA's CLUSTER mission) Planck-LFI_SED—Scientific exploitation of the Planck-LFI data (within the PLANCK mission) CFS—Growth and survival of coloured fungi in space (within ISS SURE) KEEV—Kinetic and Experimental investigation of the Earth's and Venus's plasma layers (within the VENUS EXPRESS mission) RoSpaceGRID—Romanian GRID middleware repository for Space Science Applications (within ESA-Grid) CoRoT—Romanian participation to the CoRoT mission ICRF—Improving relative positions of reference stars around ICRF radio-sources (within the GAIA mission) SOHO—Romanian contributions to the Sun-Heliosphere Studies (within the SOHO mission)
Budget (see footnote 5)	2014 ESA contribution of 22.5 million euros
Staff (see footnote 15)	64
Direct Employment in Space Manufacturing Industry	n.s.

[16] http://www.rosa.ro/index.php/en/, 30 July 2015.

Spain

Population (see footnote 1)	46.43 million
GDP (see footnote 2)	1,058,469.0 (p) million euros
Responsibility[17]	The Centre for the Development of Industrial Technology (CDTI), under the Ministry of Economy and Competitiveness, channels the funding and support applications for national and international R&D&i projects of Spanish companies, including the Spanish space activities in coordination with the Ministry of Industry, Energy and Tourism. The National Institute of Aerospace Technology (INTA) is the other important Spanish actor in the space field. INTA is an independent body related to the Ministry of Defence. INTA is responsible for organising the Spanish participation in HELIOS and is responsible for approval and certification, research, technical assistance and services in several areas related to space.
Activities (see footnote 16)	In addition to ESA and EUMETSAT programmes, Spain has several national space programmes in the field of Earth observation (SEOSAT/INGENIO), communication satellites (Hispasat, Amazonas), defence space systems (SPAINSAT, XTAR-EUR, HELIOS, SECOMSAT, Pleiades and SEOSAR/PAZ), small satellites, space exploration (MSL-REMS) and ground control stations.
Budget (see footnote 5)	2014 ESA contribution of 139.2 million euros; and 2014 National expenditure (including ESA) of approx. 200.0 million euros
Staff[18]	CDTI: app. 300
Direct Employment in Space Manufacturing Industry (see footnote 8)	3044

Sweden

Population (see footnote 1)	9.74 million
GDP (see footnote 2)	430,258.2 million euros
Responsibility[19]	The Swedish National Space Board, SNSB, is a central governmental agency under the Ministry of Education and Research. SNSB is responsible for national and international activities relating to space and remote sensing, primarily research and development.

(continued)

[17] http://www.cdti.es/index.asp?MP=7&MS=191&MN=3, 30 July 2015.

[18] http://www.esa.int/Our_Activities/Space_Engineering_Technology/Centro_para_el_Desarrollo_Tecnologico_Industrial_CDTI, 30 July 2015.

[19] http://www.snsb.se/en/Home/Home/, 30 July 2015.

	The Swedish space programme is carried out by means of extensive international cooperation, in particular through Sweden's membership of the European Space Agency, ESA. SNSB's responsibility for international activities includes the Swedish involvement in ESA as well as bilateral cooperation within space. Most of the activities funded by the Swedish National Space Board are carried out in cooperation with other countries.
Activities (see footnote 18)	The Swedish space programme is mostly carried out through international cooperation. Sweden participates in almost all the optional ESA programmes, in addition to the mandatory basic and scientific programmes. Optional programmes of special interest are the remote sensing and launcher technology programmes. Bilateral co-operation particularly concerns the co-operation with France on SPOT, Vulcain MK2 and Proteus. Sweden also has a participation in the French Pleiades programme. Other Memoranda of Understanding have been concluded as basis for co-operation with SNSB's counterparts in Austria, Brazil, China, India, Italy, Japan, Kazakhstan, Norway and Thailand. On a governmental level Sweden has framework agreement with USA and Russia for cooperative activities in the exploration and use of outer space. Sweden is also engaged in other bilateral co-operative projects carried out on ad hoc basis. Sweden is a member of the UN Committee for Peaceful Use of Outer Space, COSPAR (Committee on Space Research), Intelsat, Eutelsat, Inmarsat and Eumetsat.
Budget (see footnote 5)	2014 ESA contribution of 94.6 million euros; and 2014 National expenditure of 16.0 million euros
Staff (see footnote 18)	17
Direct Employment in Space Manufacturing Industry (see footnote 8)	760

Switzerland

Schweizerische Eidgenossenschaft
Confédération suisse
Confederazione Svizzera
Confederaziun svizra

Population (see footnote 1)	8.23 million
GDP (see footnote 2)	516,068 million euros (2013)
Responsibility[20]	The Swiss Space Office, under the authority of the State Secretariat for Education, Research and Innovation (SERI) is the administrative body in charge of planning and implementing the Swiss space policy. The Federal Commission for Space Affairs (CFAS)

(continued)

[20] http://www.sbfi.admin.ch/themen/01371/index.html?lang=en, 31 July 2015.

	• advises the Federal Council on matters relating to a coherent and forward-looking space policy by taking into consideration the European and global development and the national interests (amongst others of political, application-oriented, scientific, technological and industrial nature); • issues recommendations to the Federal Council regarding the implementation of this policy; • advises the Federal Council in terms of the evaluation of proposals for projects with important scientific or technological relevance. The SERI's Swiss Space Office serves as the secretariat for the CFAS. Interdepartmental Coordination Committee for Space Affairs (IKAR) The Federal Council has given the Interdepartmental Coordination Committee for Space Affairs (IKAR) a mandate to prepare official Swiss position papers on space on the basis of CFAS recommendations. IKAR also coordinates the activities of various federal agencies involved in space affairs. IKAR is chaired by the SERI's Swiss Space Office, which also serves as the secretariat for IKAR. Space Office, which also serves as the secretariat for IKAR.
Activities (see footnote 19)	Most of the Swiss activities are undertaken within ESA programmes (space science, Earth observation, microgravity, human spaceflight, small satellites, telecommunications, navigation, and technology).
Budget (see footnote 5)	2013 ESA contribution of 108.3 million euros; and 2013 National expenditure of 2.0 million euros.
Staff (see footnote 19)	SERI: 280
Direct Employment in Space Manufacturing Industry (see footnote 8)	874

United Kingdom

UK SPACE AGENCY

Population (see footnote 1)	64.76 million
GDP (see footnote 2)	2,222,912.2 million euros
Responsibility[21]	The UK Space Agency, an executive agency of the Department for Business, Innovation and Skills (BIS), and reporting to the Minister of State for Universities and Science, is responsible for the strategic decisions on all UK space activities. The UK Space Agency also interfaces with other departments on security and military programmes. Although

(continued)

[21] https://www.gov.uk/government/organisations/uk-space-agency, 31 July 2015.

	it does not manage these programmes directly, the UK Space Agency is kept informed and involved in decisions relating to the programmes as necessary.
Activities (see footnote 20)	UK space activities include bilateral cooperation with JAXA, India and the U.S. and within ESA in Earth observation and space exploration (Cassini-Huygens, James Webb Space Telescope, Herschel, and Planck missions).
Budget (see footnote 5)	2014 ESA contribution of 270 million euros; and 2014 National expenditure of 60.0 million euros.
Staff (see footnote 20)	UK Space Agency: 50
Direct Employment in Space Manufacturing Industry (see footnote 8)	3613

Chapter 13
Bibliography of Space Policy Publications 2014

Blandina Baranes

13.1 Monographs

Alessi EM et al (2014) Effectiveness of GNSS disposal strategies. Acta Astronaut 99:292–302

Alibay F et al (2014) Fractionated robotic architectures for planetary surface mobility systems. Acta Astronaut 95:15–29

Alvarado SC, Langston SM (2014) Establishing a new international space cooperation mechanism for protecting world heritage sites. Ann Air Space Law XXXIX

Anderson D et al (2014a) Products from NASA's in-space propulsion technology program applicable to low-cost planetary missions. Acta Astronaut 93:516–523

Anderson P, Macdonald M, Yen C-W (2014b) Novel orbits of Mercury, Venus and Mars enabled using low-thrust propulsion. Acta Astronaut 94:634–645

Armstrong D (2014) American national security and the death of space sanctuary. Astropolitics 12:69–81

Arney DC et al (2014) A space exploration strategy that promotes international and commercial participation. Acta Astronaut 94:104–115

Arnon S, Gill E (2014) The optical communication link outage probability in satellite formation flying. Acta Astronaut 95:133–140

Arnould J (2014) The explorer's complex. Space Policy 30:185–189

Asgarimehr M, Mashhadi Hossainali M (2014) Optimization of geosynchronous satellite constellation for independent regional navigation and positioning in Middle East region. Acta Astronaut 104:147–158

Bamford RA et al (2014) An exploration of the effectiveness of artificial mini-magnetospheres as a potential solar storm shelter for long term human space missions. Acta Astronaut 105:385–394

B. Baranes (✉)
ESPI, Schwarzenbergplatz 6, 1030 Vienna, Austria
e-mail: blandina.baranes@espi.or.at

C. Al-Ekabi et al. (eds.), *Yearbook on Space Policy 2014*, Yearbook on Space Policy, DOI 10.1007/978-3-7091-1899-3_13

Barger LK et al (2014) Sleep and cognitive function of crewmembers and mission controllers working 24-h shifts during a simulated 105-day spaceflight mission. Acta Astronaut 93:230–242

Becerra J (2014) Colombia's space policy: an analysis of six years of progress and challenges. Acta Astronaut 100:94–100

Bergsrud C, Straub J (2014) A space-to-space microwave wireless power transmission experiential mission using small satellites. Acta Astronaut 103:193–203

Bernasconi MC (2014) An ethical duty: let astronautical development unfold—to make the people more secure. Acta Astronaut 104:491–501

Bhat SB (2014) Application of environmental law principles for the protection of the outer space environment: a feasibility study. Ann Air Space Law XXXIX

Bonneville R (2014) What role for Europe within a future global space exploration approach? Acta Astronaut 104:487–490

Bouchey M, Delborne J (2014) Redefining safety in commercial space: understanding debates over the safety of private human spaceflight initiatives in the United States. Space Policy 30:53–61

Bowen BE (2014) Cascading crises: orbital debris and the widening of space security. Astropolitics 12:46–68

Brumbaugh Gamble K, Lightsey EG (2014) CubeSat mission design software tool for risk estimating relationships. Acta Astronaut 102:226–240

Burattini C et al (2014) A new conceptual design approach for habitative space modules. Acta Astronaut 97:1–8

Burchell MJ (2014) Human spaceflight and an asteroid redirect mission. Why? Space Policy 30:163–169

Carrelli D et al (2014) Autonomous aerobraking for low-cost interplanetary missions. Acta Astronaut 93:467–474

Chang Y-W, Chern J-S, Marec J-P (2014) Entering the 60th year of Acta Astronaut. Acta Astronaut 97:172–183

Chatterjee P (2014) Legality of anti-satellites under the space law regime. Astropolitics 12:27–45

Clark J et al (2014) Assessing the full effects of public investment in space. Space Policy 30:121–134

Clément G, Bukley AP (2014) Human space exploration—from surviving to performing. Acta Astronaut 100:101–106

De la Torre GG (2014) Toward a new cosmic consciousness: psychoeducational aspects of contact with extraterrestrial civilizations. Acta Astronaut 94:577–583

Dekens E, Engelen S, Noomen R (2014) A satellite swarm for radio astronomy. Acta Astronaut 102:321–331

Denis G, Klein S, Gueguen B (2014) Use of social networks for outreach, education and training on space applications: know-how and experience of Planete Sciences Midi-Pyrenees and CNES. Acta Astronaut 94:765–775

D'Errico M, Renga A, Graziano MD (2014) Earth observation with MEO transmitters and UAS receivers: a potential utilization of Galileo constellation. Acta Astronaut 94:93–103

Dick SJ (2014) Analogy and the societal implications of Astrobiology. Astropolitics 12:210–230

DiPaolo AJ (2014) The definition and delimitation of outer space: The present need to determine where "space activities" begins. Ann Air Space Law XXXIX

Dodge MS (2014) Earth Observation and the needs of the many: the future structure of international disaster relief law and disaster management. Ann Air Space Law XXXIX

Dougherty K, Oliver C, Fergusson J (2014) Pathways to space: a mission to foster the next generation of scientists and engineers. Acta Astronaut 99:184–192

Ehrenfreund P et al (2014) The O/OREOS mission—Astrobiology in low Earth orbit. Acta Astronaut 93:501–508

Elliott JR (2014) Search for life signatures. Acta Astronaut 105:511

Emanuelli M et al (2014) Conceptualizing an economically, legally, and politically viable active debris removal option. Acta Astronaut 104:197–205

Englert CR et al (2014) Optical orbital debris spotter. Acta Astronaut 104:99–105

Erickson AS (2014) China's space development history: a comparison of the rocket and satellite sectors. Acta Astronaut 103:142–167

Evetts SN (2014) Space life and biomedical sciences in support of the global exploration roadmap and societal development. Space Policy 30:143–145

Filatyev AS et al (2014) Advanced aviation technology for reusable launch vehicle improvement. Acta Astronaut 100:11–21

Flores Martinez CL (2014) SETI in the light of cosmic convergent evolution. Acta Astronaut 104:341–349

Fujita K et al (2014) Conceptual study and key technology development for Mars Aeroflyby sample collection. Acta Astronaut 93:84–93

Fusilli L et al (2014) Urban growth assessment around Winam Gulf of Kenya based on satellite imagery. Acta Astronaut 93:279–290

Genta G (2014) Private space exploration: a new way for starting a spacefaring society? Acta Astronaut 104:480–486

Gerzer R (2014) Salt balance: from space experiments to revolutionizing new clinical concepts on earth—a historical review. Acta Astronaut 104:378–382

Gerzer R, White R, Graef P (2014) Space medicine and physiology in the exploration era: the Cologne resolution. Acta Astronaut 104:364

Grassi L et al (2014) Satellite vulnerability to space debris—an improved 3D risk assessment methodology. Acta Astronaut 99:283–291

Grès S et al (2014) Cooperation and dialogical modeling for designing a safe human space exploration mission to Mars. Acta Astronaut 104:502–508

Grigoriev AI, Potapov AN (2014) Approaches to the development of biomedical support systems for piloted exploration missions. Acta Astronaut 94:14–16

Gruntman M (2014) Advanced degrees in astronautical engineering for the space industry. Acta Astronaut 103:92–105

Guan X et al (2014) An innovative high accuracy autonomous navigation method for the Mars rovers. Acta Astronaut 104:266–275

Guelman MM (2014) Geostationary satellites autonomous closed loop station keeping. Acta Astronaut 97:9–15

Guo J, Monas L, Gill E (2014) Statistical analysis and modelling of small satellite reliability. Acta Astronaut 98:97–110

Haeuplik-Meusburger S et al (2014) Greenhouses and their humanizing synergies. Acta Astronaut 96:138–150

Haider O, Groemer G (2014) Space Tweetup—from a participant to a Mars Tweetup organizer and a new format of space communication. Acta Astronaut 94:215–221

Hashimoto T et al (2014) Introduction to Japanese exploration study to the moon. Acta Astronaut 104:545–551

Hebert KD (2014) Regulation of space weapons: ensuring stability and continued use of outer space. Astropolitics 12:1–26

Heracleous L, Gonzalez SA (2014) Two modest proposals for propelling NASA forward. Space Policy 30:190–192

Herman JFC et al (2014) Human exploration of near earth asteroids: mission analysis for chemical and electric propulsion. Acta Astronaut 104:313–323

Herzing DL (2014a) Clicks, whistles and pulses: passive and active signal use in dolphin communication. Acta Astronaut 105:534–537

Herzing DL (2014b) Profiling nonhuman intelligence: an exercise in developing unbiased tools for describing other "types" of intelligence on earth. Acta Astronaut 94:676–680

Hlimi T (2014) The next frontier: An overview of the legal and environmental implications of near-earth asteroid mining. Ann Air Space Law XXXIX

Hobe S (2014) L'avenir du Droit de l'Espace au troisième millénaire. Zeitschrift für Luft- und Weltraumrecht 63:365–372

Hunter H, Fitzgerald R, Barlow D (2014) Improved cost monitoring and control through the Earned Value Management System. Acta Astronaut 93:497–500

Imada T, Sato N (2014) Study on JAXA elements for international lunar vicinity mission. Acta Astronaut 104:538–544

Jasper L, Schaub H (2014) Input shaped large thrust maneuver with a tethered debris object. Acta Astronaut 96:128–137

Jing Y, Chen X-Q, Chen L-H (2014) Biobjective planning of GEO debris removal mission with multiple servicing spacecrafts. Acta Astronaut 105:311–320

Kakoi M, Howell KC, Folta D (2014) Access to Mars from Earth–Moon libration point orbits: manifold and direct options. Acta Astronaut 102:269–286

Kanas N (2014) Psychosocial issues during an expedition to Mars. Acta Astronaut 103:73–80

Keller S, Collopy P, Componation P (2014) What is wrong with space system cost models? A survey and assessment of cost estimating approaches. Acta Astronaut 93:345–351

Kellman B (2014) On commercial mining of minerals in outer space: a rejoinder to Dr Ricky J. Lee. Air Space Law 39:411–420

Knapman JM, Swan PA (2014) Design concepts for the first 40 km a key step for the space elevator. Acta Astronaut 104:526–530

Korbitz A (2014) Toward understanding the active SETI debate: Insights from risk communication and perception. Acta Astronaut 105:517–520

Kramer WR (2014) Extraterrestrial environmental impact assessments—a foreseeable prerequisite for wise decisions regarding outer space exploration, research and development. Space Policy 30:215–222

Lara M et al (2014) Long-term evolution of Galileo operational orbits by canonical perturbation theory. Acta Astronaut 94:646–655

Launius RD (2014a) Power of analogies for advancing space scientific knowledge. Astropolitics 12:127–131

Launius RD (2014b) Reacting to nuclear power systems in space: American public protests over outer planetary probes since the 1980s. Acta Astronaut 96:188–200

Launius RD (2014c) The railroads and the space program revisited: historical analogues and the stimulation of commercial space operations. Astropolitics 12:167–179

Laurini KC, Gerstenmaier WH (2014) The Global Exploration Roadmap and its significance for NASA. Space Policy 30:149–155

Le Fèvre C et al (2014) Compliance of disposal orbits with the French Space Operations Act: the good practices and the STELA tool. Acta Astronaut 94:234–245

Lee J, Kim T (2014) Micro space power system using MEMS fuel cell for nanosatellites. Acta Astronaut 101:165–169

Lee RC, Jones TD, Chapman CR (2014) A decision analysis approach for risk management of near-earth objects. Acta Astronaut 103:362–369

Lladó N et al (2014) Capturing small asteroids into a Sun–Earth Lagrangian point. Acta Astronaut 95:176–188

Love SG, Harvey RP (2014) Crew autonomy for deep space exploration: lessons from the Antarctic Search for Meteorites. Acta Astronaut 94:83–92

Lowe CJ, Macdonald M (2014) Rapid model-based inter-disciplinary design of a CubeSat mission. Acta Astronaut 105:321–332

Maccone C (2014a) Evolution and history in a new "Mathematical SETI" model. Acta Astronaut 93:317–344

Maccone C (2014b) Lognormals for SETI, evolution and mass extinctions. Acta Astronaut 105:538–546

Maccone C (2014c) SETI as a part of Big History. Acta Astronaut 101:67–80

Martinez P et al (2014) Criteria for developing and testing Transparency and Confidence-Building Measures (TCBMs) for outer space activities. Space Policy 30:91–97

Masson-Zwaan T, Moro-Aguilar R, Lentsch A (2014) The future regulation of suborbital flight in Europe. Space Policy 30:75–82

Matveeva TV et al (2014) Challenges and perspectives of transport cargo vehicles utilization for performing research in free flight. Acta Astronaut 94:139–144

Mayorova VI et al (2014) Utilization of the International Space Station for education and popularization of space research. Acta Astronaut 98:147–154

Mazal L, Gurfil P (2014) Closed-loop distance-keeping for long-term satellite cluster flight. Acta Astronaut 94:73–82

McCurdy HE (2014) Reaching for higher altitudes: mountaineering analogies and the commercialization of outer space. Astropolitics 12:132–147

McKenna-Lawlor S et al (2014) Feasibility study of astronaut standardized career dose limits in LEO and the outlook for BLEO. Acta Astronaut 104:565–573

Messeri L (2014) Earth as analog: the disciplinary debate and astronaut training that took geology to the Moon. Astropolitics 12:196–209

Michaelides-Mateou S, Erotokritou C (2014) Flying into the future with UAVs: the Jetstream 31 Flight. Air Space Law 39:111–129

Michel P et al (2014) MarcoPolo-R: near-Earth Asteroid sample return mission selected for the assessment study phase of the ESA program cosmic vision. Acta Astronaut 93:530–538

Mouriaux P-F, Varnoteaux P (2014) Alexandre Ananoff (1910–1992): 30 years to promote astronautics before Sputnik. Acta Astronaut 93:266–278

Munévar G (2014) Space exploration and human survival. Space Policy 30:197–201

Munsami V (2014) South Africa's national space policy: the dawn of a new space era. Space Policy 30:115–120

Murchie S et al (2014) MERLIN: Mars-Moon exploration, reconnaissance and landed investigation. Acta Astronaut 93:475–482

Murtazin R (2014) Rendezvous missions: from ISS to lunar space station. Acta Astronaut 101:151–156

Murtazin R, Petrov N (2014) Usage of pre-flight data in short rendezvous mission of Soyuz-TMA spacecrafts. Acta Astronaut 93:71–76

Nair KK (2014) Expanding space security to contain SATCOM misuse by terrorists, narcotraffickers, criminals and other non-state actors. Ann Air Space Law XXXIX

Nazari M, Butcher EA (2014) On the stability and bifurcation analysis of dual-spin spacecraft. Acta Astronaut 93:162–175

Neal CR et al (2014) Developing the Global Exploration Roadmap: an example using the humans to the lunar surface theme. Space Policy 30:156–162

Neufeld MJ (2014) Transforming solar system exploration: the origins of the Discovery Program, 1989-1993. Space Policy 30:5–12

Newell CL (2014) Without having seen: faith, the future, and the final American frontier. Astropolitics 12:148–166

Ochiai M et al (2014) United Nations Human Space Technology Initiative (HSTI). Acta Astronaut 104:582–588

Olympio JT, Frouvelle N (2014) Space debris selection and optimal guidance for removal in the SSO with low-thrust propulsion. Acta Astronaut 99:263–275

Ongaro F (2014) From space medicine to preventive and personalized health care on earth. Acta Astronaut 104:409–411

Orgel C et al (2014) Scientific results and lessons learned from an integrated crewed Mars exploration simulation at the Rio Tinto Mars analogue site. Acta Astronaut 94:736–748

Orlov O, Belakovskiy M, Kussmaul A (2014a) Potential markets for application of space medicine achievements. Acta Astronaut 104:412–418

Orlov OI et al (2014b) Multidisciplinary Russian biomedical research in space. Acta Astronaut 101:180–187

Pace S (2014) American space strategy: choose to steer, not drift. Space Policy 30:1–4

Paikowsky D, Baram G, Ben Israel I (2014) Trends in government space activity and policy in 2013. Astropolitics 12:107–126

Paul MV et al (2014) The Penn state lunar lion: a university mission to explore the moon. Acta Astronaut 96:65–77

Peldszus R et al (2014) The perfect boring situation—addressing the experience of monotony during crewed deep space missions through habitability design. Acta Astronaut 94:262–276

Pell SJ et al (2014) Towards a cooperation between the arts, space science research and the European Space Agency—Preliminary findings of the ESA Topical Team Arts and Sciences (ETTAS). Acta Astronaut 105:181–191

Percy TK, Landrum DB (2014) Investigation of national policy shifts to impact orbital debris environments. Space Policy 30:23–33

Pezzella G, Marino G, Rufolo GC (2014) Aerodynamic database development of the ESA intermediate experimental vehicle. Acta Astronaut 94:57–72

Phipps CR (2014) L'ADROIT—a spaceborne ultraviolet laser system for space debris clearing. Acta Astronaut 104:243–255

Piattoni J, Ceruti A, Piergentili F (2014) Automated image analysis for space debris identification and astrometric measurements. Acta Astronaut 103:176–184

Preu P, Braun M (2014) German SIMBOX on Chinese mission Shenzhou-8: Europe's first bilateral cooperation utilizing China's Shenzhou programme. Acta Astronaut 94:584–591

Quantius D et al (2014) Initial design of laboratories for sustainable habitation. Acta Astronaut 94:541–562

Richard EE, Davis JR (2014) NASA Human Health and Performance Center: open innovation successes and collaborative projects. Acta Astronaut 104:383–387

Roberts SJ (2014) ENGage: the use of space and pixel art for increasing primary school children's interest in science, technology, engineering and mathematics. Acta Astronaut 93:34–44

Robinson GS (2014) Space jurisprudence and the need for a transglobal cybernation: The underlying biological dictates of humankind dispersal, migration, and settlement in near and deep space. Ann Air Space Law XXXIX

Sahara H (2014) Evaluation of a satellite constellation for active debris removal. Acta Astronaut 105:136–144

Salla ME (2014) Astropolitics and the "Exopolitics" of unacknowledged activities in outer space. Astropolitics 12:95–105

Salotti J-M (2014) New trade tree for manned mars missions. Acta Astronaut 104:574–581

Salotti J-M, Heidmann R (2014) Roadmap to a human Mars mission. Acta Astronaut 104:558–564

Salotti J-M, Suhir E (2014) Manned missions to Mars: minimizing risks of failure. Acta Astronaut 93:148–161

Santoro F et al (2014) The Italian Spacegate: study and innovative approaches to future generation transportation based on High Altitude Flight. Acta Astronaut 101:98–110

Schladebach M (2014) Die Verwendung von Satellitenbildern in nationalen Gerichtsverfahren. Zeitschrift für Luft- und Weltraumrecht 63:442–449

Schrogl K-U (2014) The new debate on the working methods of the UNCOPUOS Legal Subcommittee. Acta Astronaut 105:101–108

Schubert F (2014) The integration of remotely piloted aircraft in the air navigation services system. Ann Air Space Law XXXIX

Schwartz JSJ (2014) Prioritizing scientific exploration: a comparison of the ethical justifications for space development and for space science. Space Policy 30:202–208

Sephton MA (2014) Astrobiology can help space science, education and economy. Space Policy 30:146–148

Shakhparonov VM, Grishin DV, Karagioz O (2014) Monitoring of the space system. Acta Astronaut 104:71–76

Shelhamer M (2014) Life-sciences research opportunities in commercial suborbital space flight. Acta Astronaut 104:432–437

Shimabukuro A (2014) No deal in space: a bargaining model analysis of U.S. resistance to space arms control. Space Policy 30:13–22

Smirnova M (2014) Mars transportation vehicle concept. Acta Astronaut 103:250–256

Smith KC (2014) Manifest complexity: a foundational ethic for astrobiology? Space Policy 30:209–214

Sommariva A (2014) Motivations behind interstellar exploration and colonization. Astropolitics 12:82–94

Spiller J (2014) Scientific exploration in Antarctica as an analogy for American spaceflight. Astropolitics 12:180–195

Stansbery EK, Draper DS (2014) Role of sample return and sample science in low cost missions. Acta Astronaut 93:453–459

Straub J, Vacek J, Nordlie J (2014) Considering regulation of Small Unmanned Aerial Systems in the United States. Air Space Law 39:275–293

Su J, Lixin Z (2014) The European Union draft Code of Conduct for outer space activities: an appraisal. Space Policy 30:34–39

Summerer L (2014) Thinking tomorrows' space—research trends of the ESA advanced concepts team 2002–2012. Acta Astronaut 95:242–259

Szajnfarber Z (2014) Space science innovation: how mission sequencing interacts with technology policy. Space Policy 30:83–90

Tancredi U, Renga A, Grassi M (2014) Novel closed-loop approaches for precise relative navigation of widely separated GPS receivers in LEO. Acta Astronaut 93:243–251

Thumm T et al (2014) International space station accomplishments update: scientific discovery, advancing future exploration, and benefits brought home to earth. Acta Astronaut 103:235–242

Tronchetti F (2014a) Private property rights on asteroid resources: assessing the legality of the ASTEROIDS Act. Space Policy 30:193–196

Tronchetti F (2014b) The right of self-defence in outer space: an appraisal. Zeitschrift für Luft- und Weltraumrecht 63:92–120

Urbina DA, Charles R (2014) Symposium keynote: enduring the isolation of interplanetary travel: a personal account of the Mars500 mission. Acta Astronaut 93:374–383

Vallado DA, Finkleman D (2014) A critical assessment of satellite drag and atmospheric density modeling. Acta Astronaut 95:141–165

van der Pas N et al (2014) Target selection and comparison of mission design for space debris removal by DLR's advanced study group. Acta Astronaut 102:241–248

van der Weg WJ, Vasile M (2014) Contingency and recovery options for the European Student Moon Orbiter. Acta Astronaut 94:168–183

Vardaxis G, Wie B (2014) Near-Earth object intercept trajectory design for planetary defense. Acta Astonaut 101:1–15

Venturini K, Verbano C (2014) A systematic review of the Space technology transfer literature: research synthesis and emerging gaps. Space Policy 30:98–114

Vinokhodova AG, Gushin VI (2014) Study of values and interpersonal perception in cosmonauts on board of international space station. Acta Astronaut 93:359–365

Viscio MA et al (2014a) A methodology for innovative technologies roadmaps assessment to support strategic decisions for future space exploration. Acta Astronaut 94:813–833

Viscio MA et al (2014b) Interplanetary CubeSats system for space weather evaluations and technology demonstration. Acta Astronaut 104:516–525

Weber L, Eberg A (2014) The Cape Town Convention and its implementation in Russia and the Commonwealth of Independent States (CIS). Air Space Law 39:1–45

Weeks EE, Faiyetole AA (2014) Science, technology and imaginable social and behavioral impacts as outer space develops. Acta Astronaut 95:166–173

Weppler J (2014) Contributions of the International Space Station towards future exploration missions. Acta Astronaut 104:552–557

White AE, Lewis HG (2014) The many futures of active debris removal. Acta Astronaut 95:189–197

Wicht A, Szajnfarber Z (2014) Portfolios of promise: a review of R&D investment techniques and how they apply to technology development in space agencies. Space Policy 30:62–74

Willson D et al (2014) The performance of field scientists undertaking observations of early life fossils while in simulated space suit. Acta Astronaut 93:193–206

Wood D, Weigel A (2014) Architectures of small satellite programs in developing countries. Acta Astronaut 97:109–121

Yang C (2014) Towards a new legal framework for the use of nuclear power sources in outer space? Ann Air Space Law XXXIX

Zhao S (2014) Analysis on coverage ability of BeiDou navigation satellite system for manned spacecraft. Acta Astronaut 105:487–494

13.2 Articles

Allahdadi F, Pelton JN (2014) Handbook of cosmic hazards and planetary defense. Springer, New York

Allison AL (2014) The ITU and managing satellite orbital and spectrum resources in the 21st century. SpringerBriefs in Space Development, Springer, New York

Bainbridge Sims W (2014) The meaning and value of spaceflight: public perceptions (Space and Society). Springer, Dordrecht

Barbree J (2014) Neil Armstrong: a life of flight. Thomas Dunne Books, New York

Bizony P (2014) New space frontiers: venturing into earth orbit and beyond. Zenith Press, Minneapolis

Blottner D, Salanova M (2014) The NeuroMuscular system: from earth to space life science (SpringerBriefs in Space Life Sciences). Springer, Berlin

Boniecki DS (2014) Live TV from orbit. Apogee Prime, Burlington

Bower T (2014) Branson: behind the mask. Faber & Faber, London

Burrows WE (2014) The asteroid threat: defending our planet from deadly near-earth objects. Prometheus Books, Amherst

Capderou M (2014) Handbook of satellite orbits. From Kepler to GPS, Springer, Cham

Caroll M (2014) Living among giants: exploring and settling the outer solar system. Springer, Cham

Chaikin A (2014) John Glenn: America's astronaut. Smithsonian Books, Washington, DC

Cheung T-M (ed) (2014) Forging China's military might: a new framework for assessing innovation. Johns Hopkins University Press, Baltimore

Clegg B (2014) Final frontier: the pioneering science and technology of exploring the universe. St. Martin's Press, New York

Clément G (2014) Fundamentals of space medicine (Space Technology Library), 2nd edn. Springer, New York

Clément G, Bukley A (eds) (2014) Artificial gravity (Space Technology Library). Springer, New York

Committee on Human Spaceflight et al (2014) Pathways to exploration: rationales and approaches for a U.S. program of human space exploration. National Academies Press, Washington, DC

Consolmagno GSJ, Mueller PSJ (2014) Would you baptize an extraterrestrial?: ...and other questions from the astronomers' in-box at the Vatican Observatory. Image, New York

Couston M (2014) Droit spatial. Edition Ellipses, Paris

Crotts A (2014) The new moon: water, exploration, and future habitation. Cambridge University Press, Cambridge

Edwards KJ, Medhat Gaber M (2014) Astronomy and big data. Springer, New York
Evans B (2014) Partnership in space. The mid to late nineties. Springer, New York
Federal Research Division of Congress, NASA Headquarter (2014) Astronautics and aeronautics: A chronology, 1996-2000. CreateSpace Independent Publishing Platform, Washington, DC
Flechtner F, Sneeuw A, Schuh W-D (eds) (2014) Observation of the system Earth from space—CHAMP, GRACE, GOCE and future missions. GEOTECHNOLOGIEN science report no. 20. Springer, Berlin
Freese K (2014) The cosmic cocktail: three parts dark matter. Princeton University Press, Princeton
Gerovich S (2014) Voices of the Soviet Space Program. Cosmonauts, soldiers, and engineers who took the USSR into space (Palgrave Studies in the History of Science and Technology). Palgrave Macmillan, London
Golub L, Pasachoff JM (2014) Nearest star: the surprising science of our sun. Cambridge University Press, Cambridge
Gou F et al (2014) Space electronic reconnaissance: localization theories and methods. John Wiley & Sons, Hoboken
Habison P (2014) Weltraum-Land Österreich. Pioniere der Raumfahrt erzählen, Seifert Verlag, Wien
Hadfield C (2014) You are here: around the world in 92 minutes: photographs from the International Space Station. Little Brown and Company, New York
Harris H (2014) Challenger: an American tragedy: the inside story from launch control. Open Road Media, New York
Hartley K (2014) The political economy of aerospace industries: a key driver of growth and international competitiveness? Elgar Publications, Chelterham
Hess R (2014) International Space Station 121 success secrets: 121 most asked questions on International Space Station—what you need to know. Emero Publishing, Brisbane
Hirshfeld A (2014) Starlight detectives: how astronomers, inventors, and eccentrics discovered the modern universe. Bellevue Literary Press, New York
Hitt D, Smith H (2014) Bold they rise: the Space Shuttle early years, 1972–1986. University of Nebraska Press, Lincoln
Hobe S (ed) (2013) Pioneers of space law. Martinus Nijhoff Publishers, Leiden/Boston
Houston R, Ross J (2014) Wheels stop: the tragedies and triumphs of the Space Shuttle Program, 1986-2011 (Outward odyssey: a people's history of spaceflight). University of Nebraska Press, Lincoln
Jakhu RS, Pelton JN (2014) Small satellites and their regulation. SpringerBriefs in Space Development, Springer, New York
Jorgenson C (ed) (2014) Proceedings of the International Institute of Space Law 2013, vol 56. Eleven International Publishing, The Hague
Kahn J, Liverman CT (2014) Health standards for long duration and exploration spaceflight: ethics principles, responsibilities, and decision framework. National Academies Press, Washington, DC

Kaufman M (2014) Mars up close: inside the Curiosity Mission. National Geographic Books, Washington, DC

Lambright H (2014) Why Mars: NASA and the politics of space exploration. Johns Hopkins University Press, Baltimore

Launius RD, DeVorkin DH (eds) (2014) Hubble's legacy: reflections by those who dreamed it, built it, and observed the universe with it. Smithsonian Institution Scholarly Press, Washington, DC

Launius RD (2014) Historical analogs for the stimulation of space commerce. NASA, Washington, DC

Lele A (2014) Mission Mars. India's quest for the Red Planet (SpringerBriefs in Applied Sciences and Technology). Springer, New Delhi

Li H (2014) Geostationary satellites collocation. National Defense Industry Press, Beijing

Liang S et al. (2014) Global LAnd Surface Satellite (GLASS) products. Algorithms, validation and analysis. SpringerBriefs. Springer, Cham

Luokkala BB (2014) Exploring science through science fiction. Springer, New York

Lunan D (2014) Incoming asteroid! What could we do about it? Springer, New York

Macdonald M, Badescu V (2014) The international handbook of space technology. Springer, Berlin

Maidana CO (2014) Thermo-magnetic systems for space nuclear reactors: an introduction. SpringerBriefs in Applied Sciences and Technology. Springer, New York

Manning R, Simon WL (2014) Mars rover "Curiosity": an inside account from Curiosity's chief engineer. Smithsonian Books, Washington

Markley FL, Crassidis JL (2014) Fundamentals of spacecraft attitude determination and control. Springer, New York

Matloff G, Bangs C, Johnson L (2014) Harvesting space for a greener earth. Springer, New York

Meerman Scott D, Jurek R (2014) Marketing the moon: the selling of the Apollo Lunar Program. MIT Press, Boston

Meuser P (2014) Galina Balaschowa : Architektin des sowjetischen Raumfahrtprogramms. DOM Publications, Berlin

Mitchell E, Mahoney E (2014) Earthrise: my adventures as an Apollo 14 astronaut. Chicago Review Press, Chicago

Moltz JC (2014) Crowded orbits: conflict and cooperation in space. Columbia University Press, New York

Money S (2014) Here be dragons: the rise of Space X and the journey to Mars. Apogee Prime, Burlington

NASA (2014a) Assessment of programs in space biology and medicine, 1991. CreateSpace Independent Publishing Platform, Washington, DC

NASA (2014b) Guidelines for developing spacecraft maximum allowable concentrations for space station contaminants. CreateSpace Independent Publishing Platform, Washington, DC

NASA (2014c) Leadership in space: selected speeches of NASA administrator Michael Griffin, May 2005-October 2008. CreateSpace Independent Publishing Platform, Washington, DC

NASA (2014d) The Neurolab Spacelab Mission: neuroscience research in space: results from the STS-90 Neurolab Spacelab. MissionCreateSpace Independent Publishing Platform, Washington, DC

NASA (ed) (2014e) Space life sciences research: the importance of long-term space experiments. CreateSpace Independent Publishing Platform, Washington, DC

Neal V (2014) Discovery: the champion of the Space Shuttle fleet (Smithsonian Series). Zenith Press, Minneapolis

Neufeld M et al (eds) (2014) Milestones of space: eleven iconic objects from the Smithsonian National Air and Space Museum (Smithsonian Series). Zenith Press, Minneapolis

OECD (2014) The Space economy at a glance 2014. OECD, Paris

Pyle R (2014a) Innovation the NASA Way: harnessing the power of your organization for breakthrough success. McGraw-Hill Education, New York

Pyle R (2014b) Curiosity: an inside look at the Mars Rover Mission and the people who made it happen. Prometheus Books, New York

Sage D (2014) How outer space made America: geography, organization and the cosmic sublime. Ashgate Publishing Limited, Surrey

Scharf C (2014) The Copernicus complex: our cosmic significance in a Universe of planets and probabilities. Scientific American/Farrar, Straus and Giroux, Norwich

Seedhouse E (2014) Suborbital. Industry at the edge of space. Springer, New York

Sherr L (2014) Sally Ride: America's first woman in space. Simon & Schuster, New York

Smith M (2014) Rockets and revolution: a cultural history of early spaceflight. University of Nebraska Press, Lincoln

Soluri M (2014) Infinite worlds: the people and places of space exploration. Simon & Schuster, New York

Song Lockard E (2014) Human migration to space. Alternative technological approaches for long-term adaptation to extraterrestrial environments. Springer, Cham

Steven-Boniecki D (2014) Live TV from orbit. Apogee Prime, Burlington

Tan S-Y (2014) Meteorological satellite systems. SpringerBriefs in Space Development, Springer, New York

Tribbe MD (2014) No requiem for the space age: the Apollo moon landings and American culture. Oxford University Press, Oxford

Trotta R (2014) The edge of the sky: all you need to know about the all-there-is. Basic Books, New York

Uhlig T, Sellmaier F, Schmidhuber M (eds) (2014) Spacecraft operations. Springer, Wien

Vakoch DA (ed) (2014a) Archaeology, anthropology, and interstellar communication. NASA, Washington, DC

Vakoch DA (2014b) Extraterrestrial altruism. Springer, New York

Waltham D (2014) Lucky planet: why earth is exceptional—and what that means for life in the universe. Basic Books, New York

Weiland C (2014) Aerodynamic data of space vehicles. Springer, New York

Wells H, Whiteley SH, Kareqeannes CE (2014) The origins of NASA names. CreateSpace Independent Publishing Platform, Washington, DC

Whalen DJ (2014) The Origins of satellite communications, 1945-1965 (Smithsonian History of Aviation and Spaceflight Series). Smithsonian Institution Scholarly Press, Washington, DC

Wiesel WE (2014) Spaceflight dynamics. McGraw-Hill, New Delhi

About the Authors and Editors

Cenan Al-Ekabi joined the European Space Policy Institute in Vienna in 2011, functioning as a project manager from 2012, and a Resident Fellow from 2013. Prior to that, he obtained two Advanced LL.M. degrees in Air and Space Law, and European and International Business Law from Leiden University in the Netherlands. He also holds a U.S. JD with concentration in studies in international law from the Thomas M. Cooley law school, and a bachelor's degree in Political Science from McMaster University in Canada.

Marco Aliberti works as a Resident Fellow at the European Space Policy Institute (ESPI) in Vienna. He first joined ESPI in October 2012, after completing a Master of Advanced Studies in Space Policy and Institutions and attending the ESA/ECSL Summer Course on Space Law and Policy. Prior to that, he graduated in Oriental Languages and Cultures at the University of Rome "La Sapienza", and obtained a Master in International Relations from the Italian Diplomatic Academy (SIOI). He also pursued an Advanced Master's degree in International Asian Studies at the University of Naples "L'Orientale" and a Security Studies Program at the Institute of Global Studies—School of Government in Rome.

Blandina Baranes joined the European Space Policy Institute (ESPI) in Vienna in February 2005 and currently holds the position of ESPI Communications Manager. Prior to this, she was the chief librarian of the Jewish Studies Department of the University of Vienna. During the past years she has also worked as a documentalist and librarian for different institutions, such as the Austrian Broadcasting Corporation (ORF), the Der Spiegel magazine and others. She conducted her studies and research in Austria and Israel and graduated with a Master's Degree from Vienna University, Faculty of Philosophy, Department of Social and Cultural Anthropology. She is co-editor of the past three volumes of ESPI's "Yearbook on Space Policy", published several articles and participated as a speaker in a number of space related events on the topic of "Space and Interdisciplinarity".

Jacques Blamont, professor at the University of Paris (1957–1996) now emeritus, is an experimental physicist who turned from atomic spectroscopy to space science when, in 1959, he directed the first launches of the French Veroniques rockets that were the starting point of space science research in Europe. He was one of the few given the task of creating the French space agency CNES in 1962, as Scientific and Technical Director of the fledgling organisation. In 1972 he became Chief Scientist and advisor to the Director General of CNES. He is still advisor to the President of CNES. Involved in atmospheric research, Jacques Blamont discovered the turbopause of the Earth (1959), the interstellar wind (1970), the hydrogene envelope (or halo) of comets (1971), the polar noctilucent clouds (1973). He made the first measurement of the temperature of

C. Al-Ekabi et al. (eds.), *Yearbook on Space Policy 2014*, Yearbook on Space Policy, DOI 10.1007/978-3-7091-1899-3

the neutral atmosphere from 100 km to 500 km and the dynamic parameters of the mesopause region, developed the scientific ballooning in Europe and the Lidar technic for atmospheric probing, proposed and helped exploration of planets by balloons, with the successful flights of two balloons in the atmosphere of Venus during the soviet mission Vega (1985). He developed an image compression device which has been widely used in missions to the Moon, Mars and Titan. He was a Prime investigator on many Soviet and U.S. spacecraft (Voyager, Pioneer Venus, Vega, Phobos, Clementine).

Gerard Brachet received an Engineering degree from the Ecole Nationale Supérieure d'Aéronautique in 1967 and a Master of Sciences in Aeronautics and Astronautics from the University of Washington in 1968. From 1970 to 1982, he held various positions at Centre National d'Etudes Spatiales (CNES), the French Space Agency, before moving in 1982 to become CEO of the SPOT IMAGE company set up to develop the market for remote sensing imagery from the SPOT series of satellites. In 1995, he came back at CNES headquarters as Director for programmes, planning and industrial policy and became its Director General from 1997 to 2002. In 1997, he was Chairman of the international "Committee on Earth Observation Satellites (CEOS)". From 2003 onwards, he was consultant on space policy issues and space applications, advising industry and government institutions, in particular the European Commission and the European Space Agency. From June 2006 to June 2008, Gerard Brachet chaired the United Nations Committee for the Peaceful Uses of Outer Space (UN COPUOS). He was the French expert within the Group of Governmental Experts (GGE) set up by the Secretary General of the United Nations to develop a series of Transparency and Confidence-Building Measures (TCBMs) in Outer Space that produced its final report in 2013, endorsed by the UN General Assembly in December 2013. Gerard Brachet was President of the Air and Space Academy (Académie de l'Air et de l'Espace) from 2009 to 2012 and Vice President of the International Astronautical Federation (IAF) and chairman of the IAF Committee on Liaison with International Organizations and Developing Nations (CLIODN) during the same period. Gerard Brachet is a member of the International Academy of Astronautics and received several international awards during his career, in particular the "1985 Laurels" of Aviation Week and Space Technology, the Brock Gold Medal Award of the International Society for Photogrammetry and Remote Sensing in 1992, the Gold Medal of the British Remote Sensing Society in 1994 and the "Social Sciences Award" of the International Academy of Astronautics in 2007.

Thomas Hoerber is Professor at the EU-Asia Institute at ESSCA—School of Management. He took this Position after a lectureship at the University of Erlangen-Nuernberg, in Germany, a previous lectureship at the University of Victoria, BC, Canada and a Ph.D. from Trinity Hall, Cambridge. He has worked in the German Bundestag and the French Assemblée Nationale and retains a vivid interest in the application of his Research in concrete politics. His main field of research is European Studies for the analysis of which he applies policy fields such as energy, environment, space and security. In order to further the understanding of European Integration, Economic Theory and Integration Theory are of further interest to his research. He has published extensively an all these fields.

Peter Hulsroj pursued his legal studies at the University of Copenhagen from 1974 to 1979. After military service, Peter Hulsroj did post-graduate studies at Harvard Law School in the USA, where he earned a Master's Degree in 1981. Returning from the USA, Mr. Hulsroj joined a large commercial law firm in Copenhagen, where he became a full member of the Danish Bar. After three and half years of private practice in Copenhagen, Mr. Hulsroj became a Contracts Officer in ESTEC, and thereafter served for almost 14 years as the Head of Contracts and Legal Affairs at Eumetsat in Darmstadt, Germany. Before rejoining ESA, Mr. Hulsroj was the Legal Adviser of the Preparatory Commission for the Comprehensive Nuclear-Test-Ban Treaty Organization in Vienna. Peter Hulsroj was appointed ESA Director of Legal Affairs and External Relations (D/LEX) in April 2008, comprising four specialist areas: the Legal Department, the International Relations Department, the Communication Department and the Education and Knowledge Management Office at ESA from 2008 to 2011. Since 2011 Peter Hulsroj is Director of ESPI.

Amy Kaminski currently serves as senior policy advisor in the Office of the Chief Scientist at NASA Headquarters in Washington, DC, USA. Dr. Kaminski previously served at the White House Office of Management and Budget (OMB), where she analysed NASA programmes and made recommendations to White House policy makers about budgets and ways to improve the performance of NASA programmes. Before joining OMB, she was an analyst in the Federal Aviation Administration's Office of the Associate Administrator for Commercial Space Transportation. Dr. Kaminski holds Ph.D. and M.S. degrees in Science and Technology Studies from Virginia Tech, an M.A. in Science, Technology, and Public Policy from the George Washington University, and a B.A. from Cornell University in Earth and Planetary Sciences. She has written several articles and book chapters on space policy and history topics. Her current research focuses on NASA's relationships with publics beyond traditional government, industry, and academic space policy developers.

Stephen D. Krasner is the Graham H. Stuart Professor of International Relations at Stanford, and a Senior Fellow at the Freeman Spogli Institute, the Hoover Institution, and at SIEPR. From February 2005 to April 2007 he was Director of the Policy Planning Staff at the Department of State. In 2002 he served as Director for Governance and Development at the National Security Council. He is a member of the Board of Directors of the United States Institute of Peace, and was a member of the Foreign Policy Advisory Board of the Department of State from 2012 to 2014. He was a fellow at the Center for Advanced Studies in the Behavioral Science in 1987–1988 and at the Wissenschaftskolleg Berlin I 2000–2001. He has been a Marcator Fellow at the Free University Berlin since 2014. Professor Krasner is a Fellow of the American Academy of Arts and Sciences and a Member of the Council on Foreign Relations.

Arne Lahcen is Resident Fellow at the European Space Policy Institute in Vienna, Austria. His research interests include Earth observation governance, international cooperation schemes, innovation management, and the relationship between space and society. Prior to joining ESPI in 2011, he obtained an Advanced Master's degree in Space Studies with a specialisation in space law, policy, business and management at the Faculty of Science of the Catholic University of Leuven, Belgium. He also holds a Bachelor's and Master's degree in Social-Economic Sciences pursued at the Faculty of Applied Economics, University of Antwerp, Belgium. In addition to ESPI research activities, he is editor-in-chief of the ESPI Perspectives series and co-responsible for the follow-up of the ESPI Autumn Conference and the editing of the ESPI Yearbook.

Xavier Li-Wen Liao is a Ph.D. Candidate of the Ghent Institute for International Studies (GIIS), Belgium. He earned his B.A. and M.A. degrees on political science from Brussels' Facultés Universitaires Saint Louis, and Université Libre de Bruxelles respectively. His doctoral research is related to global astropolitics and global space governance on the use of outer space. His recently published 'The Growing Regionalization of Global Space Regime Complex' (2015) The Aviation & Space Journal XIV (1) (January/March 2015): 23–35, and 'Global Space Governance in the 21st Century: A Regime Complex perspective' (2014) Astrosociology Insights 3(2): 3–5. Xavier has extensive interests in other research fields, such as Chinese official discourse analysis. He published his MA thesis 'A case study on China's (re-)securitization discourse along the institutionalization of the Shanghai Cooperation Organization (SCO) in The Role of Discourse as the Interface between various Disciplines studying Chinese Society, (2012) (Eds.) L. Lams & X. Liao published by the Belgian Royal Academies for Science and the Arts (KVAB). He also co-wrote 'Tracing Taiwanization processes in the Taiwanese presidential statements in times of cross-Strait rapprochement' with L. Lams published in the Journal of Chinese Current Affairs, 2011(1): 63–98.

Rajeswari Pillai Rajagopalan is Senior Fellow at the Observer Research Foundation (ORF), New Delhi. Dr. Rajagopalan joined ORF after an almost 5-year stint at the National Security Council Secretariat (NSCS), Government of India (2003– 2007), where she was Assistant Director. Prior to joining the NSCS, she was Research Officer at the Institute for Defence Studies and Analyses, New Delhi. She was also a Visiting Professor at the Graduate Institute of International Politics,

National Chung Tsing University, Taichung, Taiwan, in early 2012. She is the author of three books: Clashing Titans: Military Strategy and Insecurity Among Asian Great Powers; The Dragon's Fire: Chinese Military Strategy and Its Implications for Asia; and Uncertain Eagle: US Military Strategy in Asia. She has also co-authored two books. She has co-authored and edited a number of books and her research articles have appeared in edited volumes, and in India Review, Bulletin of Atomic Scientists, The National Interest, Indian Foreign Affairs Journal, Strategic Studies Quarterly, Strategic Analysis, and CLAWS Journal. Other writings have appeared in the Journal of Strategic Studies, Journal of Peace Research and Contemporary South Asia and she has also contributed essays to newspapers such as the Wall Street Journal, The Times of India, The Hindustan Times, The Economic Times and The Pioneer. Her areas of research interests include Indian and Asian foreign policy and security issues, US foreign policy, nuclear and space security, and other arms control issues. She can be contacted at rajeswarirajagopalan@gmail.com; rpr@orfonline.org.

Emmanuel Sigalas will be joining the Czech Institute of International Relations as Senior Research Fellow. Dr. Sigalas was visiting professor of EU and European Studies at the Institute of European, Russian and Eurasian Studies of Carleton University in Canada. Previously he worked as assistant professor at the Institute for Advanced Studies and at the Institute of European Integration Research of the Austrian Academy of Sciences. He holds a Ph.D. from the University of Reading, a Master's from the University of Essex and a Bachelor from the Athens University of Economics and Business. He has published in scientific journals, such as Space Policy, European Union Politics and West European Politics. Currently, he is working as co-editor on the Theorizing Space Policy book (Lexington) and on the European Space Policy entry in the Oxford Research Encyclopedia.

Kazuto Suzuki is an expert in the Panel of Experts for Iranian Sanction Committee under the United Nations Security Council. He is also professor of International Politics at Graduate School of Law of Hokkaido University, Japan. He graduated Department of International Relations, Ritsumeikan University, and received Ph.D. from Sussex European Institute, University of Sussex, England. He has worked in the University of Tsukuba from 2000 to 2008, and moved to Hokkaido University. He also spent 1 year at Woodrow Wilson School of Public and International Affairs at Princeton University from 2012 to 2013 as visiting researcher. He has conducted researches from International Political Economy perspective in Space Policy, together with nuclear energy policy (contributed to Independent Investigation Commission of Fukushima nuclear accident), export control policy, science and technology policy, counter-terrorism, non-proliferation, export control and policies on market regulation. He has contributed to the drafting the Basic Space Law of Japan, and served as space expert in number of governmental activities.

CPSIA information can be obtained at www.ICGtesting.com
Printed in the USA
BVOW10*0212090216

436061BV00001B/6/P

9 783709 118986